Smart Sensors, Measurement and Instrumentation

Volume 39

More information about this series at http://www.springer.com/series/10617

Olfa Kanoun · Nabil Derbel
Editors

Advanced Systems for Biomedical Applications

Springer

Editors
Olfa Kanoun
Fakultät für Elektrotechnik und Informationstechnik
TU Chemnitz
Chemnitz, Germany

Nabil Derbel
Department of Electrical Engineering
Ecole Nationale d'ingenieurs de Sfax
University of Sfax
Sfax, Tunisia

ISSN 2194-8402 ISSN 2194-8410 (electronic)
Smart Sensors, Measurement and Instrumentation
ISBN 978-3-030-71220-4 ISBN 978-3-030-71221-1 (eBook)
https://doi.org/10.1007/978-3-030-71221-1

This Springer imprint is published by the registered company Springer Nature Switzerland AG
The registered company address is: Gewerbestrasse 11, 6330 Cham, Switzerland

Preface

In this book, we present recent developments in the field of biomedical systems covering a wide range of methods, systems, and instrumentation techniques for diagnosis, monitoring, treatment, and assistance.

It provides insights into theory, applications, and perspectives relevant to the field of biomedical engineering, as well as the general paradigms and methodologies behind them. Each chapter provides advances in a research topic along with a balanced treatment of the relevant theories, methods, or applications. It reports on the latest advances achieved in the corresponding field of biomedical engineering.

This book is a valuable reference for graduate students, researchers, educators, engineers, and scientists and includes 12 chapters in total, structured into three parts as follows:

The first part of this book focuses on methods and technologies for medical systems, and comprises four chapters:

- The first chapter entitled "Potential of Impedance Spectroscopy as a Manifold Non-invasive Method for Medical Applications" addresses the importance and the applicability of impedance spectroscopy as a non-invasive measurement method in the medical field and classifies the corresponding applications in a review form.
- The second chapter entitled "Electrode Design for Reproducible Study of Tissues Impedance in Medical Applications" discusses theoretical and experimental investigations of the influence of electrode geometry with the aim to reduce the effect of the anisotropy effect.
- The third chapter entitled "Resonant Inductive Coupling for Wirelessly Powering Active Implants: Current Issues, Proposed Solutions and Future Technological attempts" concerns the Wireless Power Transfer (WPT) as a promising solution to overcome any health or side complications for medical implants, which provide a substantial improvement in health care by helping to manage diseases and save innumerable lives of patients.
- The fourth chapter entitled "Fault Diagnosis for Nonlinear Biological Processes Based on Machine Learning Models" addresses the use of learning techniques to monitor and detect faults in biological systems resulting in a high computation cost in case of too large sets of training data.

The second part concerns systems for the Covid-19 context, and comprises three chapters:

- The fifth chapter entitled "Prospects of Internet of Things (IoT) and Machine Learning to Fight Against COVID-19" discusses different aspects of IoT and machine learning, for automatic thermal monitoring and for measuring and real-time monitoring of heart rate with wearable IoT devices, in order to help healthcare systems for detecting and monitoring Coronavirus patients.
- The sixth chapter entitled "Development of an IoT-Based System for Training in Cardiopulmonary Resuscitation" applies guidelines from the American Heart Association to build and design a Dummy-based system for Cardiopulmonary Resuscitation (CPR), for the implementation of the method of cardiopulmonary resuscitation, as well as to make them aware of the risks of cardiac arrest.
- The seventh chapter entitled "Portable Cardiopulmonary Resuscitation and Ventilator Device: Design and Implementation" describes a mechanical ventilator in the first stage to apply the concepts of reverse engineering to the design and construction of a low-cost portable device as well as a cardiopulmonary resuscitation system, based on wireless sensor network technologies in order to access the device and adjust its main parameters by the specialist according to the patient's condition.

The third part reports on medical systems for persons with disabilities and comprises five chapters:

- The eighth chapter entitled "Virtual Reality and Augmented Reality Technologies for Smart Physical Rehabilitation" considers wearable, environment embedded or remote sensing solutions together with different type of IoT physical rehabilitation architectures, based on different virtual reality and augmented reality technologies, which has been implemented using unity game development software for different computation platform and considering different types of interfaces based on new technologies, including wearable smart sensors, environment embedded sensors, mobile devices, and big displays.
- The ninth chapter entitled "Control of Lower Limb Exoskeletons for Gait Rehabilitation Purposes" presents rehabilitation methods to increase the mobility deficiency for children with Cerebral Palsy, using robotic systems based gait training therapy as effective key tools to compensate and rehabilitate their functional skills.
- The tenth chapter entitled "Indoor Scene Simplification for Safe Navigation Using Saliency Map for the Benefit of Visually Impaired People" proposes an approach that simplifies observed scenes while ensuring autonomous navigation, involving two main steps (i) the highlight of salient regions in order to suppress background objects, and (ii) the employment of morphology operations to simplify scene interpretation.
- The eleventh chapter entitled "Towards Intelligent Control of Electric Wheelchairs for Physically Challenged People" deals with the use of soft computing techniques for solving the mobility problems of physically handicapped people using available signals such as face directional gesture, voice, brain, and electromyogram

signals, depending on the type and degree of the handicap, in order to classify commands required to drive a wheelchair.
- The twelfth chapter entitled "Fuzzy Control of an Intelligent Electric Wheelchair Using an EMOTIV EPOC Headset" reports on the design and the implementation of an intelligent control method of an electric wheelchair for handicapped persons using technologies derived from mobile robotics, improving the performance of the wheelchair controlled by the brain without any requirement and feedback from the user, and avoiding obstacles based on the ultrasonic sensor detectors.

Chemnitz, Germany
Sfax, Tunisia
May 2021

Olfa Kanoun
Nabil Derbel

About This Book

The book highlights recent developments in the field of biomedical systems covering a wide range of technological aspects, methods, systems and instrumentation techniques for diagnosis, monitoring, treatment, and assistance.

Biomedical systems are becoming increasingly important in medicine and in special areas of application such as supporting people with disabilities and under pandemic conditions. They provide a solid basis for supporting people and improving their health care.

As such, the book offers a key reference guide about novel medical systems for students, engineers, designers, and technicians.

Contents

Potential of Impedance Spectroscopy as a Manifold Non-invasive Method for Medical Applications

Dhouha Bouchaala, Hanen Nouri, Bilel Ben Atitallah, Nabil Derbel, and Olfa Kanoun

Abstract Electrical Bioimpedance Spectroscopy (BIS) is an interesting method for assessing the state or composition of human organs and various types of biological tissues in-vivo in both clinical and research applications as well as in-vitro. The non-invasiveness of this method, its manifold applicability and adaptability make it suitable for usability as a part of health monitoring systems as portable, wearable or implantable systems. The foremost applications in personal health and biological tissues monitoring are fluid and body mass changes, tissue and cell characterization and dynamically variable biological systems assessment such as cardiac and lung function monitoring. In this contribution, we provide an overview of BIS and its applications for fluid monitoring and body cell mass changes in the second section, tissue characterization and cell growth monitoring in the third section and dynamically variable biological systems monitoring in the last section.

Keywords Impedance spectroscopy · Fluid montoring · Tissue characterisation · Disease prevention

---

D. Bouchaala (✉) · H. Nouri · B. Ben Atitallah · N. Derbel
Control and Energy Management laboratory (CEM Lab), National School of Engineers of Sfax (ENIS), University of Sfax, Sfax, Tunisia
e-mail: bouchaala.dhouha@gmail.com

H. Nouri
e-mail: hanen.nouri@etit.tu-chemnitz.de

B. Ben Atitallah
e-mail: bilel.ben-atitallah@s2019.tu-chemnitz.de

N. Derbel
e-mail: nabil.derbel@enis.tn

Digital Research Centre of Sfax, University of Sfax, Sfax, Tunisia

H. Nouri · B. Ben Atitallah · O. Kanoun
Chair for Measurement and Sensor Technology, Technische Universität Chemnitz, Chemnitz, Germany
e-mail: olfa.kanoun@etit.tu-chemnitz.de

O. Kanoun and N. Derbel (eds.), *Advanced Systems for Biomedical Applications*, Smart Sensors, Measurement and Instrumentation 39,
https://doi.org/10.1007/978-3-030-71221-1_1

1 Introduction

To fulfil a cost-effective and good quality healthcare, long-term and continuous personal health monitoring noninvasively is required. Several methods such X-ray and Magnetic Resonance Imaging (MRI) are able to perform such a task. Nonetheless, they are lethal and raise question to their general safety. This is added to their lab-exclusivity and medical knowledge requirement, which makes them off-limits to the general end-users. Bioimpedance spectroscopy (BIS) is an alternative enabling in-vitro biological tissue state monitoring (Guermazi 2016; Guermazi et al. 2014; Bouchaala et al. 2015) and a point-of-care monitoring during patients' daily routine heath evaluations for medical diagnosis (Barsoukov and Macdonald 2005; Cheung et al. 2005; Tränkler et al. 2007; Grimnes and Martinsen 2000), whose gap to end-users is shortened thanks to the development of portable and low cost embedded systems, which offer fast response and reliable results (Tränkler et al. 2007; Gordon et al. 2005; Pliquett et al. 2000; Min et al. 2004; Fu et al. 2018; Lisdat and Schäfer 2008). Several bioimpedance devices are designed for a wide frequency range (Munjal et al. 2019; Bouchaala et al. 2013, 2015). While designing a multifrequency bioimpedance device, the voltage controlled current source is added to ensure the safety of patients. The latter provides a stable of the current amplitude over different frequencies and different biological tissues (Bouchaala et al. 2016). This technique has been widely used on in-vivo measurements in both clinical and research applications. The foremost applications in personal health and biological tissues monitoring are fluid and body mass changes, tissue and cell characterization and dynamically variable biological systems assessment such as cardiac and lung function monitoring (Fig. 1).

Bioimpedance spectroscopy is performed by exciting a constant amplitude signal wave voltage or current and measuring the current or voltage at each frequency (Grimnes and Martinsen 2014). The spectra are determined either by obtaining the impedance real and imaginary parts, or by directly obtaining its modulus and phase. Current flows through a tissue. At high frequencies, the cell membrane is short circuited making it invisible to the current. Therefore, current proceeds in straight lines. In the other hand, at low frequencies, the membrane impedance is very high that direct the current to flow only through the extracellular fluid. The main constituent of

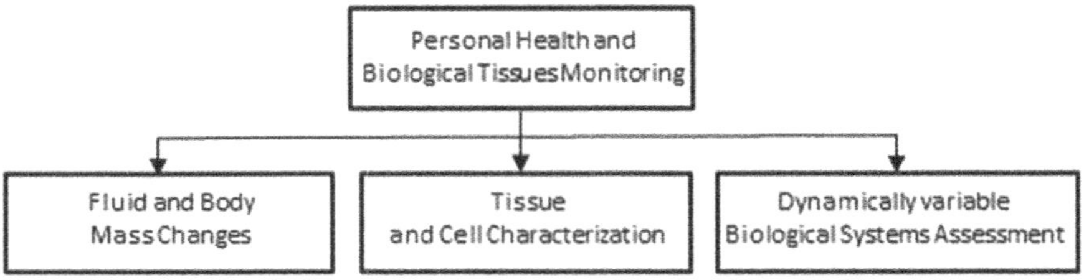

Fig. 1 Impedance spectroscopy for biological tissues monitoring

tissue are the cells comprising an outer membrane, which separates the intracellular fluids and the cell nucleus from its external environment. The intracellular fluids together with the nucleus fluids make up about 95% of the tissue contents, which make the bulk tissue a good electrical conductor. The membrane is composed of two layers of lipids into which proteins are inserted. The lipid is an insulator, which makes the membrane a poor electrical conductor.

Studying the health of the cells infers the study of the outer cells plus the inner part. This is roughly translated to a frequency sweep from Hz range to GHz range, which are often split into three overlapped regions, each expresses a different dispersion mechanism: α region, β region and γ region (Grimnes and Martinsen 2010; Schwan 1957, 1994; Simini and Bertemes-Filho 2018). For tissues cells, α region could also be correlated with the Sarcoplasmic reticulum. When skin is involved (e.g. measured), Stratum corneum's impedance dominates the spectrum (Li et al. 2019), which involves both living and dry tissues therein, as referred (Grimnes and Martinsen 2000). The β region corresponds to 100 kHz to MHz range (1 MHz to 100 MHz (Kuang et al. 1998)). It is influenced by the cellular structure of tissues. In the Impedance Spectroscopy and according to Schwan (1994), the cellular structures' relaxation times depend on the capacitance of the cell membrane and radius, in addition to the fluid, modelled as a resistance. Finally, the γ region corresponds to 100 MHz to GHz range and is mainly correlated to Debye equation, the water relaxation inside the membrane cells, plus several other contributions from the amino acids and various polar proteins (Schwan 1994). The GHz frequency range associated AC signals could also penetrate the tissues. While bones and organs' impedance is spread throughout the spectrum (Bronzino 1999), when external electrodes are used, they would not be visible outside gamma or higher frequency range. Therefore, bioimpedance in tissue could be seen in two different perspectives: Macroscopically and in the case of external electrodes, the signals' wavelength is correlated with the depth of the penetration. Therefore, surface or penetrating electrodes can be used to investigate organs such as liver, ageing of muscles (Dastjerdi et al. 2016; Guermazi et al. 2013). In order to be able to bypass the skin, a frequency in the β region is necessary. On the other hand, in the molecular level, in the case of penetrability of the signals in the tissue level, each of regions is able to extract an information about the tissue' cells using outer-cellular or intracellular or a combined frequency range.

2 Monitoring of Fluid and Body Cell Mass Changes

2.1 Body Composition Analysis

For body composition estimation, bioimpedance analysis (BIA) and bioimpedance spectroscopy (BIS) are most used methods based respectively on single frequency and multifrequency impedance measurement (Smith et al. 2009). BIA presumes that the measured impedance at a single frequency correlates with the intracellular fluid

(ICF) which is related to the fat free mass (FFM). This latter is divided into the extracellular mass (ECM), including extracellular water (ECW) and the structural bone matrix, and body cell mass (BCM) surrounding the intracellular water (ICW) (Andreoli et al. 2011). While, BIS presumes that the measured impedance at several frequencies correlates with the extracellular fluid (ECF) which is related to total body water (TBW) (Ivorra 2003; Jaffrin et al. 2009). This leads to the determination of fat free mass (FFM) and fat mass (FM). Therefore, bioimpedance spectroscopy has various advantages over bioimpedance analysis. It is widely used as a non-invasive and rapid technique in body composition analysis of healthy patients (Kalvoy et al. 2008; Matthie 2008; Lichtenbelt et al. 1994; Barbosa-Silva and Barros 2005) and nutritional status assessment particularly in cancer patients (Wieskotten et al. 2008; Sarhill et al. 2003; Thibault et al. 2011). In fact, body composition assessment is a key factor for nutritional status assessment and malnourished patients' identification.

2.2 Assessment of Dry Weight and Body Hydration States

Bioimpedance spectroscopy is a potential method for fluid changes monitoring in patients after gastric bypass surgery (Mager et al. 2008) and for patients with disease progression in cardiac failure and end-stage renal disease (Davenport et al. 2013; Machek et al. 2010; Onofriescu et al. 2014; Baek et al. 2014; Su et al. 2011; Oh et al. 2018; Parmentier et al. 2013; Sipahi et al. 2004; Davison et al. 2009; Yilmaz et al. 2014; Luo et al. 2004; Biesen and Biesen 2011; Aguiar et al. 2015; Devolder et al. 2010). For end-stage renal patients, medical diagnosis based on impedance spectroscopy are used in order to estimate clinical hydration states in patients under renal replacement therapy in both hemodialysis and peritoneal dialysis. Many studies are performed to evaluate the normal hydration states (dry weight) for hemodialysis (HD) patients (Wystrychowski et al. 2007; Jaeger and Mehta 1999; Chamney et al. 2002; Passauer et al. 2010) and to estimate body hydration states for peritoneal dialysis (PD) patients (Zhu et al. 2008; Arroyo et al. 2015).

2.3 Blood, Glucose and Uric Acid Monitoring

Impedance spectroscopy is used also to characterize serum and blood (Dai et al. 2009; Pradhan et al. 2012; Addabbo et al. 2019; Bohli et al. 2017) and to assess blood glucose for diabetes patients (Andersen et al. 2019; Pedro et al. 2020). Researchers from China University focused on developing an electrochemical analyser using smartphone for real-time uric acid characterization (Guo et al. 2016). A drop of 3 μL of whole blood is applied on the electrochemical sensor for uric acid evaluation which is also compared with sophisticated biochemical analyser. Results show a good correlation between the biochemical analyser and medical smartphone with correlation coefficient of 0.969 varying from 100 to 600 μM/L which can be considered as reliable and accurate results. However, the blood cells can influence the results as it affects the mass transfer of UA molecules from the bulky solution to the electrode surface. This device can ensure not only precise and accurate results but also patient friendly device that can be also connected online with the doctor. The same device is associated with test strip to characterize blood ketone to evaluate the diabetic ketoacidosis (DK) and diabetic ketosis acid (DKA) (Guo 2017). The results using smartphone were compared with biochemical analyser and the evaluation shows good accuracy for blood ketone assessments. Both applications show a good performance of the miniaturized electrochemical analyser. Furthermore, it is possible to store and upload the data which can be evaluated by doctors for better recovery. Moreover, researchers enhanced their sensor for simultaneous detection of glucose and uric acid which can be used for patients having both gout and diabetes (Guo and Ma 2017). The developed sensor has two channels, the first channel located above the substrate to detect glucose in blood and the second channel located on the bottom of the substrate to detect the uric acid. These results were also compared with biochemical analyser and it proves a good accuracy. Therefore, pain less and patient friendly device is successfully developed with comparable results of the clinical biochemical analyser. More investigation should be done in order to design a complete portable monitoring device comparable to the existing healthcare products in the market.

3 Tissue Characterization and Cell Growth Monitoring

3.1 Tissue Characterization

Tissue characterization was firstly investigated based on bioimpedance spectroscopy by Herman P. Schwan in 1984. In this study (Schwan 1984), several topics are addressed: Dielectric properties of tissues from low frequencies up to 20 GHz, identification of the responsible mechanism of dielectric relaxation such us Maxwell-Wagner effect and relaxation of water molecules, ultrasonic properties of biopolymers and influence of electromagnetic field are also studied (Table 1).

In Gersing (1998), ischemia can be studied in a canine heart muscle and a porcine liver using bioimpedance spectroscopy because the accumulation of metabolic components and cell swelling can easily have detected by this method. The extracellular pathways become smaller because of the gap junction and the cell swelling and this phenomenon can be seen for frequencies up to 1 kHz. At high frequencies higher than 1 MHz, the ions content can be detected. Results show that the organ remains revival until the increase of slope phase in resistance but when the plateau after the slope is reached the recovery is not effective and the revival is no more possible. Human and animal tissues are studied from 10 Hz to 20 GHz to extract dielectric properties (Gabriel et al. 1996). The electrode polarization has significant effect on

Table 1 Impedance spectroscopy for tissue characterization

Reference	Frequency range	Studied parameters	Subjects	Remarks	Application
Schwan (1984)	DC – 20 GHz	• Resistance • Capacitance • Dielectric constant • Complex dielectric constant • Conductivity	–	–	Tissue charact.
Gabriel et al. (1996)	10 Hz–20 GHz	• Permittivity • Conductivity	• Human • Animals	Open-ended co-axial probes	Tissue diagnostic: • Heart Muscle • Kidney • Tone • Liver
Gersing (1998)	1 Hz–10 MHz	• Impedance • Real part of impedance • Imaginary part of the impedance	• Canine heart muscle • Porcine liver	Samples are placed on surface and electrodes on the floor of the chambers	Study of ischemia

the measurements at low frequencies especially lower than 100 Hz. After correction of data in order to eliminate the polarization effect, the results were more correlated to the literature results.

3.2 Muscle State Assessment

Numerous studies choose single frequency or multifrequency bioimpedance analysis as a non-invasive technique to examine muscle injury and health health (Table 2). It has successfully used for muscle injure and recovery (Nescolarde 2011, 2014). In Nescolarde (2011), segmental and whole-body impedance are measured for 14 athletes' half of them are from basketball team and the rest are from football team. All measurements are taken before the training for one frequency of 50 kHz. Bioelectrical impedance vector analysis (BIVA) method is selected to assess the whole-body impedance and 6 segmental impedance vectors of the main muscular groups in the lower- limbs in order to analyse the tolerance ellipses of the athletes. The impedance vector is standardized by the height H, Z/H, for each segment. In this study, three

Table 2 Impedance spectroscopy for muscle assessment

Reference	Frequency range	Studied parameters	Subjects	Tissue type	Application
Nescolarde (2011, 2013, 2014, 2015)	50 kHz	• Resistance • Reactance • Phase angle of impedance • Standardized impedance (by height)	Injured athletes of basketball and football	• Lower limb • Quadriceps • Hamstring • Calf muscles	Assessment of muscle injury and its symmetry: • Before injury • After injury • During the recovery • Classification of injuries
Bartels et al. (2015)	Multi-frequency 10 kHz–10 MHz	• Impedance • Resistance • Reactance • Membrane capacitance • Centre frequency • Phase angle • Extracellular resistance • Intracellular resistance	8 Individuals	Gastrocnemius region of each leg	Classification of three levels of skeletal muscles: • Anatomical • Physiological • Metabolic state
Bartels et al. (2019)	4 kHz–1 MHz	• Impedance • Resistance • Reactance • Membrane capacitance • Centre frequency • Phase angle • Centre frequency • Phase angle • Extracellular resistance • Intracellular resistance	Fifty healthy subjects aged 20–69 years: • 25 men • 25 women	• Lower extremities • Upper extremities • Upper back	Classification of muscle health with gender and age

(continued)

Table 2 (continued)

Reference	Frequency range	Studied parameters	Subjects	Tissue type	Application
Harrison et al. (2015)	4 kHz–1 MHz	• Impedance • Resistance • Reactance • Membrane capacitance • Centre frequency • Phase angle • Centre frequency • Phase angle • Extracellular resistance • Intracellular resistance	10 Horses	• Hip • Back • Neck	Four levels of horse's muscles health: • Healthy muscle • Injured muscle • Training • Re-training
Freeborn and Fu (2018)	10 kHz–100 kHz	• Equivalent model parameters: R_1, R_∞, f_p	–	Biceps	Prior and post fatigue exercise

hypotheses are considered. Firstly, it is estimated that the bioimpedance vector does not have the same pattern for all sports. Secondly, professional and well trained athletes have symmetrical muscle groups for right and left side. Finally, a change in the bioimpedance vector is acquired between normal and injured muscle. Hoteling's T2 test is used for the results evaluation and comparison between different complex vectors. A significant difference in bioimpedance vectors ($p < 0.05$) is observed according to the team sport and also between normal and injured muscle which produces hyper-hydration. Furthermore, segmental bioelectrical impedance vector analysis could be a promised method in the assessment of muscle injury and its symmetry of lower-limbs muscles could detect the changes in hydration and muscular structure. Moreover, localized bioimpedance analysis (BIA) is used as technique for the assessment of injuries to lower limb muscles which are common injuries for football players in Nescolarde (2013), the aim of the measurements is the assessment of cell membrane integrity and soft tissue hydration in non-invasive way. The measurements are performed for single frequency of 50 kHz on three football players before injury, after injury and during the recovery period. Three parameters, resistance (R), reactance (X_c) and phase angle (P_A), are considered to study the change in BIA in three different muscles, quadriceps, hamstring and calf muscles, with different injury degree. All parameters decrease with increasing the severity of muscle injury. Thus, the decrease on the resistance (R) can be associated to the accumulation of localized fluid, and the decrease of the reactance (X_c) and phase angle (P_A) reflect the perturbation of cellular membrane integrity and injury. Researchers conclude the potential of localized BIA measurements on the detection of soft tissue injury and also its severity. The localized bioimpedance is also used in another study three types of injuries, myositis ossificans, intramuscular seroma and trochanteric (hip) bursitis, for four football players (Nescolarde 2014). Results shows that the relative variations of resistance (R) and reactance (X_c) at the injury time is overlapped with the non-injured values. On myositis ossificans, localized BIA shows a 7–8 % decrease in Xc but negligible change of R which can be explained by the accumulation of fluid in the soft tissue cavity. Furthermore, a significant change on both resistance and reactance of the localized BIA acquires between post-injury and non-injured side due to the soft tissue destruction and interstitial fluid accumulation. This technique can be useful not only for classification of soft tissue injuries but also the understanding of pathophysiology and structural impairments of other kind of injuries. In Nescolarde (2014), three degree of injuries according to severity, grade I, II and III, are classified for the lower limb using tetrapolar localized bioimpedance analysis at 50 kHz and magnetic resonance imaging (MRI). The measurements are performed after 24 h of the injury in 21 injuries in the quadriceps and hamstring and calf. The aim is to establish a pattern between the change in BIA parameters, resistance (R) as indicator of fluid and reactance (X_c) and phase angle (P_A) as cell structure integrity, and MRI results. After 24 h, only reactance Xc has a significant decrease ($p < 0.001$) especially between no injured contralateral muscle and grade III injuries. However, the fluid distribution is not proportional to the severity degree which explains the small change in the resistance R. Only the reactance Xc indicates the disruption of soft tissue structure and it is proportional to the severity of the injury. In previous studies, only one frequency is extracted from muscle which may limit the analysis of the muscle in different states.

A multi frequency bioimpedance analysis from 10 kHz to 10 MHz is used for the assessment of particular muscle at the fibre level, body mass and muscular health assessment (Bartels et al. 2015). In this work, three levels of skeletal muscles, anatomical, physiological, and metabolic state are extracted using the bioimpedance measurements. The measurements are performed on eight healthy individuals and three of them are used for further investigation in training in sport. The following bioimpedance parameters are extracted from the measurement: Impedance, resistance, reactance, membrane capacitance, centre frequency, phase angle, and both extracellular and intracellular resistance. The combination of different parameters can be used as indicator of contracted state, cell metabolic activity and contralateral muscle loss. In Bartels et al. (2019), several relaxed muscle categories are analysed and compared according to the age and gender using impedance measurements to create a data set. The measurements are performed from 4 kHz to 1 MHz from the muscles of the hand, the upper back and the lower and upper extremities on fifty healthy individuals half of the men and half women aged from 20 to 60 years old. The extracted parameters to be analysed are resistance, reactance, impedance, centre frequency, phase angle, membrane capacitance and external and internal resistance. Results shows no significant change with age but significant change with gender where women have higher centre frequency and intracellular and extracellular resistance but lower membrane capacitance. In Harrison et al. (2015), researchers focus on four levels of a horse's muscles health: Healthy muscle, injured muscle, training and re-training. Same parameters extracted by (Bartels et al. 2019, 2015) are used in this study. Four points impedance measurements are carried out from 4 kHz to 1 MHz on 10 horses where 5 of them with known anamnesis. Multi frequency bioimpedance analysis is a suitable tool from muscle diagnosis and also effects of treatment monitoring. In Freeborn and Fu (2018), different approach is considered compared to the other works about muscle assessment. In order to simplify the data taken from 10 kHz to 100 kHz, an equivalent model of the impedance spectrum, known as Cole-Cole model, is considered. The data are collected from biceps tissue in two levels immediately before and after completion of a fatiguing exercise protocol. All parameters of the equivalent model, R_1, R_∞, f_p, show significant change between prior and post fatigue exercise.

3.3 Cerebral Monitoring

Bioimpedance spectroscopy has been utilized in the detection and monitoring of stroke, cerebral ischemia and haemorrhage (Table 3). Hypoxia/ischaemia is one of the most common diseases of brain damage in infants. This disease consists of the alteration on the concentration of intracellular and extracellular ions which modifies the electrical properties of the tissue.

Therefore, multifrequency bioimpedance is identified as a non-invasive method to diagnose the poor neurodevelopmental by identifying cerebral oedema. In Lingwood et al. (2003), the measurements are carried out from 4 to 1012 kHz on piglets in order to identify a relationship between the severity of a hypoxic/ischemic episode, and cerebral bioelectrical impedance. Based on the results, individuals can be classified into two categories: Individuals suffering of a mild, acute hypoxic episode and individuals suffer from a severe hypoxic episode. More impedance data are considered in Seoane et al. (2005).

Measurements are taken from 20 to 750 kHz and resistance and reactance are used to study the effects of the hypoxia on the perinatal brain in comparison with simulation results of a suspension of cells during cell swelling. The resistance is more sensitive at low frequencies than at high frequency but the reactance considerably changes in the whole frequency range during hypoxia. Furthermore, coherent results are found compared to the simulation results which leads to the development of clinical non-invasive and portable tool based on bioimpedance spectroscopy. Some limitations should be considered such as the effect of frequency on sensitivity of measurements, the effect of skull on the surface measurement and the suitable frequency points that should be considered. Stroke is another type of cerebral disease causing death or permanent disability. The latter is mainly diagnosed using neuroimaging and monitored in the pre-hospital stage using a portable bioimpedance measurement device [Atefi2015]. Finite Element Method (FEM) is used for preliminary study to check the electrical potential distribution on the head. The simulation results prove

Table 3 Impedance spectroscopy for cerebral monitoring

Reference	Frequency range	Studied parameters	Subjects	Remarks	Application
Lingwood et al. (2003)	4 kHz–1012 kHz	• Impedance	Piglets	• Shaved scalp of the piglet	Classification of hypoxia episode:
				• 4 mm either side of the midline	• Control
				• One pair above the eyes	• Mild
				• One pair in the occipital region	• Severe
Seoane et al. (2005)	20 kHz–750 kHz	• Resistance	New born piglets	• 2 Electrodes on the right side of brain	Comparison between electrical parameters and simulation of a suspension of cells during cell swelling
		• Reactance		• 2 Electrodes on the left side of brain	
Atefi (2007)	3 kHz-1 MHz	• Resistance	10 Patients:	• 12 electrodes on the head	Classification of 3 phases of stroke:
			• 5 females		• Healthy
			• 5 males		• Haemorrhagic
					• Ischemic stroke

that Intra-Cranial Haemorrhage (ICH) creates larger left-right potential compared to healthy model. Only the resistance of the bioimpedance measurement is considered in this work. These parameters have smaller amplitude of haemorrhagic stroke patient than in healthy patients while ischemic stroke patients have a larger resistance. The bioimpedance measurements have the potential to classify stroke into 3 groups: Healthy, haemorrhagic and ischemic stroke. However, age and sex are influencing factors that should be considered while analysing the impedance data.

3.4 Cancer Diagnosis

Bioimpedance spectroscopy is used to analyse and differentiate between skin cancer and benign nevi (Aberg et al. 2004, 2005). In β-dispersion range, up to 10 MHz, the current flows in the extracellular and intracellular fluids and also the insulating membrane. Thus, any change in the fluid is reflected in impedance measurement. The electrical properties are affected by several aspects of cell such as the cell shape, cell structure and cell membrane composition but can be differentiated from the impedance of cancer cells.

Typically, the surface bioimpedance measurements are performed by taken cancerous skin and normal skin as a reference (Table 4). 100 skin cancers and 511 benign nevi are used in this study and the reference is measured ipsi-laterally to the lesions. A specificity higher than 75 % is found to distinguish no melanoma skin cancer and malignant melanoma from benign nevi. The results are promising to deliver more sensitive results than conventional visual screening. However, a high care should be addressed because the impedance varies with the electrode location, gender and age. In Gersing et al. (2003), the study of tumours on rats during treatment with photodynamic therapy (PDT), localized hyperthermia (HT) and combination of both are investigated using impedance spectroscopy from 37 Hz to 3.7 MHz. Results reveal that HT generates an increase in the conductance specially at high frequencies caused by the increase of extracellular fluid volume.

Moreover, with PDT, the conductivity at high frequency increases with the start of irradiation and a decrease in extracellular space index proves the development of intracellular oedema. Colorectal cancer is also diagnosed with bioimpedance spectroscopy (Gupta et al. 2008). The aim is to compare the phase angle derived from the impedance measurement and Subjective Global Assessment (SGA), which is an indicator of nutritional status, in advanced colorectal cancer. The evaluation is made by collecting data from 73 patients classified into two categories, well-nourished or malnourished, using the SGA. The Spearman correlation coefficient is used for the evaluation. The results show higher phase angles with better nutritional status. The phase angle can be a potential indicator in colorectal cancer. However, further investigations are needed to find the optimum cut-off level of bioimpedance phase angle revealing better nutritional evaluation and monitoring. Multifrequency impedance, from 10 to 100 kHz, is examined in rats for normal liver tissue and liver cancer tissues (Salazar-Anguiano et al. 2008). Hepatocellular carcinoma (HCC) is associated to the real part of impedance and has no significant change on imaginary part of impedance. The Cole model parameters are also considered to analyse the relationship between the cancer and multi-frequency impedance.

Table 4 Impedance spectroscopy for cancer assessment

Reference	Frequency range	Studied parameters	Subjects	Electrodes positions	Application
Salazar-Anguiano et al. (2008)	10–100 kHz	• Cole parameters • Real part of impedance • Imaginary part of impedance	Rat liver	• Tetrapolar electrode • Placement in the liver	Liver cancer
Aberg et al. (2004)	1 kHz–1 MHz	• Magnitude of the impedance • Phase of the impedance	374 Patients	• Non-invasive probe with 4 concentric electrodes • Placement over the center of the lesion	Differentiation between: • Skin cancer • Benign lesion
Gersing et al. (2003)	37 Hz–3.7 MHz	• Real part of impedance • Imaginary part of impedance	Rats	• 2 Stainless steel needles • Insertion of electrodes into the tumour	Study of tumors on rats: • During treatment with photodynamic therapy • Localized hyperthermia

3.5 Monitoring of Cell Growth

Biological tissues are an association of microstructures with non-linear signal behaviour. The optical methods are commonly chosen to extract cells information as it is sensitive and efficient method. Besides the existing biotechnological processes, a reliable and real-time measurement should be developed for the operation of bioreactors. Real time monitoring of growth, cell density and physiological state can be developed based on impedance of interdigitated electrode structures (IDES) (Ehert et al. 1997).

The influence of serum components and the toxicity from heavy metals can be detected using this method (Table 5). In this work, the cells were cultured for 14 days on the electrode. The main factor of impedance change is the insulting property of the cell membrane. Moreover, using IDE, it is possible to study more complex factors such the shape and morphology of cells. However, a complete information of cells needs pH measurement with the impedance.

Table 5 Impedance spectroscopy for cell growth monitoring

Reference	Frequency range	Studied parameters	Subjects	Electrodes	Application
Ehert et al. (1997)	100 Hz	Admittance	CV- 1 cell from the kidney of a male adult African green monkey	• Interdigitated electrodes (IDES)	Detection of serum component and heavy metals
	1 kHz	• Capacitance			
	10 kHz				
	100 kHz				
Soley et al. (2005)	10 kHz–10 MHz	• Impedance	Yeast	• Teflon tubular electrode	Estimation of cell concentration
		• Capacitance	Saccharomycs cerevisiae		
Cheng et al. (2007)	100 Hz	• Impedance	Blood mononuclear cells	• Platinium probe	Enumerate immobilized cells
	1 MHz	• Conductance			

More frequency range is considered in Soley et al. (2005) to study the relative variation of measured impedance at 10 kHz and 10 MHz of the growth monetarization of a Saccharomyces cerevisiae culture. The aim is to characterize the influence of chemical and physical components on the impedance measurement with in-situ and ex-situ probe. The in-situ probe delivers good results but sensitive to agitation which can be overcomes with ex-situ probe to have more robust measurements. Results shows a good correlation between measured and estimated values with coefficient of 0.99. Thus, this technique has the potential to estimate the cell concentration, however it is sensitive to agitation. The authors suggest ex-situ probe or external flow-through system to overcome this problem. The change of conductivity provides information about the number of cells due to the change of the surrounding medium (Cheng et al. 2007). Results shows a linear increase on the conductance of bulk solution with the number of isolated cells with an accuracy of 20 cell μL^{-1}. We can successfully isolate specific cell types from a complex fluid but also quantifying them with impedance measurement at one frequency of 760 Hz. This approach is the most sensitive non-optical approach to count the immobilized cells

4 Monitoring of Dynamically Variable Biological Systems

Respiratory and cardiac motion artefacts work together to ensure the suitable blood circulation debit and the necessary oxygen quantity throughout the body, which human needs for biological cells to live and function properly. A lot of diseases e.g Heart Failure, Abnormal Heart Rhythms, Chronic Bronchitis, Pneumonia, can affect the biological system. In this direction, bio-impedance measurements have been studied in the detection of the internal motion of organs due to respiration and cardiac contractions, which can lead to better diagnosis of the mentioned diseases. However, the time-varying bio-impedance is recorded based on single or multi-frequency measurements as for example, impedance cardiography (Kubicek et al. 1966), impedance pneumography (Koivumäki et al. 2012) or pleural effusion in the lungs (Giovinazzo et al. 2011).

4.1 Cardiovascular Diseases Prevention

4.1.1 Early Detection of Congestive Heart Failure

Currently, cardiovascular diseases have become one of the major threats to human life. Based on the last statistical analysis carried out in 2016 by the World Health Organization (WHO), cardiovascular diseases are causing 31% of all global deaths, which is estimated at 17.9 million people. According to the related investigation, 85% of these patients' death is caused by heart failure and stroke. Hence, these cardiovascular diseases are preferably avoided in the early stage. Fu et al. (2018) presented a portable, reliable and low-cost electrochemical analyser integrated into a smartphone to be a blood lipid monitoring device. The proposed system can provide an accurate evaluation of the patient's blood lipid. However, the bio-impedance change during the electrochemical reaction caused by the blood, precisely the total cholesterol and the developed cell, is measured. The produced cell current was calculated from the resolved impedance during the biochemical reaction and to be mapped into a blood total cholesterol quantity. The measured biochemical parameter was stored into the smartphone and be updated with the users' personalized health data centre. Point of care monitoring the blood lipid level is capable of making a positive contribution to the prevention of cardiovascular disease. The proposed medical smartphone system is a promising platform as a device for blood total cholesterol monitoring, which can be applied for long-term prevention of cardiovascular disease due to its internet-based medical data interaction. Yu et al. (2005) present a reliable method for heat failure prediction based on the measured intrathoracic impedance to identify potential fluid overload before hospitalization. Based on this study, regular monitoring of the intrathoracic impedance may provide an early warning of a heart failure occurrence. In the same context, Darling et al. (2017) proposed a novel, non-invasive, wearable fluid accumulation vest for measuring the real-time transthoracic bio-impedance change. Bluetooth communication is established for the transmission and recording of the data within a smartphone. 106 participants were asked to wear the proposed system and transmit the recorded bio-impedance for 45 consecutive days to be treated later using a developed algorithm as a predictor of heart failure events. The analysing thoracic bio-impedance algorithm shows 72% accuracy and 87% sensitivity for identifying recurrent heart failure events. Moreover, Dovancescu et al. (2014) proposed a wearable monitoring device for improved home-based disease management in congestive heart failure. The assessment of the thoracic fluid status is followed from the bioimpedance spectroscopy measurements.

4.1.2 Determination of Myocardial State

Impedance is a useful parameter for determining the properties of biological tissues due to their electrical conductivity. Based on the extracted parameters, physiological events can be analyzed and investigated, in particular those related to the myocardial state (Bragos et al. 1996). Since biological tissue cells are equivalent to electronic components, some functional lesions of human organs and changes in biological cell activity can lead to changes in the electrical properties of biological tissues. Yufera et al. (2005) proposed a microelectronic system for biological impedance measurements and tissue characterization. Tests are performed in vivo conditions showed the feasibility of the system as a useful integrated circuit for impedance measurements. Mathematical methods can be used to adjust the measured impedance data to cole-cole models and study the main parameters of ischemia in different organs, especially in the heart muscle level. Casas et al. (1999) demonstrate that the best Cole parameters to characterize the ischemia are R0 and fc. Moreover, electrical bioimpedance shows the potential to determine whether ischemic and reperfusion damage in cardia surgery (Mellert et al. 2010). In the same context, Warren et al. (2000) prove that bioimpedance measurements allow identification of areas of healed myocardial infarction by measuring myocardial electrical impedance with an intracavitary contact electro catheter, which helps to detect the reversible myocardial ischemic injury. Sanchez et al. (2011) are investigating myocardium tissue impedance to recognize the myocardial state using the electrical impedance spectroscopy approach. A broadband signal is developed to collect simultaneous impedance data with a multi-sine excitation signal. To investigate the dynamic time-varying properties of in-vivo and ex-vivo tissue, a novel approach is presented for the impedance-frequency response estimation based on the local polynomial method (LPM). One important aspect that should be mentioned, is the high complexity structure of the myocardial fibre orientation and the distribution of gap junctions, which can lead to inaccurately measured impedance spectra. Salazar et al. (2000) carried out a comparative study of two measurement methods to investigate the differentiation among normal, ischemic, and infarcted myocardial tissue. Using both methods the transmural and non-transmural measurements using respectively an intracavitary catheter and a four-needle probe inserted into the epicardium, the in-situ provides a specific impedance spectrum (magnitude and phase angle) of normal, ischemic, and infarcted pig myocardium tissue from 1 kHz to 1 MHz.

4.2 *Lung Tissue Characterization*

The lungs are the primary organs of the respiratory system in humans, which make it necessary to investigate deeply the abnormal changes commonly known as the pleural effusion using electrical impedance spectroscopy. A pleural effusion is an excess fluid that accumulates in the pleural cavity, the fluid-filled space that surrounds the lungs. This excess fluid can impair breathing by limiting the expansion of the lungs. Various kinds of pleural effusion, depending on the nature of the fluid and what caused its entry into the pleural space, are hydrothorax (serous fluid), hemothorax (blood), urinothorax (urine), chylothorax (chyle), or pyothorax (pus) (Sanchez et al. 2013). They proved that the bioelectrical impedance is changing in correlation with the thoracic fluid level for patients with pleural effusion after thoracentesis (Zink et al. 2015). A significant increase in bioelectrical impedance is detected for 45 participants during the low-frequency domain for the transthoracic vector after thoracentesis. Besides, Zerahn et al. (1999) demonstrate that the increase of the baseline impedance is always correlating with the increase of the lung volume after thoracentesis but it is depending on the primary disease and reflecting the overall fluid content of the thorax. Where, fifteen patients with pleural effusion due to malignant or cardiac diseases were asked to perform measurements. Fein et al. (1979) evaluate the effect of pulmonary oedema disease using transthoracic electrical impedance. A comparison study is carried, where 27 normal subjects and 33 patients were selected. Analysis results show that single measurements of impedance are useless in diagnosing pulmonary oedema.

5 Conclusion

The potential of bioimpedance spectroscopy is illustrated by various applications in medical field. It represents a low cost, non-invasive and safe technique. It is widely used in body composition analysis of healthy patients, fluid changes monitoring in patients after gastric bypass surgery and for patients with disease progression in cardiac failure and end-stage renal disease and nutritional status assessment particularly in cancer patients. For end-stage renal patients, medical diagnosis based on impedance spectroscopy are used in order to estimate clinical hydration states in patients under renal replacement therapy in both haemodialysis and peritoneal dialysis. Furthermore, impedance spectroscopy is used also to characterize tissues, serum and blood and to assess blood glucose for diabetes patients, examine muscle injury and health and detect cerebral ischemia and haemorrhage. Also, bio-impedance measurements have been studied in the detection of the internal motion of organs due to respiration and cardiac contractions, which can lead to better diagnosis of different diseases such as heart failure, abnormal heart rhythms, chronic bronchitis, pneumonia, can affect the biological system.

References

Aberg, P., Nicander, I., Hansson, J., Geladi, P., Holmgren, U., & Ollmar, S. (2004). Skin cancer identification using multifrequency electrical impedance-a potential screening tool. *IEEE Transactions on Biomedical Engineering, 51*(12), 2097–2102.

Aberg, P., Geladi, P., Nicander, I., Hansson, J., Holmgren, U., & Ollmar, S. (2005). Non-invasive and microinvasive electrical impedance spectra of skin cancer- a comparison between two techniques. *Skin Research and Technology, 11*(4), 281–286.

Addabbo, T., Fort, A., Mugnaini, M., Parri, L., Pinzi, M., Vignoli, V., et al. (2019). On the suitability of low-cost compact instrumentation for blood impedance measurements. *IEEE Transactions on Instrumentation and Measurement, 68,* 2412–2424.

Aguiar FN: P. V. , et al. (2015). Overhydration prevalence in peritoneal dialysis - A 2-year longitudinal analysis. *Nefrologia, 35,* 189–196.

Andersen, J. H., et al. (2019). Bioimpedance and NIR for non-invasive assessment of blood glucose. *Journal of Electrical Bioimpedance, 10,* 133–138.

Andreoli, A., De Lorenzo, A., Cadeddu, F., Iacopino, L., & Grande, M. (2011). New trends in nutritional status assessment of cancer patients. *European Journal of Medical and Pharmaceutical Science, 15,* 469–480.

Arroyo, D., et al. (2015). Intraperitoneal fluid overestimates hydration status assessment by bioimpedance spectroscopy. *Peritoneal Dialysis International, 35,* 85–89.

Atefi, S. R. (2007). *Electrical bioimpedance cerebral monitoring: From hypothesis and simulation to first experimental evidence in stroke patients*. Ph.D. Thesis, Royal Institute of Technology KTH, Stockholm, Sweden.

Baek, S. H., et al. (2014). Control of fluid balance guided by body composition monitoring in patients on peritoneal dialysis (COMPASS): Study protocol for a randomized controlled trial. *Trials, 15,* 432.

Barbosa-Silva, M. C., & Barros, A. J. (2005). Bioelectrical impedance analysis in clinical practice: A new perspective on its use beyond body composition equations. *Current Opinion in Clinical Nutrition and Metabolic Care, 8,* 311–317.

Barsoukov, E., & Macdonald, J. R. (2005). *Impedance spectroscopy: Theory, experiment, and applications* (p. 2005). New York: Wiley.

Bartels, E. M., Sorensen, E. R., & Harrison, A. P. (2015). Multi-frequency bioimpedance in human muscle assessment. *Physiological Reports*.

Bartels, E. M., Andersen, E. L., Olsen, J. K., Kristensen, L. E., Bliddal, H., Danneskiold-Samsoe, B., & Harrison, A. P. (2019). Muscle assessment using multi-frequency bioimpedance in a healthy Danish population aged 20–69 years: A powerful non-invasive tool in sports and in the clinic. *Physiological Reports, 7*(11), 14109.

Van Biesen, W., et al. (2011). Fluid status in peritoneal dialysis patients: The European Body Composition Monitoring (EuroBCM) study cohort. *PLoS One, 6,* 17148.

Bohli, N., Chammem, H., Meilhac, O., Mora, L., & Abdelghani, A. (2017). Electrochemical impedance spectroscopy on interdigitated gold microelectrodes for glycosylated human serum albumin characterization. *IEEE Transactions on NanoBioscience, 16,* 676–681.

Bouchaala, D., Guermazi, M., Kanoun, O., & Derbel, N. (2015). Portable device design for in-vitro muscle tissue monitoring. *tm-Technisches Messen., 82*(10), 485–494.

Bragos, R., Riu, P. J., Warren, M., Tresilnchez, M., Carreiio, A., & Cinca, J. (1996). Changes in myocardial Impedance spectrum during acute ischemia in the in-situ pig heart. In *IEEE EMBS conference, Amsterdam.*

Bronzino, J. D. (1999). *Biomedical engineering handbook* (Vol. 2). CRC Press.

Casas, O., Bragos, R., Riu, P., Rosell, J., Tresilnchez, M., Warren, M., et al. (1999). In vivo and in situ ischemic tissue characterization using electrical impedance spectroscopy. *Annals of the New York Academy of Sciences, 873,* 51–58.

Chamney, P. W., Kramer, M., Rode, C., Kleinekofort, W., & Wizemann, V. A. (2002). New technique for establishing dry weight in hemodialysis patients via whole body bioimpedance. *Kidney International, 61,* 2250–2258.

Cheung, K., Gawad, S., & Renaud, P. (2005). Impedance spectroscopy flow cytometry: On-chip label-free cell differentiation. *Cytometry, 65A,* 124–132.

Cheng, X., Liu, Y., Irimia, D., Demirci, U., Yang, L., Zamir, L., et al. (2007). Cell detection and counting through cell lysate impedance spectroscopy in microfluidic devices. *Lab on a Chip, 7*(6), 746–755.

Dai, T., & Adler, A. (2009). In vivo blood characterization from bioimpedance spectroscopy of blood pooling. *IEEE Transactions on Instrumentation and Measurement, 58*(11), 3831–3838.

Darling, C. E., Dovancescu, S., Saczynski, J. S., Riistama, J., Kuniyoshi, F. S., Rock, J., et al. (2017). Bioimpedance-based heart failure deterioration prediction using a prototype fluid accumulation vest-mobile phone dyad: An observational study. *JMIR Cardio, 1*(1).

Davenport, A., et al. (2013). Can non-invasive measurements aid clinical assessment of volume in patients with cirrhosis? *World Journal of Hepatology, 5,* 433–438.

Davison, S. N., Jhangri, G. S., Jindal, K., & Pannu, N. (2009). Comparison of volume overload with cycler-assisted versus continuous ambulatory peritoneal dialysis. *Clinical Journal of the American Society of Nephrology, 4,* 1044–1050.

Devolder, I., Verleysen, A., Vijt, D., Vanholder, R., & Van Biesen, W. (2010). Body composition, hydration, and related parameters in hemodialysis versus peritoneal dialysis patients. *Peritoneal Dialysis International, 30,* 208–214.

Dovancescu et al., S. (2014). Detection of electrocardiographic and respiratory signals from transthoracic bioimpedance spectroscopy measurements with a wearable monitor for improved home-based disease management in congestive heart failure. In *IEEE computing in cardiology conference (CinC)* (pp. 985–988).

Ehret, R., Baumann, W., Brischwein, M., Schwinde, A., Stegbauer, K., & Wolf, B. (1997). Monitoring of cellular behaviour by impedance measurements on interdigitated electrode structures. *Biosensors and Bioelectronics, 12*(1), 29–41.

Fein, A., Grossman, R. F., Jones, J. G., Goodman, P., & Murray, J. (1979). Evaluation of transthoracic electrical impedance in the diagnosis of pulmonary edema. *Circulation, 60*(5), 1156–1160.

Freeborn, T., & Fu, B. (2018). Fatigue-induced cole electrical impedance model changes of biceps tissue bioimpedance. *Fractal Fractional, 2,* 27.

Fu, Y., & Guo, J. (2018). fBlood cholesterol monitoring with smartphone as miniaturized electrochemical analyzer or cardiovascular disease prevention. *IEEE Transactions on Biomedical Circuits and Systems, 12*(4), 784–790.

Gabriel, S., Lau, R. W., & Gabriel, C. (1996). The dielectric properties of biological tissues: II. Measurements in the frequency range 10 Hz to 20 GHz. *Physics in Medicine and Biology, 41,* 2251–2269.

Gersing, E., Kelleher, D. K., & Vaupel, P. (2003). Tumour tissue monitoring during photodynamic and hyperthermic treatment using bioimpedance spectroscopy. *Physiological Measurement, 24,* 625–637.

Gersing, E. (1998). Impedance spectroscopy on living tissue for determination of the state of organs. *Bioelectrochemistry and Bioenergetics, 45,* 145–149.

Giovinazzo, G., Ribas, N., Cinca, J., & Rosell-Ferrer, J. (2011). The feasibility of transoesophageal bioimpedance measurements for the detection of heart graft rejection. *Physiological Measurement, 32,* 867–876.

Gordon, R., Land, R., Min, M., Parve, T., & R. W. Salo (2005). A virtual system for simultaneous multi-frequency measurement of electrical bioimpedance. *International Journal of Bioelectromagnetism, 7*, 243–246.

Grimnes, S., & Martinsen, O. G. (2000). *Bioimpedance and bioelectricity basics* (1st ed., p. 2000). Cambridge: Academic.

Grimnes, S., & Martinsen, O. G. (2010). Alpha-dispersion in human tissue. *Journal of Physics: Conference Series. IOP Publishing, 224*, 012073.

Grimnes, S., & Martinsen, O. G. (2014). *Bioimpedance and bioelectricity basics*. Cambridge: Academic.

Guermazi, M. (2016). In-vitro biological tissue state monitoring based on impedance spectroscopy. Ph.D. Thesis, Chemnitz University of Technology (Vol. 3).

Guermazi, M., Kanoun, O., & Derbel, N. (2014). Investigation of long time beef and veal meat behaviour by bioimpedance spectroscopy for meat monitoring. *IEEE Sensors Journal, 14*(10), 3624–3630.

Guo, J. (2017). Smartphone-powered electrochemical dongle for point-of care monitoring of blood SS-ketone. *Analytical Chemistry, 89*(17), 8609–8613.

Guo, J., & Ma, X. (2017). Simultaneous monitoring of glucose and uric acid on a single test strip with dual channels. *Biosensors and Bioelectronics, 94*, 415–419.

Guo, J. (2016). Uric acid monitoring with a smartphone as the electrochemical analyzer. *Analytical Chemistry, 88*(24), 11986–11989.

Gupta, D., Lis, C. G., Dahlk, S. L., King, J., Vashi, P. G., Grutsch, J. F., & Lammersfeld, C. A. (2008). The relationship between bioelectrical impedance phase angle and subjective global assessment in advanced colorectal cancer. *Nutrition Journal, 7*, 19.

Harrison, A. P., Elbrond, V. S., Riis-Olesen, K., & Bartels, E. M. (2015). Multifrequency bioimpedance in equine muscle assessment. *Physiological Measurement, 36*(3), 453–464.

Ivorra, A. (2003). Bioimpedance monitoring for physicians: An overview. Centre National de Microelectronica, Barcelona. Diploma thesis, Hospital Clínic de Barcelona.

Jaeger, J. Q., & Mehta, R. L. (1999). Assessment of dry weight in hemodialysis: An overview. *Journal of the American Society of Nephrology, 10*, 392–403.

Jaffrin, M. Y. (2009). Body composition determination by bioimpedance: An update. *Current Opinion in Clinical Nutrition and Metabolic Care, 12*(5), 482–486.

Kalvoy, H., Martinsen, O. G., & Grimnes, S. (2008). Determination of tissue type surrounding a needle tip by electrical bioimpedance. In *Conference proceedings: Annual international conference of the IEEE engineering in medicine and biology society* (pp. 2285– 2286).

Koivumäki, T., Vauhkonen, M., Kuikka, J. T., & Hakulinen, M. A. (2012). Bioimpedance-based measurement method for simultaneous acquisition of respiratory and cardiac gating signals. *Physiological Measurement, 33*, 1323–1334.

Kuang, W., & Nelson, S. (1998). Low-frequency dielectric properties of biological tissues: A review with some new insights. *Transactions of the ASAE, 41*(1), 1998.

Kubicek, W. G., Karnegis, J. N., Patterson, R. P., Witsoe, D. A., & Mattson, R. H. (1966). Development and evaluation of an impedance cardiac output system. *Aerospace Medicine, 37*, 1208–1212.

Li, Y., Ma, R., Wang, X., Jin, J., Wang, H., Liu, Z., & Yin, T. (2019). Tissue coefficient of bioimpedance spectrometry as an index to discriminate different tissues in vivo. *Biocybernetics and Biomedical Engineering, 39*, 923–936.

Van Marken Lichtenbelt, W. D., Westerterp, K. R., Wouters, L., & Luijendijk, S. (1994). Validation of bioelectrical-impedance measurements as a method to estimate body-water compartments. *The American Journal of Clinical Nutrition, 60*(2), 159–166.

Lisdat, F., & Schäfer, D. (2008). The use of electrochemical impedance spectroscopy for biosensing. *Analytical and Bioanalytical Chemistry, 391*(5), 1555–1567.

Lingwood, B. E., Dunster, K. R., Healy, G. N., Ward, L. C., & Colditz, P. B. (2003). Cerebral impedance and neurological outcome following a mild or severe hypoxic/ischemic episode in neonatal piglets. *Brain Research, 969*(1), 160–167.

Luo, Y. J., Lu, X. H., Woods, F., & Wang, T. (2011). Volume control in peritoneal dialysis patients guided by bioimpedance spectroscopy assessment. *Blood Purification, 31,* 296–302.

Machek, P., Jirka, T., Moissl, U., Chamney, P., & Wabel, P. (2010). Guided optimization of fluid status in haemodialysis patients. *Nephrology, Dialysis, Transplantation: Off. Publ. European Dialysis and Transplant Association - European Renal Association, 25,* 538–544.

Mager, J. R., Sibley, S. D., Beckman, T. R., Kellogg, T. A., & Earthman, C. P. (2008). Multifrequency bioelectrical impedance analysis and bioimpedance spectroscopy for monitoring fluid and body cell mass changes after gastric bypass surgery. *Clinical Nutrition, 27,* 832–841.

Matthie, J. R. (2008). Bioimpedance measurements of human body composition: Critical analysis and outlook. *Expert Review of Medical Devices, 5*(2), 239–261.

Mellert, F., Winkler, K., Schneider, C., Dudykevych, T., Welz, A., Osypka, M., et al. (2010). Detection of (reversible) myocardial ischemic injury by means of electrical bioimpedance. *IEEE Transactions on Biomedical Engineering, 58,* 1511–1518.

Min, M., Kink, A., Land, R., Parve, T., & Rätsep, I. (2004). Modification of pulse wave signals in electrical bioimpedance analyzer for implantable medical devices. In *Proceedings of the 26th IEEE International Conference on Engineering in Biology and Medicine (San Francisco)* (pp. 2263–2266).

Nescolarde, L., Yanguas, J., Medina, D., Rodas, G., & Rosell-Ferrer, J. (2011). Assessment and follow-up of muscle injuries in athletes by bioimpedance: preliminary results. In *Conference proceedings: Annual international conference of the IEEE engineering in medicine and biology society* (pp. 1137–1140).

Nescolarde, L., Yanguas, J., Lukaski, H., Alomar, X., Rosell-Ferrer, J., & Rodas, G. (2013). Localized bioimpedance to assess muscle injury. *Physiological Measurement, 34*(2), 237–245.

Nescolarde, L., Yanguas, J., Lukaski, H., Rodas, G., & Rosell-Ferrer, J. (2014). Localized BIA identifies structural and pathophysiological changes in soft tissue after post-traumatic injuries in soccer players. In *Conference proceedings: Annual international conference of the IEEE engineering in medicine and biology society* (pp. 3743–3746).

Nescolarde, L., Yanguas, J., Lukaski, H., Alomar, X., Rosell-Ferrer, J., & Rodas, G. (2015). Effects of muscle injury severity on localized bioimpedance measurements. *Physiological Measurement, 36*(1), 27–42.

Oh, K. H., et al. (2018). Does routine bioimpedance-guided fluid management provide additional benefit to non-anuric peritoneal dialysis patients? results from compass clinical trial. *Peritoneal Dialysis International, 38,* 131–138.

Onofriescu, M., et al. (2014). Bioimpedance-guided fluid management in maintenance hemodialysis: A pilot randomized controlled trial. *American Journal of Kidney Disease, 64,* 111–118.

Parmentier, S. P., et al. (2013). Influence of peritoneal dialysis solution on measurements of fluid status by bioimpedance spectroscopy. *International Urology and Nephrology, 45,* 229–232.

Passauer, J., Petrov, H., Schleser, A., Leicht, J., & Pucalka, K. (2010). Evaluation of clinical dry weight assessment in haemodialysis patients using bioimpedance spectroscopy: A cross-sectional study. *Nephrology Dialysis Transplantation, 25,* 545–551.

Pedro, B. G., Marcôndes, D. W. C., & Bertemes-Filho, P. (2020). Analytical model for blood glucose detection using electrical impedance spectroscopy. *Sensors, 20*(23), 6928.

Pliquett, U., Gersing, E., & Pliquett, F. (2000). Evaluation of time-domain based impedance measurements on biological tissue. *Biomedical Technology, 45,* 6–13.

Pradhan, R., Mitra, A., & Das, S. (2012). Impedimetric characterization of human blood using three-electrode based ECIS devices. *Journal of Electrical Bioimpedance, 3,* 12–19.

Salazar, Y., Bragos, R., Casas, O., Cinca, J., & Rosell, J. (2000). Transmural versus nontransmural in situ electrical impedance spectrum for healthy, ischemic, and healed myocardium. *IEEE Transactions on Biomedical Engineering, 51,* 1421–1427.

Salazar-Anguiano, J., Chávez-López, M. G., Zúñiga-García, V., Camacho, J., & Elías-Viñas, D. (2018). Resistive part of impedance as a possible indicator of hepatocellular carcinoma. *Archives of Medical Research, 49,* 89–93.

Sanchez, B., Vandersteen, G., Rosell-Ferrer, J., Cinca, J., & Bragos, R. (2011). In-cycle myocardium tissue electrical impedance monitoring using broadband impedance spectroscopy. In *Annual international conference of the IEEE engineering in medicine and biology society (EMBC), 2011*.

Sanchez, B., Vandersteen, G., Martin, I., Castillo, D., Torrego, A., Riu, P. J., Schoukens, J., & Bragos, R. (2013). In vivo electrical bioimpedance characterization of human lung tissue during the bronchoscopy procedure. A feasibility study. *Medical Engineering and Physics, 35*(7), 949–957.

Sarhill, N., Mahmoud, F. A., Christie, R., & Tahir, R. (2003). Assessment of nutritional status and fluid deficits in advanced cancer. *American Journal of Hospice and Palliative Medicine, 20,* 465–473.

Schwan, H. P. (1984). Electrical and acoustic properties of biological materials and biomedical applications. *IEEE Transactions on Biomedical Engineering, 31*(12), 872–878.

Schwan, H. P. (1957). Electrical properties of tissue and cell suspensions. *Advances in Biological and Medical Physics, 5,* 147–209.

Schwan, H. P. (1994). Electrical properties of tissues and cell suspensions: Mechanisms and models. In *Proceedings of 16th annual international conference of the IEEE engineering in medicine and biology society* (Vol. 1, pp. A70–A71).

Seoane, F., Lindecrantz, K., Olsson, T., Kjellmer, I., Flisberg, A., & Bågenholm, R. (2005). Spectroscopy study of the dynamics of the transencephalic electrical impedance in the perinatal brain during hypoxia. *Physiological Measurement, 26,* 849.

Simini, F., & Bertemes-Filho, P. (2018). *Bioimpedance in biomedical applications and research*. Berlin: Springer.

Sipahi, S., et al. (2011). Body composition monitor measurement technique for the detection of volume status in peritoneal dialysis patients: The effect of abdominal fullness. *International Urology and Nephrology, 43,* 1195–1199.

Smith, D., Johnson, M., & Nagy, T. (2009). Precision and accuracy of bioimpedance spectroscopy for determination of in vivo body composition in rats. *International Journal of Body Composition Research, 7,* 21–26.

Soley, A., Lecina, M., Gamez, X., Cairo, J., Riu, P., & Rosell, X., et al. (2005). Online monitoring of yeast cell growth by impedance spectroscopy. *Journal of Biotechnology, 118*(4), 398–405.

Su, W. S., et al. (2011). The fluid study protocol: A randomized controlled study on the effects of bioimpedance analysis and vitamin D on left ventricular mass in peritoneal dialysis patients. *Peritoneal Dialysis International: Journal of International Society of Peritoneal Dialysis, 31,* 529–536.

Thibault, R., Cano, N., & Pichard, C. (2011). Quantification of lean tissue losses during cancer and HIV infection/AIDS. *Current Opinion in Clinical Nutrition and Metabolic Care, 14*(3), 261–267.

Tränkler, H. R., Kanoun, O., Min, M., & Rist, M. (2007). Smart sensor systems using impedance spectroscopy. *Proceedings of the Estonian Academy of Science and Engineering, 13,* 455–478.

Warren, M., Bragos, R., Casas, O., Rodriguez-Sinovas, A., Rosell, J., Anivarro, I., & Cinca, J. (2000). Percutaneous electrocatheter technique for on-line detection of healed transmural myocardial infarction. *Pacing and Clinical Electrophysiology, 23,* 1283–1287.

Wieskotten, S., Heinke, S., Wabel, P., Moissl, U., Becker, J., Pirlich, M., et al. (2008). Bioimpedance-based identification of malnutrition using fuzzy logic. *Physiological Measurement, 29,* 639–654.

Wystrychowski, G., & Levin, N. W. (2007). Dry weight: Sine qua non of adequate dialysis. *Advances in Chronic Kidney Disease, 14,* 10–16.

Yilmaz, Z., et al. (2014). Evaluation of fluid status related parameters in hemodialysis and peritoneal dialysis patients: Clinical usefulness of bioimpedance analysis. *Medicina, 50,* 269–274.

Yu, C. M., Wang, L., et al. (2005). Intrathoracic impedance monitoring in patients with heart failure: correlation with fluid status and feasibility of early warning preceding hospitalization. *Circulation, 112*(6), 841–848.

Yufera, A., Rueda, A., Munoz, J. M., Doldan, R., Leger, G., & Rodriguez-Villegas, E. O. (2005). A tissue impedance measurement chip for myocardial ischemia detection. *IEEE Transactions on Circuits System. I, Regular Papers, 52*(12), 2620–2628.

Zerahn, B., Jensen, B. V., Olsen, F., Petersen, J. R., & Kanstrup, I. L. (1999). The effect of thoracentesis on lung function and transthoracic electrical bioimpedance. *Respiratory Medicine, 93*(3), 196–201.

Zhu, F., Kuhlmann, M. K., Kotanko, P., Seibert, E., Leonard, E. F., & Levin, N. W. (2008). A method for the estimation of hydration state during hemodialysis using a calf bioimpedance technique. *Physiological Measurement, 29,* 503–516.

Zink, M. D., Weyer, S., Pauly, K., Napp, A., Dreher, M., Leonhardt, S., et al. (2015). Feasibility of bioelectrical impedance spectroscopy measurement before and after thoracentesis. *BioMed Research International.*

Munjal, R., Wendler, F., & Kanoun, O. (2019). Embedded wideband measurement system for fast impedance spectroscopy using undersampling. *IEEE Transactions on Instrumentation and Measurement, 69*(6), 3461–3469.

Bouchaala, D., Kanoun, O., & Derbel, N. (2016). High accurate and wideband current excitation for bioimpedance health monitoring systems. *Measurement, 79,* 339–348.

Bouchaala, D., Kanoun, O., & Derbel, N. (2013). Portable bioimpedance spectrometer for total frequency range of β-dispersion. *Technisches Messen, 80*(11), 373–378.

Bouchaala, D., Guermazi, M., Derbel, N., & Kanoun, O. (2015). Portable device design for in-vitro muscle tissue monitoring. *Technisches Messen, 82*(10), 485–494.

Heidary Dastjerdi, M., Kanoun, O., & Himmel, J. (2016). Method to adjust gradiometer for medical applications. *Technisches Messen, 83*(5), 247–256.

Guermazi, M., Kanoun, O., & Derbel, N. (2013). Reduction of anisotropy influence and contacting effects in in-vitro bioimpedance measurements. *Journal of Physics: Conference Series, 434*(012058).

Electrode Design for Reproducible Study of Tissues Impedance in Medical Applications

Mahdi Guermazi, Hanen Nouri, and Olfa Kanoun

Abstract Anisotropy significantly influences impedance spectra of biological tissues and needs therefore to be considered. A theoretical and experimental investigation of the influence of electrode geometry was carried out with the aim to reduce the effect of the anisotropy. The results confirm the existence of a bad contact between electrodes and biological tissue for non-penetrating probes because of the rough surface. We propose the use of penetrating electrodes with a cylindrical geometry and a sufficient space between the electrodes to reduce the influence of anisotropy, allow a good probe contacting and permit a measurement in the volume of the tissue. Measurements show a good reproducibility and allow therefore a reliable study of tissue impedance. We demonstrate that the proposed design can be adopted for medical applications by miniaturization and by using acupuncture needles as electrodes.

Keywords Electrode design · Biological tissues · Anisotropy · Bad contact · Reproducible measurements · Penetrating electrodes · Cylindrical geometry · Acupuncture needles

1 Introduction

Bioimpedance measurements are gaining increasingly importance in medicine (Kanoun 2019), ambient assisted living and wearable sensors in general (Barioul et al. 2019). The reproducibility of experimental procedure is very important for in-

M. Guermazi (✉) · H. Nouri · O. Kanoun
Chair for Measurement and Sensor Technology, Technische Universität Chemnitz, Chemnitz, Germany
e-mail: mahdi.guermazi@polytecsousse.tn

H. Nouri
e-mail: hanen.nouri@etit.tu-chemnitz.de

M. Guermazi · H. Nouri
Control and Energy Management Laboratory (CEM Lab), National School of Engineers of Sfax (ENIS), University of Sfax, Sfax, Tunisia

O. Kanoun and N. Derbel (eds.), *Advanced Systems for Biomedical Applications*, Smart Sensors, Measurement and Instrumentation 39,
https://doi.org/10.1007/978-3-030-71221-1_2

vitro bio-impedance measurements. Especially the electrodes are decisive because they are the intermediate element between the patient and the measurement device (Naranjo-Hernández et al. 2019). Electrode structures have attracted more and more attention in the monitoring in the physiological electrical signals, including electrode-skin interface impedance (EII), electromyography (EMG) and electrocardiography (ECG), different types of electrodes have been proposed, such as the micro-needle electrodes (MEs) developed by (Chen et al. 2019). Different electrodes are used for in-vitro bio-impedance measurements such as needle electrodes and non-penetrating circular electrodes. Penetrating electrodes have been excessively used for several bio-impedance measurement systems.

Theoretical and experimental investigations of the influence of electrode geometry have been carried out to identify a suitable geometrical design of the electrodes. (Lepetit et al. 2002) have developed different geometries of electrodes working with or without penetration to muscle. They are like plates electrodes made consisting of two electrodes realized by two rows of five cylindrical bull nosed stainless steel probes with 20 mm between the two rows of probes and with the same electrical potential on each row. They also have developed bars electrodes made of two parallel, rectangular cross-sectional bars of stainless steel with 10 mm between the two bars. Other geometries of needles electrodes are made of two rows of nine cylindrical needles and with 10 mm between the two rows of needles, with the same electrical potential of each row. It has been reported that needles electrodes have shown a big disadvantage considering reproducibility in anisotropic materials. (Li et al. 2016) applied CNT-modified acupuncture needles for a real time monitoring of serotonin (5-HT) in vivo and state that for the first time it could be directly probed into rat body for real time monitoring of 5-HT in vivo and it shows a great potential for better understanding the mechanism of acupuncture treatment. (Park et al. 2018) improve the biopsy-based early diagnosis of cancer, which increases the complete recovery rate of the cancer patients using biopsy needles with an electrical impedance sensor array based on stainless steel microelectrodes that was developed for real-time four electrode measurement and multi-spot sensing of tissues during the biopsy process.

Needle electrodes insure a very good contact but have a big disadvantage considering reproducibility in anisotropic materials. Biological tissues have an anisotropic structure, so that impedance varies according to whether the current runs parallel or perpendicular to muscle fibers (Swatland 1980). The conductivity of biological tissue increases in the case of measurement in parallel to the muscle fibers. The current is better conducted by the fibers than in other cases where it is obliged to cross them. Different authors propose the use of tetra polar electrodes, this type of electrodes shows the same problematic of reproducibility. Anisotropy dependency of bioimpedance measurements was reported by (Swatland 1991) and (Bodakian and Hart 1994). They classify three directions for the placement of electrodes for the measurement of electrical properties of biological tissue that are across the fibers and parallel to their long axis, then across the fibers and perpendicular to their long axis and the third is along the fibers and perpendicular to their long axis.

To avoid the influence of anisotropy, (Damez et al. 2008) proposed an experimental solution consisting of the measurement longitudinally then transversally to the fiber direction to allow diametrically opposite electrodes successively commutated to measure the impedance in a radial direction. (García-Breijo et al. 2008) have proposed electrodes with a cylindrical geometry. It consists of a hollow needle made of stainless steel that acts as the outer electrode inside, which is a wire made of steel and plays the role of inner electrode and with a dielectric material, the epoxy resin between both electrodes. The needle has an outer diameter equal to 0.46 mm.

(Altmann and Pliquett 2006) have used different directions of probe insertion for the prediction of intramuscular fat by impedance spectroscopy. The probe is inserted three times with 5, 7, and 9 cm from dorsal midline. The used penetration depth of the field lines is 0.8 mm for an average with a frequency range between 500 Hz and 100 kHz. They confirm that the increase in penetration depth by increasing the electrode gap would compromise the local resolution. The stroke of the electrode system is 120 mm and it is fully inserted into the muscle prior to the measurement. This work deals with experimental investigations for designing a bioimpedance measurement procedure by designing electrodes and defining measurement parameters allowing tissue diagnostic. Measurements with high reproducibility represent thereby an important challenge. This is why we look for methods reducing the contact influences and eliminating the influence of anisotropy effects of the biological tissues.

2 Investigations of the Influence of Anisotropy and Contacting Effects

The influence of anisotropy and contacting effects is carried out using a measurement setup consisting of a laboratory impedance analyzer (Agilent 4294 A) connected to a personal computer for data acquisition. The spectrometer measures impedance within a frequency band ranging from 40 Hz to 110 MHz. Electrodes are connected with a sample in series. A calibration of the electrode is implemented to eliminate influencing effects, especially cable effects, before every measurement.

2.1 Definition of Measurement Parameters

Electrochemical impedance is normally measured using a small excitation signal, which can be a current or a voltage having a small amplitude. For in-vivo measurements, a current excitation fulfilling the medical security rules is preferred (Bouchaala et al. 2016). For in-vitro measurements a voltage excitation can be adopted. The use of a voltage amplitude of 500 mV provides a good signal to noise ratio then 5 mV (Fig. 1). For in-vitro measurement (Bouchaala et al. 2015), this excitation voltage is adopted in all bioimpedance measurement in this study. This value corresponds to the maximum value of the voltage fulfilling the linearity condition, which is necessary in impedance measurements (500 mV). At this voltage amplitude, the current

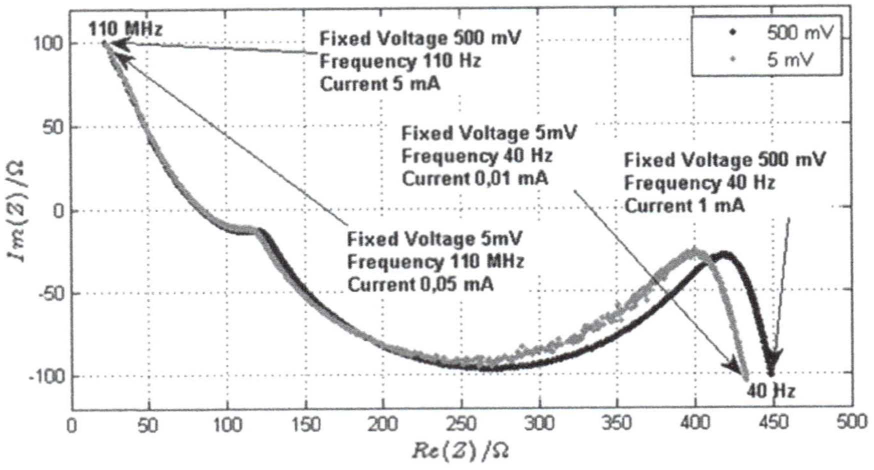

Fig. 1 Influence of the amplitude of excitation in the frequency range between 40 Hz and 110 MHz

is increasing from 1 mA at low frequency (40 Hz) to 5 mA at high frequency (110 MHz). At 5 mV, the current increases from 0.01 mA (40 Hz) to 0.05 mA at 110 MHz.

2.2 *Influence of Anisotropy*

For the anisotropy dependency, a first experiment is realized using a probe composed by 36 needle electrodes in 18 angles with a step of 10° (Fig. 2). For example, for an

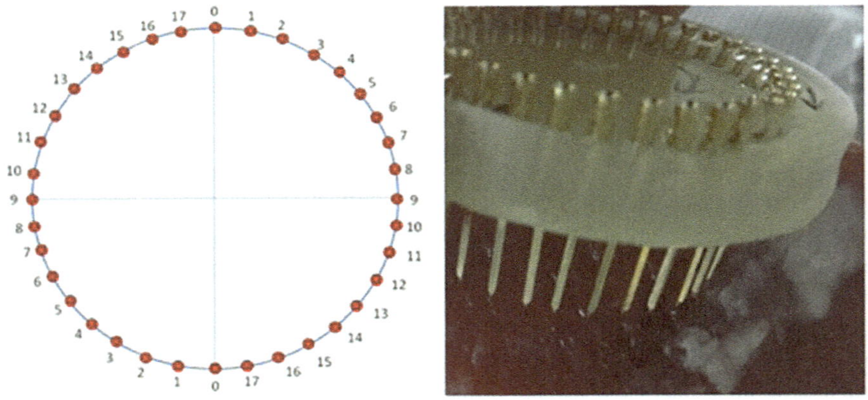

Fig. 2 Anisotropy investigation probe

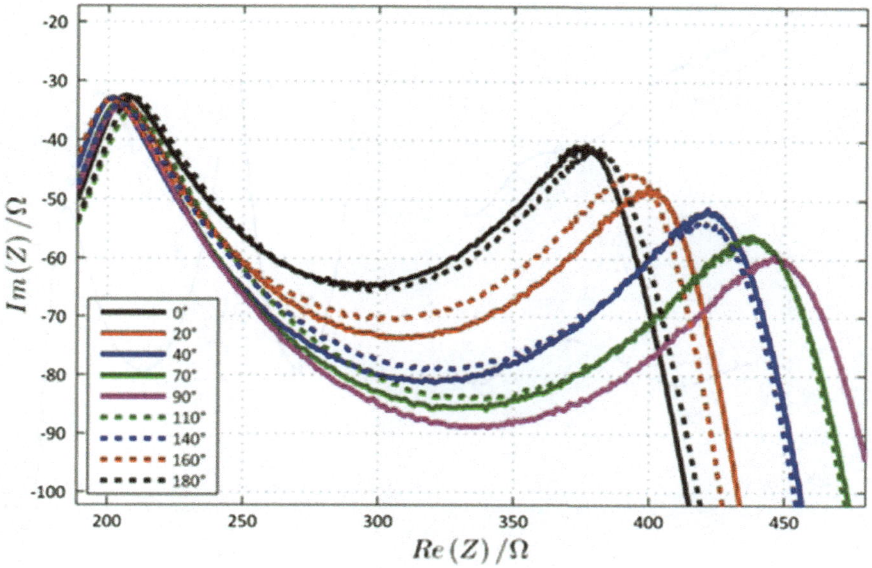

Fig. 3 Nyquist plot for different angles to fibers direction

angle of 20° the needle number 16 are chosen. For the angle 160° the needles number 2 are chosen. Experimental results in Fig. 3 show an almost linear distribution of the spectrum behavior with the angle. Small deviations are observed between 0° and 180° due to the mechanical stress during the experiment and due to rest deviations. Nevertheless, these changes are very small and therefore negligible. Some research groups suggested to measure in parallel or perpendicular to the direction of fiber direction. This deviation shows that this would be not possible. On the one hand, it is difficult to exactly detect the direction of the fiber of muscles. On the other hand, small changes of the direction of electrodes lead to significant changes of the impedance spectrum.

Results of the anisotropy investigations are analyzed deeply by evaluating the characteristic points in the impedance related to the dispersion β that is suitable for biological tissue characterization (Altmann et al. 2004) especially the characteristic points 1 and 2 (CP1 and CP2) (Fig. 4). CP1 corresponds to maximum of the spectrum in the transition region between α and β dispersion and CP2 corresponds to the minimum of the spectrum near the predominant critical frequency of the β dispersion. Characteristic points can be easily determined as maxima and minima of the imaginary part in the corresponding frequency ranges of the impedance as explained in Fig. 4.

Figure 5 shows the angular dependency of characteristic points CP1 and CP2. For spectral results in the region of the β dispersion, biological tissue conductivity varies in dependence of the angle. When the angle to fiber is close to 0°, corresponding to electrodes parallel to the muscle fibers, the conductivity increases, in this case the

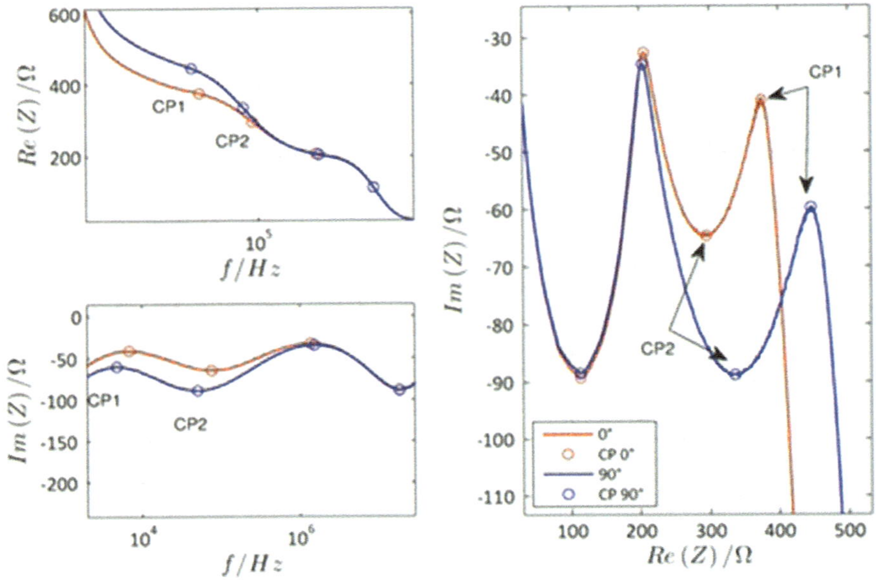

Fig. 4 Fiber direction dependency of characteristic points

current is better conducted by the fibers than in other cases where it is obliged to cross them.

$\Re e(Z)$ value is minimal at the angles 0° and 180° that correspond to the case parallel to the fiber direction, approximately 380 Ω for the CP1 and 300 for the CP2. Re (Z) increases according to the angle value with approximately 400 Ω for the angle 20° and 160° for the CP1 and approximately 310 Ω for the CP2. The impedance reaches a peak value at the angle close to 90° that corresponds to the position perpendicular to fibers. The behavior of the critical points of the spectrum in dependence of the angle is better predictable than impedance points at specific frequencies. This means, that the critical points CP1 and CP2 are very good features for describing the behavior of the impedance spectrum.

2.3 Influence of Contacting Effects

In this section, we investigate the contact between the electrodes and the biological tissue for non-penetrating probes. For this purpose, measurements of biological tissue is realized using a cylindrical non penetrating electrode with and without applied force (Fig. 6). Different beef muscles (*N* '*neck*', *LD* '*longisimus dorsi*', *RA* '*Rectus Abdominis*' and *SM* '*Semi Membranous*') are subjected to the experiment. Results show changes of the impedance according to the force applied (around 5 N) on the

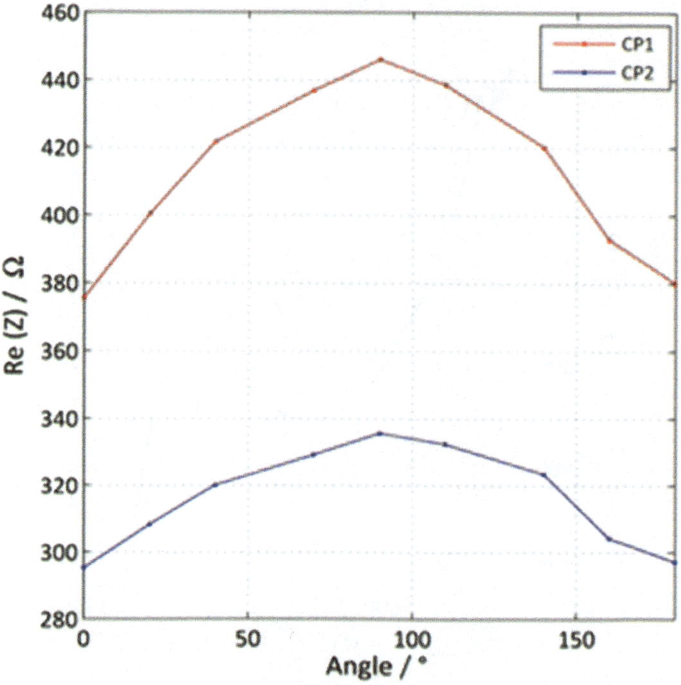

Fig. 5 Angle dependency of the considered characteristics points CP1 and CP2

electrode (Guermazi et al. 2013). This effect becomes more serious if the biological tissue surface is drying, e.g. because of aging.

The investigation results show that non-penetrating circular electrodes overcome the anisotropy effect in two dimensions. However, it has a bad contact to the rough surface of the biological tissue and needs therefore a sufficient force applied to get acceptable results. We decide to take penetrating electrodes for realizing measurement reliability.

3 Design of Electrodes

Based on the previous investigations, we propose electrodes having a cylindrical geometry to eliminate the anisotropy effect in two dimensions and combining cylindrical non penetrating electrode and needle electrodes. The objective is thereby to develop electrodes that can guarantee reproducible and consistent measurements despite these effects. Different theoretical and experimental investigations are provided for the proposed electrode configuration in order to ensure a reduction of the inadequate electrode contact and the influence of anisotropy at the same time.

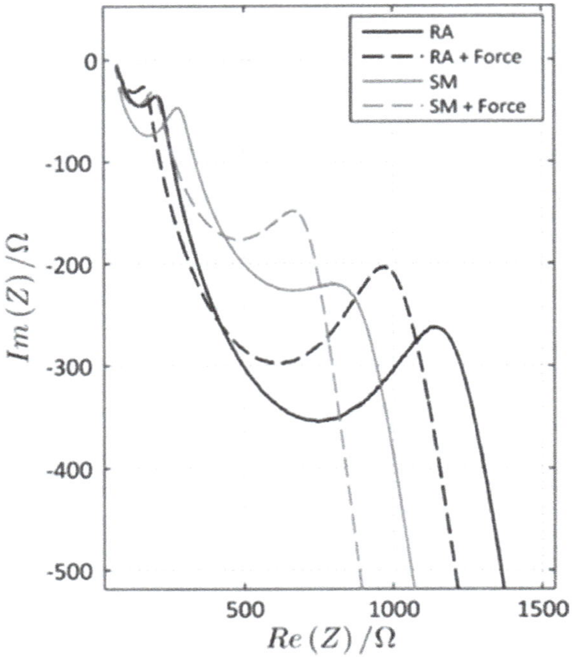

Fig. 6 Impedance spectra dependency to an applied force

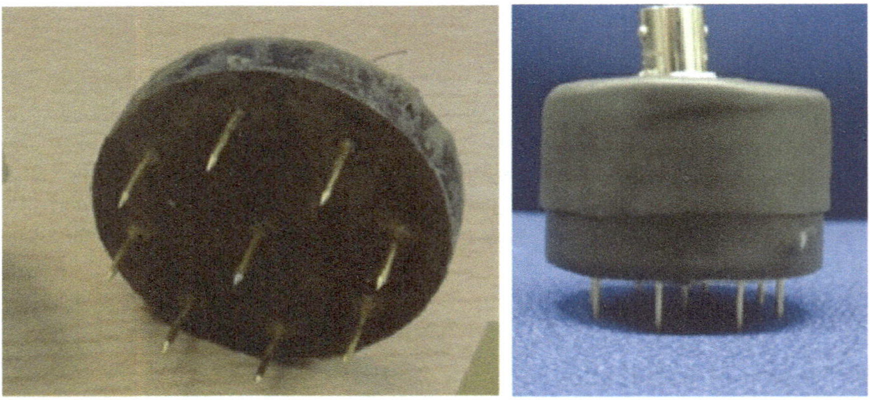

Fig. 7 Preliminary prototype of electrode design

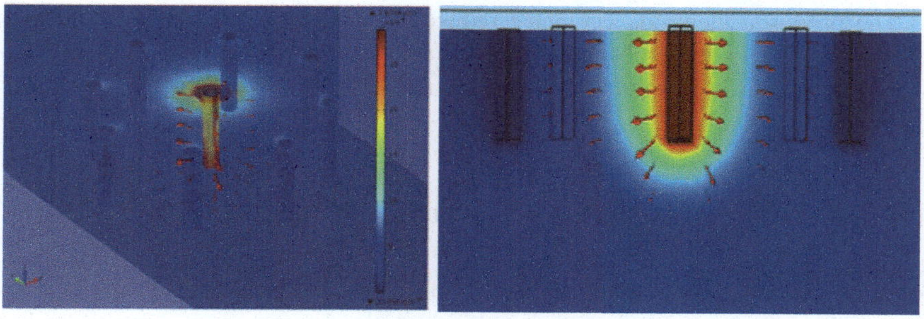

Fig. 8 Electric displacement field in 3D and in YZ view

A preliminary prototype of the electrode design is developed with a cylindrical geometry consisting of 9 steel needle electrodes well-spaced and coated with gold (Altmann and Pliquett 2006; Guermazi et al. 2014) (Fig. 7). To ensure a measurement in a big representative volume of biological tissue, we choose that the inner electrode has a distance of 15 mm to the outer electrodes. Electrodes deeply inserted into the sample and located close to each other show the most stable values according to different tests and according to the literature. This proves the importance of needle electrodes that can be inserted in the biological tissue. They do not destroy the biological tissue structure a lot, as a length of only 10 mm is chosen.

In order to provide information about the distribution of the electric field in the biological tissue sample, an electrode configuration with nine needles in cylindrical configuration is investigated by Finite Element Methods (Fig. 8). The phantom in the simulation is chosen according to the following geometric parameters: Length 80 mm, width 80 mm, height 40 mm. The conductivity has been chosen equal to 0.3 S/m and the relative permittivity has been chosen equal to 100 as mentioned in (Andretzko 2007; Gabriel et al. 1996a, b).

In presence of an electric field, the phantom is subjected to a displacement of free charges (conduction current) and a movement of charges (displacement current) depending on the electrical conductivity σ and the dielectric permittivity ε respectively. The displacement field corresponds to the following equation, where E is the electric field, ε_0 is the vacuum permittivity and P is the polarization density.

$$D = \varepsilon_0 E + P \tag{1}$$

The simulation shows that the electric displacement field D varies in the frequency range [10 kHz – 10 MHz] between $2.87 \times 10^{-7}\text{Cm}^{-2}$ to $2.1 \times 10^{-16}\text{Cm}^{-2}$ and propagates in the whole volume of the phantom in all possible directions. The choice of 9 electrodes with a diameter of 1 mm and 10 mm depth represents an accurate configuration. The preliminary prototype of electrode design is tested for different experiments with sufficiently big biological tissue samples.

The proposed electrode geometry reduces the influence of anisotropy, allows simultaneously a good probe contacting, showing a great potential, permits a measurement in the whole volume of the tissue and shows a good reproducibility over time.

4 Electrode Optimization for Medical Applications

Keeping the proposed electrode geometry and electrode penetration to ensure a reduction of the influence of anisotropy at the same time, we propose a multi-needles electrode using acupuncture needles used in Chinese medicine.

The used electrodes are highly polished and safe for ex-vivo and in-vivo applications as they can be inserted into the skin safely and without significant damage, as acupuncture needles are generally thin. The chosen needles are usually used for ears and can be inserted up to one week in the ear because of their shape as a dart as shown in Fig. 9a. They are golden needles composed of three main parts and the total length is 2.8 mm. The electrode geometry proposed in last section is miniaturized by keeping the cylindrical geometry. It is composed of 9 needles, where the outer circle radius is 7.5 mm and the inner circle radius is 1.25 mm (Fig. 9b).

In order to study the performance of the electrode, simulations using Finite Element Method (FEM) are considered. An agar-agar gelatin phantom has been prepared with different NaCl concentrations. It has a conductivity from 1.1 to 3.7 and permittivity from 70 to 75 (Duan et al. 2014). The study is done in the frequency range from 40 Hz to 110 MHz. The results show that the electric field displacement changes from 0.52 to 0.15 μC m^{-2} by increasing the NaCl concentration from 0.9 to 3.6%.

The simulation results show, that even by decreasing the length of the needle and the diameter of the electrode, the electrical field propagates well between the inner and the outer electrode. Thus, the miniaturization of the electrode can provide an accurate measurement as the circular shape can solve the anisotropic problem and the small length of needles can ensure that the electrical field does not propagate so much in the surrounding tissue and the field is only concentrated the aimed tissue.

(a) (b)

Fig. 9 **a** Acupuncture needles. **b** Miniaturized circular multi-needles electrode

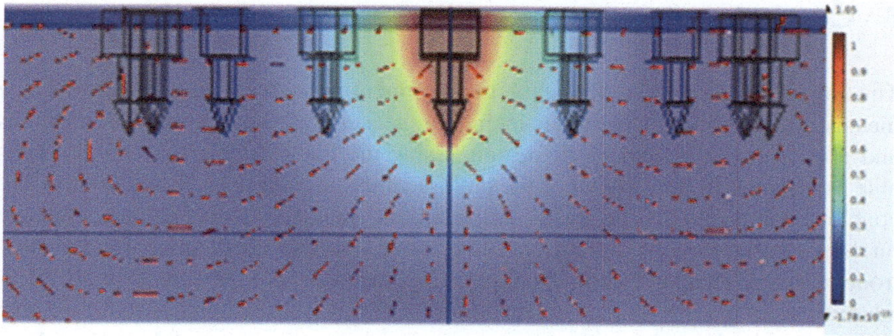

Fig. 10 Simulation of multi-needle electrode using Finite Element Method (FEM)

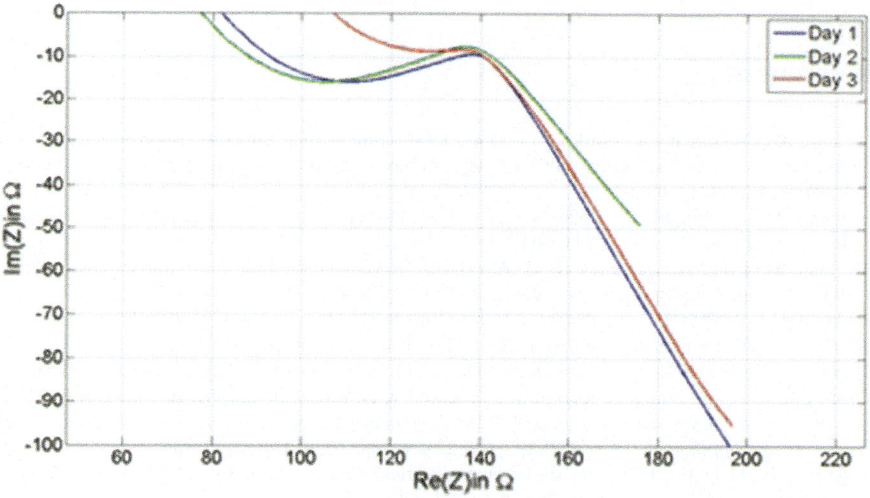

Fig. 11 Experimental results with chicken breast for three consecutive days

The electrode, in Fig. 10, has been applied for preliminary experiments on chicken breast. The measurements are performed using the impedance analyzer Agilent 4294A with an excitation signal of 500 mV for 3 consecutive days. The experimental results are shown in Fig. 11.

Using the electrode with acupuncture needles, the results show that breast tissue clearly exhibits its α and β dispersions. α dispersion is varying from 40 Hz to 10 kHz, β dispersion is found from 10 kHz to 100 MHz and γ dispersion is mainly at high frequency up to 100 MHz. The electrode shows therefore a great potential to be applied in medical applications. The acupuncture needles are safe for human body and the association of them in circular shape will allow accurate information about the tissue.

5 Conclusion

The aim of this work is to define the measurement procedure for the bioimpedance measurement of biological tissues including measurement set-up, excitation signals and probe contacting and probe design. Using the Finite Element Method, a suitable assessment of the volume of the biological tissue sample has been defined. We conclude that the use of electrodes with a circular geometry allows an independency on the anisotropy in two dimensions. The use of small needles realizes a good electrode electric contact to biological tissue. The electrode geometry can be realized by adopting acupuncture needles, which are thin, highly polished, and safe for ex-vivo and in-vivo applications as they can be inserted into the skin safely and without significant damage.

References

Alcaniz, M., Masot, R., & Fraile, R. (2008). Development of a puncture electronic device for electrical conductivity measurements throughout biological tissue salting. *Sensors and Actuators A: Physical, 148*(1), 63–67.

Altmann, M., & Pliquett, U. (2006). Prediction of intramuscular fat by impedance spectroscopy. *Biological Tissue Science, 72*(4), 666–671.

Altmann, M., Pliquett, U., Suess, R., & Von Borell, E. (2004). Prediction of lamb carcass composition by impedance spectroscopy. *Journal of Animal Science, 82*(3), 816–825.

Andretzko, J. P. (2007). *Contribution à l'optimisation et à la modèlisation d'un banc de mesure CEM* (p. I27). Ph.D. instrumentation and microelectronic, University of Nancy.

Barioul, R., Fakhfakh Ghribi, S., Ben Jemaa Derbel, H., & Kanoun, O. (2020). Four sensors bracelet for American sign language recognition based on wrist force myography. In *25th annual IEEE international conference on computational intelligence and virtual environments for measurement systems and applications, CIVEMSA, 2020, 22–24 June 2020*. Tunisia: Tunis.

Bodakian, B., Hart, F. (1994). The dielectric properties of biological tissue. *IEEE Transactions on Dielectrics and Electrical Insulation, 1*(2), 181–187.

Bouchaala, D., Guermazi, M., Derbel, N., & Kanoun, O. (2015). Portable device design for in-vitro muscle tissue monitoring. *tm–Technisches Messen, 82*(10), 485–494.

Bouchaala, D., Kanoun, O., & Derbel, N. (2016). High accurate and wideband current excitation for bioimpedance health monitoring systems. *Measurements, 79*, 339–348.

Chen, K., Ren, L., Chen, Z., Pan, C., Zhou, W., & Jiang, L. (2016). Fabrication of micro-needle electrodes for bio-signal recording by a magnetization-induced self-assembly method. *Sensors, 16*, 1533.

Damez, J. L., Clerjon, S., Abouelkaram, S., & Lepetit, J. (2008). Electrical impedance probing of the muscle food anisotropy for biological tissue ageing control. *Journal Food Control, 19*(10), 931–39.

Duan, Q., Duyn, J. H., Gudino, N., de Zwart, J.A., Van Gelderen, P., Sodickson, D. K., & Brown, R. (2014). Characterization of a dielectric phantom for high-field magnetic resonance imaging applications. *Medical Physics, 41*(10), 102303.

Gabriel, C., Gabriel, S., & Corthout, E. (1996). The dielectric properties of biological tissues I. Literature survey. *Physics in Medicine and Biology, 41*(11), 2231–2349.

Gabriel, S., Lau, R.W., & Gabriel, C. (1996). The dielectric properties of biological tissues: II. Measurements in the frequency range 10 Hz to 20 GHz. *Physics in Medicine and Biology, 41*(11), 2251–2269.

Guermazi, M., Kanoun, O., & Derbel, N. (2013). Reduction of anisotropy influence and contacting effects in in-vitro bioimpedance measurements. *Journal of Physics: Conference Series*, *434*, 1–4.

Guermazi, M., Kanoun, O., & Derbel, N. (2014). Investigation of long time beef and veal biologi-cal tissue behaviour by bio-impedance spectroscopy for biological tissue monitoring. *IEEE Sensors Journal*, *14*(10), 3624–3630.

Kanoun, O. (2019). Impedance spectroscopy advances and future trends: A comprehensive review. In *Impedance Spectroscopy* (pp. 1–22). De Gruyter.

Lepetit, J., Salé, P., Favier, R., & Dalle, R. (2002). Electrical impedance and tenderisation in bovine biological tissue. *Biological tissue Science*, *60*(1), 51–62.

Li, Y., Tang, L., Ning, Y., Shu, Q., Liang, F., Wang, H., et al. (2016). In vivo monitoring of serotonin by nanomaterial functionalized acupuncture needle. *Scientific Reports*, *6*, 28018. https://doi.org/10.1038/srep28018.

Naranjo-Hernández, D., Reina-Tosina, J., & Min, M. (2019). Fundamentals, recent advances, and future challenges in bioimpedance devices for healthcare applications. *Hindawi Journal of Sensors*, Article ID 9210258, 42 pages.

Park, J., Choi, W., Kim, K., Jeong, W., Seo, J., & Park, I. (2018). Biopsy: Needle integrated with electrical impedance sensing microelectrode array towards realtime needle guidance and tissue discrimination. *Scientific Reports*, *8*, 264. https://doi.org/10.1038/s41598-017-18360-4.

Swatland, H. J. (1980). Anisotropy and postmortem changes in the electrical resistivity and capacitance of skeletal-muscle. *Journal of Animal Science*, *50*, 67–74.

Swatland, H. J. (1991). Post-mortem changes in pork using parallel needles for both light scattering and low frequency electrical properties. *Food Research International*, *30*(3–4), 293–298.

Resonant Inductive Coupling for Wirelessly Powering Active Implants: Current Issues, Proposed Solutions and Future Technological attempts

Yosra Ben Fadhel, Aref Trigui, Salem Rahmani, and Kamal Al-Haddad

Abstract Since the 20th century, portable electronic devices such as laptops, cell phones, tablets, and medical implants have emerged in our daily life. All these devices are mainly power supplied by batteries. However, their lifespan is a major disadvantage. This challenge becomes more critical when it comes to Implantable Medical Devices (IMDs). Therefore, Wireless Power Transfer (WPT) is considered as a promising alternative to overcome this challenge. The near-field magnetic WPT approach is a common-used way to power-up active implants wirelessly. Although many improvements have made on the implant's design and their remote powering approaches, stringent requirements and several parameters should be considered during the design of the WPT system, such as the power transfer efficiency, the transfer distance, the implant size, and its biocompatibility. In this paper, a comprehensive study of the commonly proposed approaches to power-up IMDs have been reviewed. In particular, various advanced mechanisms and technical solutions related to the Near-field magnetic WPT approach have been discussed. These solutions are carried out to optimize transcutaneous WPT transfer efficiency, maintaining robustness and safety against many factors such as coil misalignments and load variations. Human safety concerns and exposure limits to the electromagnetic field by respecting international guidelines are also explored in this review. Finally, a summary of relevant information and directions for future research investigations are provided.

Y. Ben Fadhel (✉) · S. Rahmani
Research Laboratory of Biophysics and Medical Technology (BMT), High Institute of Medical Technologies, University of Tunis El-Manar, Tunis, Tunisia
e-mail: benfadhelyosra@gmail.com

S. Rahmani
e-mail: rsalem02@gmail.com

A. Trigui
Department of Electrical Engineering, Polytechnique Montreal, Montreal QC H3T 1J4, Canada
e-mail: aref.trigui@gmail.com

K. Al-Haddad
Canada Research Chair in Energy Conversion and Power Electronics CRC-ECPE Ecole de Technologie Supérieure, Montréal QC H3C 1K3, Canada
e-mail: kamal@ele.etsmtl.ca

O. Kanoun and N. Derbel (eds.), *Advanced Systems for Biomedical Applications*, Smart Sensors, Measurement and Instrumentation 39,
https://doi.org/10.1007/978-3-030-71221-1_3

Keywords Wireless power transfer (WPT) · Active implant (AI) · Near-field · Human safety · Resonant inductive coupling (RIC) · Converters · Specific absorption rate (SAR) · Biocompatibility

1 Introduction

In-body medical devices provide a substantial improvement in healthcare by helping to manage diseases and save innumerable lives of patients. In particular, implants may replace an organ's function or treat associated diseases, such as brain or cardiac troubles. To date, researchers and engineers have created varieties of in-body devices helping to manage diseases and saving innumerable lives of patients. These devices may be divided into three categories: implantable, ingestible and injectable based on how they are inserted inside the human body (Kiourti et al. 2017). Implantable devices can also be classified as passive or active. Passive devices do not need electrical power for their operation. They are used for support or mobility, such as a simple screw or artificial valves. However, active implants need power for their operation. As reported by the Council Directive of the European Union 1990 (90/385/EEC), active implants are employed to replace a total or partial organ's function. They may collect and transmit useful physiological signals (heartbeat, blood pressure or temperature) and treat an associated disease such as brain or cardiac troubles. Mostly, the power requirements of all these implants range from a few microwatts to a few tens of milliwatts as illustrated in Fig. 1, and that depends on their specific applications (Bihr al. 2014; Council Directive of 20 1990). For the sake of simplicity and standardization, this type of devices is called Implantable Medical Devices (IMDs).

Powering IMDs is still a subject very addressed by researchers in order to meet their evolving requirements. Conducting wires and cables are nowadays standard mediums used for transmitting electrical power. Since, they ensure a flow of power in a precise and efficient manner (Wen 2012). However, wires are inappropriate to power-up IMDs. They restrict patient mobility. They also may corrode and cause tissue infections due to patient motion (Ben Amor et al. 2015). Alternative solutions consist of using an internal battery integrated into the implantable device. It is also possible to employ external power sources to wirelessly powering these devices.

In the literature, several methods are proposed to develop reliable WPT systems for powering IMDs, such as optical coupling (Ben Fadhel et al. 2016; Cruciani et al. 2014; Mashhadi et al. 2019), ultrasonic coupling (Bihr al. 2014; Maleki et al. 2011), capacitive coupling (Dai et al. 2015b), and inductive coupling (Dai et al. 2015a; Ozeri et al. 2010). Until now, inductive coupling is the most adopted technique in transcutaneous applications, due to its simplicity, reliability, and safety (Li et al. 2015). Table 1 illustrates the important proposed techniques to power-up active implants and it presents their main disadvantages, whereas each of these methods may be the subject of further research.

Nowadays, batteries and inductive coupling are the most used techniques to power-up commercial IMDs. Hence, these two supplying power methods will be investigated

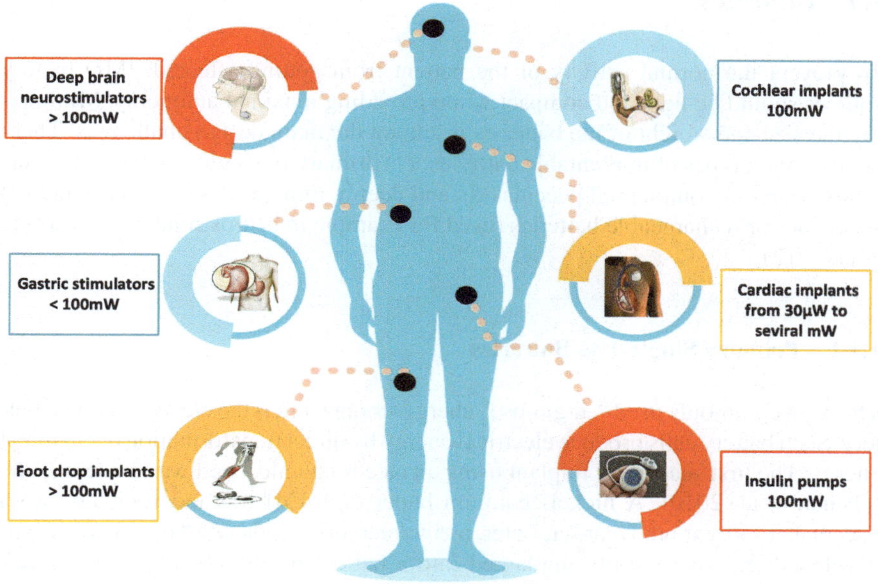

Fig. 1 Power requirement of commonly used active IMDs

Table 1 Major disadvantages of WPT techniques proposed for transcutaneous applications

Proposed WPT techniques	Disadvantages
Optical coupling	High absorption of the electromagnetic (EM) radiation by the body tissue and potential health hazards at sufficiently high-power density
Ultrasonic coupling	High reflection in the air-tissue interface and the need for a matching layer (as a coupling gel) between the transducer and the skin
Capacitive coupling	Large coupling area needed when high power is required, very sensitive to the distance variations

in the following sections. First, single-use and rechargeable batteries meant for IMDs have been reviewed. Bio-batteries and specific supercapacitor power sources adding complementary features are then presented. The concept of inductive coupling, its benefits and its limitations for implantable biomedical applications have also been discussed.

1.1 Batteries

To prevent the normal activity of the patient from getting affected, IMD battery sources should be light and compact while providing maximal autonomy. Minimizing the size and weight of the batteries is among the main design challenges. There are two main types of implantable batteries: (1) primary or single-use batteries (currently found in commercial pacemakers and deeply implanted stimulators) and (2) secondary or rechargeable batteries (used for example in neurostimulators) (Bazaka et al. 2012).

1.1.1 Primary Single-Use Batteries

The most commonly used "single-use" energy storage unit is the electrochemical battery. Such battery cells produce electrical energy by direct transformation of chemical energy. The first successful implant using an electrochemical cell was made in 1960 (Osman et al. 2010). A nickel-cadmium battery (Ni-Cd) was incorporated into a pacemaker to treat bradycardia. Later, a zinc/mercuric oxide (Zn/HgO) battery was developed and successfully implanted into a patient by the Greatbatch-Chardack team (Biel et al. 2017). Afterward, the invention of lithium-based power cells in the late 60s led to an attractive alternative to the first electrochemical cells. With their high energy density (three times the energy of the Zn/HgO battery) lithium batteries are still, to date, the most used type of batteries. The lithium/iodine-polyvinylpyridine (PVP) cell was the first lithium-based battery implanted into a pacemaker. Over several decades, it has proved its safety and reliability for clinical uses. Therefore, with their high power density (milliampere range), low self-discharge rate, and longevity, Lithium/Manganese Dioxide (Li-MnO2), Lithium/Carbon monofluoride (Li/CFx), lithium/thionyl chloride, and other varieties of lithium batteries were used to power neurostimulators, drug delivery systems, and pacemakers with additional functionalities (Ben Fadhel et al. 2016). Implantable Cardioverter Defibrillators (ICDs) pacing at fast rates require high power rate batteries. These batteries must be capable of delivering high current pulses of 2–3A to rapidly charge the capacitors of the device that subsequently shocks the heart to stop ventricular fibrillation (Wang et al. 2001). Lithium/silver vanadium oxide (Li/SVO) and Lithium/Manganese Dioxide batteries are commonly used in most of the ICDs today due to their high-power capabilities. These single-use lithium-based batteries establish advanced powering solutions, however, they require replacement surgery when depleted. In the 1970s, another type of battery was introduced in the IMDs industry to extend the service lifetime, which is the nuclear battery. Some manufacturers of IMDs, including Medtronic, have introduced nuclear models (such as Laurens-Alcatel model 9000) in their product lines. These devices have given patients the opportunity to have a lasting pacemaker. Nuclear batteries have been proved safe and effective. Their output power is stable and immune to external factors such as temperature, pressure, and electromagnetic fields. However, nuclear batteries are costly as their internal materials, particularly the radionuclide, are very difficult to extract and use.

1.1.2 Secondary Rechargeable Batteries

The rechargeable battery is an electrochemical generator that can be "revived" when depleted. In rechargeable batteries, energy is generated by a chemical reaction between an electrolyte and two metallic electrodes (Li et al. 2007). When there is not enough reactive material in the electrodes, the energy storage reduces, and the battery must be recharged. To do so, a DC-voltage is applied to the battery. This reverses the direction of the chemical process and replenishes the starting reactants thus allowing the battery to be used several times for many discharging/charging cycles. However, these cycles cannot be done indefinitely. After several charging/discharging cycles the battery needs to be replaced. The first rechargeable implantable pacemaker used a new kind of Ni-Cad battery (Biel et al. 2017). Later pacemakers were commercialized by Pacesetter Systems Inc (became then a division of St. Jude Medical) in 1973. The sale of these rechargeable pacemakers was not successful. Pacemakers integrating single-use lithium batteries were also introduced in the market. They were smaller and offered longer service life without requiring the battery recharging equipment. Targeting higher energy capabilities and other improvements, such as smaller size and longer lifetime, rechargeable lithium-ion cells have emerged with great promises of compactness, safety, reliability, and durability for many applications. Therefore, some commercial IMDs, such as Medtronic neurostimulators (ex. the ACTIVA' RC Neurostimulator for deep brain stimulation) were equipped with lithium-ion rechargeable cells (Al Idrus 2016). Currently, the use of the lithium-ion rechargeable battery is intended for several IMDs under development such as neurostimulators to cure epilepsy and Parkinson's diseases, cochlear implants, Left Ventricular Assist Devices (LVADs), and artificial hearts. The Lithium-ion battery is not complicated to be manufactured, however it requires elaborated tests to ensure safety. Actually, these secondary batteries should be well sealed and monitored all the time to prevent overheating and damage. Lithium-ion batteries are commonly recharged via an inductive link.

1.2 Other Types of Embedded Power Sources

1.2.1 Supercapacitors

Although batteries are the primary choice for IMDs on-board powering, there are some specific applications that do require special power sources. In fact, some IMDs need to have a very small volume or extremely high-power density. Supercapacitors also named ultra-capacitors provide these features. They are of smaller sizes and have higher-power densities compared to electrochemical batteries. They can generate

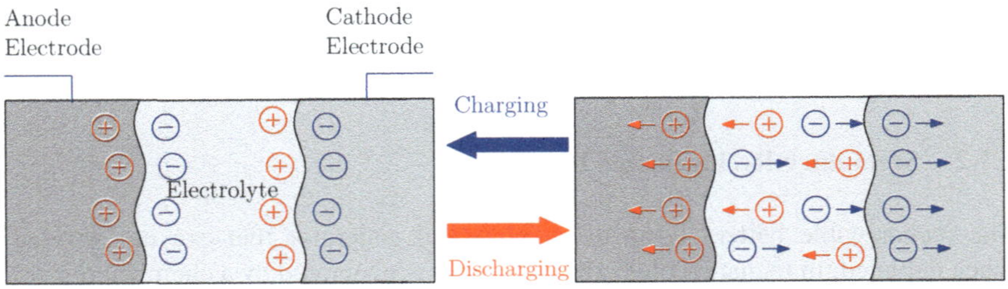

Fig. 2 Electrical double-layer capacitor

short pulses of relatively high power (Mendoza-Ponce et al. 2018). They have faster-charging rates (several milliseconds to seconds vs several hours for batteries), longer lifespan (up to 1 million charge/discharge cycles vs hundreds to thousands of cycles), larger operating temperature ranges, and they are not subject to overcharge, which makes them safer (Pandey et al. 2011). Also known as an electric double-layer capacitor (EDLC), a supercapacitor consists of two disjoined electrodes (separated by an ion-permeable membrane) immersed in an electrolyte that carries positive and negative ions, as shown in Fig. 2. The energy can be stored in EDLC by applying two storage principles: (i) electrostatic storage, which is an accumulation of ions on the electrode/electrolyte interfaces; and (ii) electrochemical storage (faradic pseudo-capacitance), which is a chemical reaction in which an exchange of electrons occurs at the surface or at the near-surface of the electrodes. Compared to conventional electrolytic capacitors, supercapacitors have a much higher capacitance (50 times greater) and a relatively smaller size (Kosage et al. 2016). However, the use of supercapacitors in IMDs is still limited due to its low energy density (5 % of lithium batteries), fast self-discharge rate and low cell voltage (limited to 2.5V or 2.75V).

The rapid technological progress in the field of miniaturized portable electronic devices led to the fabrication of on-chip micro-scale supercapacitors. These micro-supercapacitors provide ultra-fast charge/discharge rates and higher specific energy than conventional supercapacitors (2 times greater). Special attention has been given to the structures, the dimensions and the materials of the micro-supercapacitor elements to improve their performances. However, several challenges need to be addressed to increase the capacitance value and energy storage performances (Liu et al. 2017). There is also a trend supporting the combination of the supercapacitor with the lithium-ion battery resulting in a hybrid energy storage system. These hybrid elements offer the advantages of both the supercapacitor (fast-charge rate, high-power density) and the rechargeable battery (high-energy density). However, the latter combination adds size and complexity.

1.2.2 Bio-Fuel Cells

Another power source alternative to standard battery sources is the bio batteries for IMDs. In the 1970s, many studies were carried on the enzyme-based biofuel cells (using glucose or lactate as fuel and oxygen as oxidizer) in order to build new batteries for IMDs. These biological molecules are present in physiological fluids such as blood. In addition, the body pH, temperature, and pressure are adequate for efficient power generation. Since 1984, many improvements have been made on these biofuel cells as they proved quite advantageous for IMDs. However, most biofuel cells proposed till date are unable to meet all the IMDs' requirements. This is due to their short lifespan and inability to generate sufficient power values (microwatt level) (Osman et al. 2010).

2 Wireless Power Transfer

Wireless Power Transfer (WPT) solutions are based on external source units. Two types of WPT can be distinguished:

- Far-field or radiative techniques where the power is transferred by electromagnetic waves (RF and optical link) or acoustic waves (ultrasonic link).
- Near-field or non-radiative techniques where the power is transferred over short distances by electrical field (capacitive link) or magnetic field (inductive link).

Far-field transmission range is up to several kilometers and its frequency bands are extremely high, up to several GHz. Unfortunately, the far-field topology engenders health complications such as tissue warming up. This aspect is an important reason to make it inappropriate for powering IMDs.

Near-field WPT is to transfer power over short distances approximatively for few centimeters. This method uses reactive coupling technique such as the capacitive and the inductive coupling (Ben Fadhel et al. 2019). The inductive coupling results from the magnetic coupling, whereas capacitive coupling is from the electrical coupling. Near-field approaches are considered the most suitable for powering IMDs compared to far-field approaches since they bring several benefits. Firstly, they experience less power absorption in a high loosely dielectric medium. Secondly, they may transfer high amount of power while meeting the Specific Absorption Rate (SAR)limit. Finally, the magnetic field is biocompatible with the human body and it is controlled by the human body exposure limits to the EM field (Mohamad Jawad et al. 2017). Actually, near-field techniques have distinct advantages over the other conventional techniques for powering implants, since they can efficiently deliver power without causing harmful effects to the human body and they have a simple implementation. Nonetheless, near-field approaches also have a few limitations, such as low power transfer efficiency, short transmission range, and misalignment sensitivity (Kanoun et al. 2021). In fact, the short transmission range of the magnetic systems may be

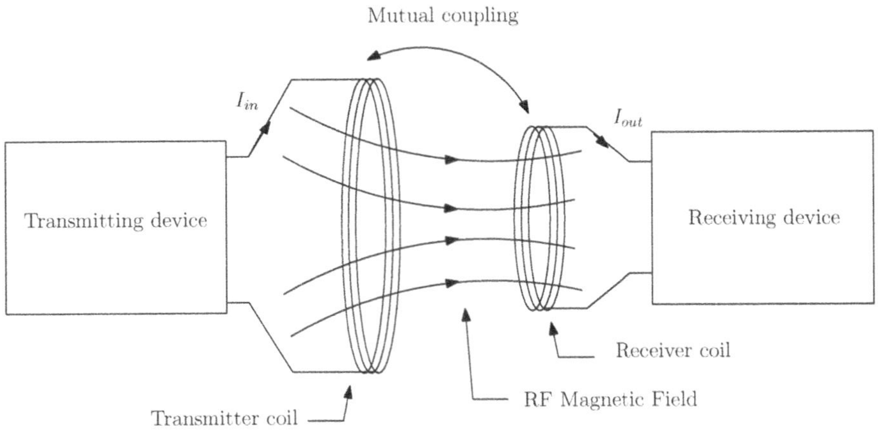

Fig. 3 Inductive coupling between two circular coils

of benefit from a security and interference immunity standpoint. However, those issues need to be solved to apply near-field technologies to various future applications, which may require higher power, longer transmission distance, and mobility support.

2.1 Inductive Coupling WPT

Figure 3 illustrates the simplest case of the inductive coupling phenomenon. It is implemented using two coils positioned in proximity to each other. When an AC current flows through the primary coil, an AC magnetic flux is created in its vicinity. Part of this flux is picked up by the secondary coil and converted to an ac voltage across it.

Where the magnetic flux couples the transmitter coil (T_X) and the receiver coil (R_X). It relies on the use of a magnetic field generated by an AC voltage, in order to induce a voltage in a second conductor. This effect occurring in the electromagnetic near-field, with the secondary circuit is in proximity to the primary circuit (Trigui et al. 2015b).

2.2 Resonant Inductive Coupling WPT

The resonant inductive coupling is an improved version of the inductive coupling.It uses additional capacitors (C_1 and C_2) that may be connected to the coupled coils either in series or in parallel. The transmitter circuit (L_1, C_1) and the receiver cir-

Table 2 Far-field and near-field characteristics

	Far-field	Near-field	
		Inductive coupling	Resonant inductive coupling
Transmission range	Far	Close: less than 1/5 of the size	Close: between 1 to 3 times the size
Produced power	Medium or large	Low: power below 5 watts.	Average: can reach a hundred Watts
Human body safety	High power level can be harmful	Safe for high power level	
System complexity	–Average without monitoring mechanism. –Complex with a monitoring mechanism	Simple	Simple
Transmission mechanism	Electromagnetic radiation	Magnetic induction	Magnetic induction
TX/RX medium	Antennas	Coils	Coils
Antenna interaction	No interaction	Medium interaction	Strong interaction
Efficiency	Low (10–50)	Medium (30–60)	High (70–90)
Commercial applications	No	Yes	Yes

cuit (L_2, C_2) are tuned to resonate at the same frequency as the driving frequency. Table 2 summarizes the transcutaneous energy transfer methods. Briefly, it can be noticed from literature and from this table that the the most suitable method is the resonant inductive coupling owing to its extended transmission range, its relatively high efficiency, its ability to scale over a range of output power levels milliwatts to tens of watts, and its safety (less exposure to radiofrequency radiation) (Dai et al. 2015a; Lazzi et al. 2005).

2.2.1 Fundamentals of Resonant Inductive Coupling WPT

Power transfer efficiency (PTE) is a crucial metric to evaluate the WPT system performances. It is the ratio of the power delivered to the load from the transmitter unit. Many parameters contribute to limit it such as coil losses (eddy current effects and proximity effects), weak coupling coils (large separations, misalignments), and near-field absorption in layered biological tissue (Jiang et al. 2014). For IMDs applications, high coupling efficiency is desirable as it reduces power losses and heat dissipations. The first step towards increasing the PTE is the use of resonant tuning. In fact, when the on-body and the in-body circuits are in the resonance state i.e. the TX/RX LC circuits operate at the same resonant frequency ($f_{res} = f_1 = f_2$), the PTE

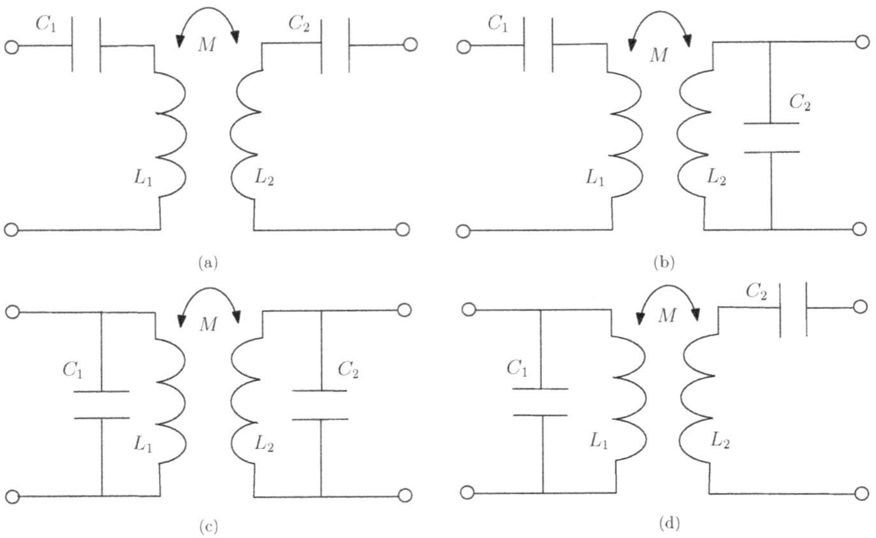

Fig. 4 Possible RIC topologies

reaches its maximum. To get the resonance between the two circuits, four resonance topologies are possible as shown Fig. 4: Series-to-Series (SS), Series-to-Parallel (SP), Parallel-to-Parallel (PP), and Parallel-to-Series (PS) (Wang et al. 2019). Each of the configuration arrangements in Fig. 4 can be considered as a 2-port network, where the transmitter coil (L_1) is driven by a power source and the receiver coil (L_2) is connected to a load. The 2-port representation is illustrated in Fig. 5, where (Z_{link}) is the impedance looking into the link, and will depend on the topology chosen from Fig. 4. Z_{out} is the impedance connected to the output port of the link. For good input matching, a series resonant primary Fig. 4a, b is driven by a voltage source, and a parallel resonant primary Fig. 4c, d is driven by a current source. Using the circuit analysis presented by Terman (Na et al. 2015), parameters such as the link gain and impedance can be determined for any of the link configurations in Fig. 4.

The gain is considered either as a unitless voltage gain (V_{out}/V_{in}), or as a transimpedance (V_{out}/I_{in}), depending on the driver. This analysis is developed further by Van Schuylenbergh and Puers, where equations for link efficiency and optimization procedures are developed and presented (Christ et al. 2013). Combining these analyses is sufficient to perform rudimentary analysis of an inductive link in terms of the electrical circuit parameters, extracting parameters such as the power delivered to the load (PDL) and the power transfer efficiency (PTE). WPT between the coils is ensured by the Mutual inductance (M) expressed by:

$$M = k\sqrt{L_1 L_2} \tag{1}$$

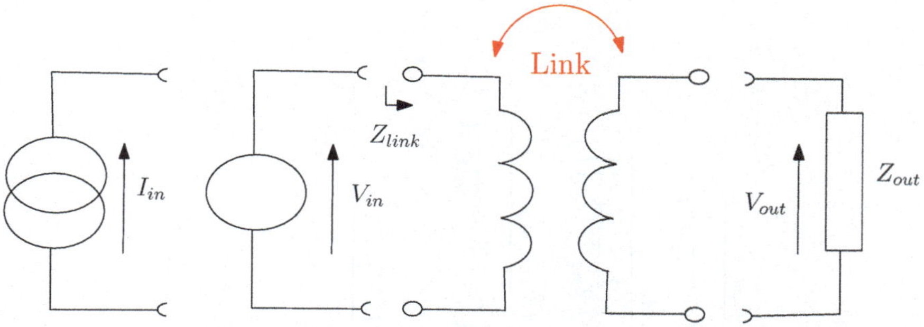

Fig. 5 General 2-port model: an appropriate source should be chosen depending on the resonant configuration

where L_1 and L_2 are the self-inductance of the primary and the secondary coils, respectively:

$$L_1 = \frac{\mu_0 \, \mu_r \, N_1^2 \, A}{l} \tag{2}$$

$$L_2 = \frac{\mu_0 \, \mu_r \, N_2^2 \, A}{l} \tag{3}$$

with: L_1 and L_2 are in Henry, $\mu_0 = 4\pi \; 10^{-7}$ the permeability of free space, N the number of turns, $A = \pi \, r^2$ the Inner core Area in m^2, l the length of the coil in meters, μ_r the relative permeability of the iron core.

Another parameter that link the two coils is the coupling factor k expressed as follows:

$$k = \frac{M}{\sqrt{L_1 \, L_2}} \tag{4}$$

The following table presents the link impedance, the link gain, and the link efficiency related to Figs. 4 and 5. Mattew et al. (2018) explain in details how to obtain the tabulated equations (Table 3).

For IMDs applications, the SP topology is the most employed for three main reasons listed below:

- The SP topology ensures power transfer efficiency greater than the other topologies in WPT applications (Wang et al. 2019).
- The serial topology in the transmitter circuit is appropriate for the class-E power amplifier topology (if it is used). Since a class-E power amplifier has a tuning capacitor in parallel and the second one in a series (Ben Fadhel et al. 2019), which reduces the system complexity.
- Usually, a rectifier is added to the receiver unit which makes the parallel topology the most adequate.

Table 3 Link impedance, gain and efficiency for different topology (Schormans et al. 2018)

Topology	Link impedance	Link gain
SS	$Z_{L_1}+\frac{1}{j\omega C_1}+\frac{(\omega M)^2}{Z_{L_2}+Z_{out}+\frac{1}{j\omega C_2}}$	$-\frac{jM\omega Z_{out}}{Z_{link}\left(Z_{L_2}+Z_{out}+\frac{1}{j\omega C_2}\right)}$
SP	$Z_{L_1}+\frac{1}{j\omega C_1}+\frac{(\omega M)^2}{Z_{L_2}+\frac{1}{j\omega C_2+\frac{1}{Z_{out}}}}$	$-\frac{jM\omega}{Z_{link}\left(1+Z_{L_2}\left[\frac{1}{Z_{out}}+j\omega C_2\right]\right)}$
PS	$\frac{1}{j\omega C_1+\frac{1}{Z_{L_1}+\frac{(\omega M)^2}{Z_{L_2}+Z_{out}+\frac{1}{j\omega C_2}}}}$	$-\frac{jM\omega\left(Z_{out}Z_{link}\right)}{\left(Z_{L_1}+Z_{ref}\right)\left(Z_{L_2}+Z_{out}+\frac{1}{j\omega C_2}\right)}$
PP	$\frac{1}{j\omega C_1+\frac{1}{Z_{L_1}+\frac{(\omega M)^2}{Z_{L_2}+\frac{1}{j\omega C_2+\frac{1}{Z_{out}}}}}}$	$-\frac{jM\omega\left(Z_{out}Z_{link}\right)}{\left(Z_{L_1}+Z_{ref}\right)\left(1+Z_{L_2}\left[\frac{1}{Z_{out}}+j\omega C_2\right]\right)}$

Link efficiency: $\eta_{link}=\frac{P_{out}}{P_{in}}=\frac{|V_{out}|^2}{|i_{in}|^2}\times\frac{\Re e\left(Z_{out}\right)}{|Z_{out}|^2\,\Re e\left(Z_{link}\right)}$

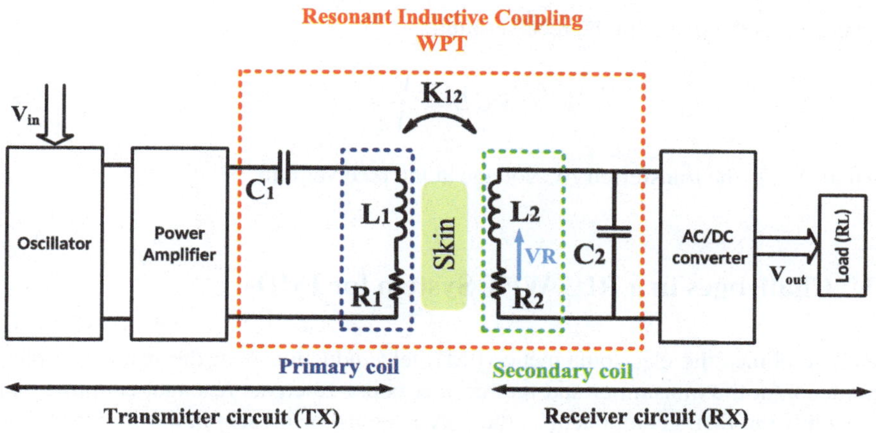

Fig. 6 Block diagram of a resonant wireless power transfer system with Series-Parallel toplogy

The resonance at the receiver circuit can be tuned in serial or in parallel. However, the resonant serial topology does so by using a large voltage and small current. Since rectifiers work better at a large voltage and low current, which make the parallel topology as the leading choice in biomedical transcutaneous applications (Ben Fadhel et al. 2019).

Figure 6 illustrates a conventional block diagram of a RIC WPT system with SP topology between two coils, it contains a DC/AC converter, a wireless coupling link, and an AC/DC converter. Firstly, the DC/AC converter is composed of a DC power supply, an oscillator, a gate driver and a power amplifier (PA). Secondly, the wireless coupling link consists of series-parallel LC tanks (C_1 L_1 and C_2 L_2). Finally, the AC/DC converter is composed of a rectifier and a regulator, in order to convert the AC voltage (VR) at the receiver coil into DC voltage V_{out} for a giving load (R_L).

In the aim to compute the overall RIC WPT system efficiency, four parameters ensure it. Firstly, the power delivered to the load (PDL). Secondly, the power transmission efficiency (PTE). Thirdly, the power conversion efficiency (PCE) in the receiver unit. Finally, the voltage conversion efficiency (VCE). All these parameters are determined respectively as follows:

$$PDL = \frac{V_{out}^2}{R_L} \tag{5}$$

$$PTE = \frac{PDL}{P_T} \tag{6}$$

where P_T is the power in the transmitter unit.

$$PCE = \frac{PDL}{P_R} \tag{7}$$

where P_R is the power in the receiver unit.

$$VCE = \frac{V_{out}}{V_R} \tag{8}$$

where V_R, is the transferred AC voltage at the receiver unit.

3 Challenges in a RIC WPT System for IMDs

RIC combines the electromagnetic (EM) field induction with the resonance phenomenon at the transmitter and the receiver coils. To create resonant coupling and to supply the IMDs components, other circuits must be present in the complete system. In addition, to generate an alternative signal at a precise amplitude, shape, and frequency, an oscillator and a Power Amplifier (PA) are essential (Cai et al. 2014). For a long time, the class-E PA is widely used in the transmitter circuit in transcutaneous systems, because of its simplicity, its high efficiency (100%) and its operating frequency range (3 MHz until 10 GHz) (Peña-Eguiluzet et al. 2020). Overall system efficiency is quantified by considering specific metrics such as the power transfer efficiency (PTE) and the transfer air gap. During system operation power losses may occur and PTE decreases. Therefore many research projects have investigated these issues and they have proposed technical solutions to minimize power losses as much as possible. Some researchers have interest in maintaining the resonance state in the transmitter circuit despite the inductive coupling parameters fluctuation, hence they suggested servo loops and control systems. Others groups, try to increase the transfer efficiency by optimizing the coil geometry. The optimisation of the power conversion and management circuits in the receiver unit is a subject of major interest for other researches.

Improving system efficiency is possible by increasing the magnetic strength (increase the current), the rate of the magnetic field change (increase the frequency), and the flux linkage between the two coils (reduce misalignment and distance transfer). Nevertheless, that can engender health complications if the human body exposure to the EM field exceeds certain limits and it can be incompatible with the new implants design requirements (big size) .However, it should be remembered that the main goal in this area is to transfer power wirelessly with high efficiency and power stability in safe conditions. To achieve this objective, a lot of challenges should be overcome. Firstly, the receiver's coil form factor, as it is implanted inside the human body, should be considerably small. However, this limits the system PTE. Secondly, increasing the operating frequency implies an increase of PTE. However, energy losses in the tissue may also increase, which increases tissue temperature and threatens human body safety. Thirdly, the power transfer efficiency is inversely proportional to the separation distance between the coils. Finally, meeting the power need of implants is challenging especially for hight power transcutaneous applications. The main issue is how to transfer power with high efficiency since the latter decrease with the power level. On the other hand, high power applications present another challenge that is how to avoid the thermal issue and how to reduce at maximum the EM power absorption by the human body. Thereby, ensuring patient safety, minimizing device size and reducing power consumption are always considered as significant challenges in modern IMDs design. In recent years, RIC WPT systems for IMDs have undergone a remarkable improvement thanks to research efforts. Nevertheless, practical difficulties and stringent design requirements still present many limitations and challenges to overcome.

4 Proposed Solutions to Improve the Overall RIC WPT System Efficiency

Resonant inductive coupling parameters such as the operating frequency, the input voltage, coils misalignment, and load variations may affect the overall system efficiency. For transcutaneous applications, wireless power transfer efficiency should be important and the load power supply should be well-regulated. Moreover, in the RIC WPT systems that consider coil misalignments, load and coupling distance variations, the regulation become more important. Hence, a poor system regulation deeply decrease the PTE. In the aim to increase the wireless coupling efficiency, and provide a strong regulation in the transmitter circuit (DC/AC converter + transmitter coil) and in the receiver circuit (receiver coil + AC/DC converter). Proposed RIC WPT systems can be classified into two types: systems modeled in tissue and others modeled in the air medium. Figure 7 illustrates the main proposed alternatives to increase the overall RIC WPT system efficiency. In the following, we provide a comprehensive investigation of these alternatives.

4.1 Transmitter Circuit Optimization

Transmitter circuit (TX) optimization is a very important step, hence it controls the transmitter power, and maintains the resonant state in the transmitter circuit. As we will explain in this section, TX optimization is possible by controlling the PA supply voltage, the resonant frequency, or the power switch control.

4.1.1 PA Supply Voltage Control

Li et al. have proposed a 13.56MHz wireless power transfer system with a $1X/2X$ reconfigurable resonant regulating (R^3) rectifier in which the load condition in the receiver side can be feed-back to the transmitter side (Li et al. 2015). The rectifier works as a bridge rectifier in $1X$ mode ($mode = 0$) with an output equal to V_{1X},

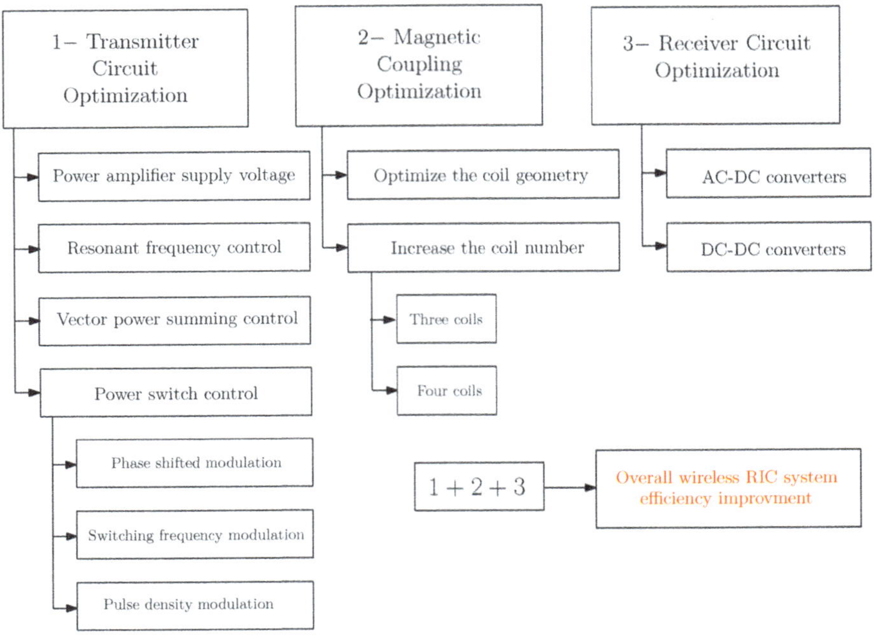

Fig. 7 Main proposed alternatives to increase the overall RIC WPT system efficiency

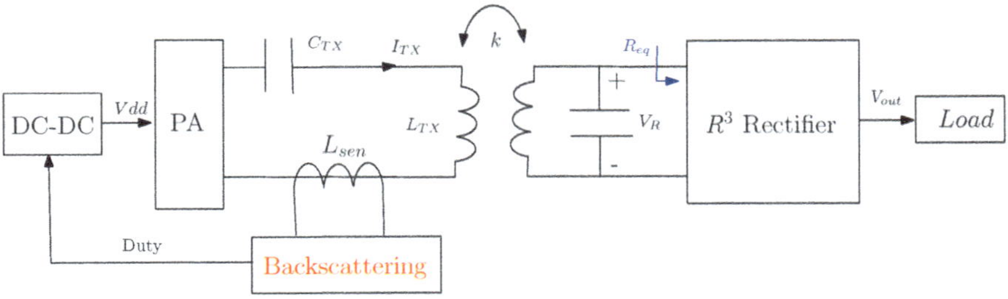

Fig. 8 Block diagram of the transmitter coil (T_X) regulation based on the supply voltage control

and as a voltage doubler in $2X$ mode ($mode = 1$) with an output equal to V_{2X}. The voltage window $[V_{1X}, V_{2X}]$ depends on the transmitter power and load current (resistance). The system is useful to extend the output range in the receiver circuit, where $V_{2X} < V_{1X}$ and thus false regulation may occur under certain load conditions. To avoid this, the output power in the transmitter circuit was increasing during the 2X mode to ensure that V_{2X} is always greater than V_{1X}. Figure 8 illustrates the proposed system, it shows that the feedback is ensured by a global control loop, and the load status is sent by a backscattering approach. Later, the difference of Req between 1X mode and 2X mode is sent to the transmitter side. An additional coil coupling L_{sen} sends the courant difference I_{TX} between the two mods to the transmitter coil T_X.

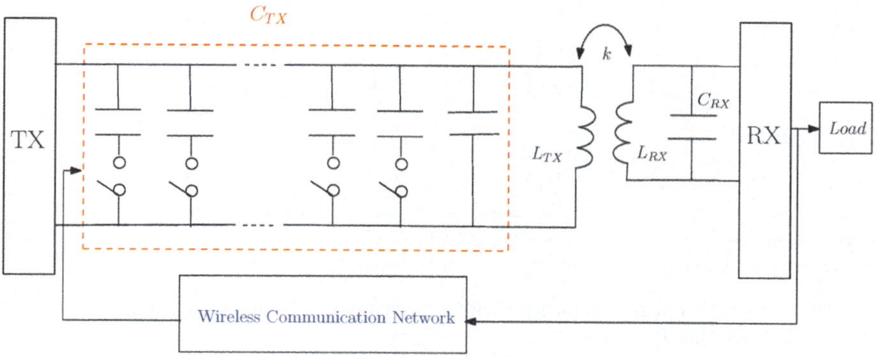

Fig. 9 Transmitter circuit optimization based on resonant frequency control using a weighted-capacitor array

4.1.2 Resonant Frequency Control

Transmitter circuit optimization can be achieved by adjusting the resonant frequency f_0. Si et al. proposed a resonant frequency control system to maintain the resonant state (Si et al. 2008). Figure 9 illustrates the proposed TX regulation system, that adjust the the resonant frequency by controlling the capacitor C_{TX} value in the resonant TX tank. The capacitor value depends on the load condition of the receiver side. It is controlled through a wireless communication network. Measurements results showed that the system regulates effectively the power delivered to the load over a wide load range. In addition, it ensures a power conversion efficiency of 80% for $10mm$ distance transfer and $15W$ of transferred power.

To maintain the resonant state in RIC WPT system for transcutaneous applications by controlling the resonant frequency, Trigui et al. proposed a new technique to compensate for the sensitivity of the inductive link to the coupling factor variations between the transmitter and the receiver coils (Trigui et al. 2015a). The system maintains the resonant state at the transmitter circuit and ensures a better wireless power transfer despite these variations. Figure 10 presents the proposed inductive power transmitter circuit optimization. The system contains an oscillator followed by a gate driver circuit that drive the class E PA switch at 13.56 MHz. The primary resonant frequency was maintained by tuning the series capacitor. The self-calibrated system includes:

- Sensing unit to measure the positive peak of the primary coil AC voltage.
- Control unit to control the high-resolution stepper motor.
- Stepper motor attached to a trimmer capacitor to automatically adjusts the capacitance value according to the sensed peak voltage.
- The trimmer capacitor is tunable from 25pF to 500pF, which allows a large immunity against large coupling variations.

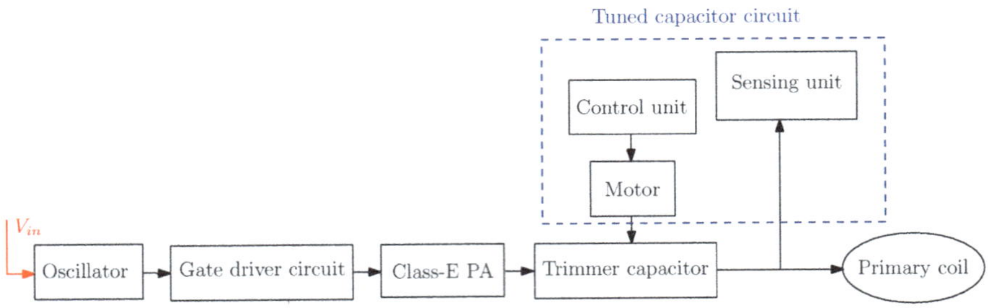

Fig. 10 Transmitter circuit optimization with a mechatronic module that dynamically tunes the primary resonant capacitor value

The proposed approach allows the RIC WPT system to reach an efficiency improvement of 66.5% compared to a typical system. Further details about the functional principles are provided in Trigui et al. (2015a).

4.1.3 Vector Power Summing Control

Usually, transmitter circuit optimization is achieved only with a single transmitter coil, while it can also be by using multiple coils in the transmitter side. Kosage et al. proposed a WPT system with two transmitter coils (Kosage et al. 2016). As Fig. 11 illustrates, the system contains a vector power summing control technique to regulate the received power P_{Total}. The two power vectors implemented by two separated transmitter coils, are then summed at a single receiver coil R_X. The two coils are driven by two switching clocks (CLK_1 and CLK_2) with different phase θ under the same switching frequency ($F_{sw} = 1/T_{sw}$). Consequently, the transmitted powers, P_0 and $P_0\ e^{i\theta}$, are achieved in the overlapped T_X coils L_{TX1} and L_{TX2}, respectively. The combined transmitted power is expressed as follows:

$$P_{Total} = P_0\ (1 + \cos\theta) \tag{9}$$

Equation 9 illustrates that it is possible to regulate the transferred power by controlling θ using a delay locked loop (DLL) and phase control circuit based on global feedback from the R_X.

4.1.4 Power Switch Control

Transmitter circuit optimization may also be achieved using a power switch control approach based on phase-shifted modulation, switching frequency modulation, and pulse density modulation. These approaches control the resonant state in the transmitter side by keeping a constant resonant frequency (f_0). These designs are

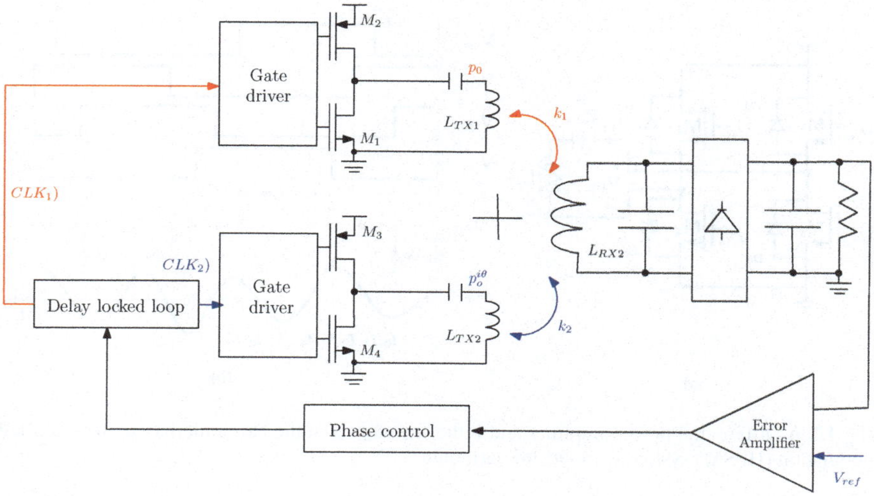

Fig. 11 Transmitter coil regulation based on the vector power summing control technique, where the output power is summed over two-transmitted powers

presented in the following paragraphs.

(a) Phase-Shifted Modulation

The output voltage V_{PA} of a differential power amplifier is regulated by varying the phase-shifted angle α of the gate signals $V_{G1\sim4}$. However switching loss increases as α appraoches the 180° which results in low efficiency. To address this, Cai et al. proposed a transmitter circuit optimization with harmonic-based-shifted modulation (HPSM) (Cai et al. 2014). This approach applies the nth harmonic component of the switching frequency to adjust the transmitted power. Figure 12 illustrates the example of the $3rd$ HPSM.

According to Fig. 12b, a half-switching period is subdivided into six stages delimited by the time interval between t_i and t_{i+1}, with i=[1; 6]. These half-switching modes acts on the the primary LC tank voltage and current V_{PA} and I_{TX}. The 3rd HPSM approach allows the current I_{TX} circulates in the transmitter circuit three times during one switching period which reduces switching loss. Measured results show that the proposed approach improves the power amplifier efficiency by 10% compared to fundamental-based phase shifted modulation.

(b) Switching Frequency Modulation

Figure 13 presents the block diagram of the developed wireless power delivery system. It consists of a resonant transmitter circuit (TX), a receiver secondary side resonator circuit (RX), and a rectifier. The resonant frequency f_{res} is tuned at 13.56

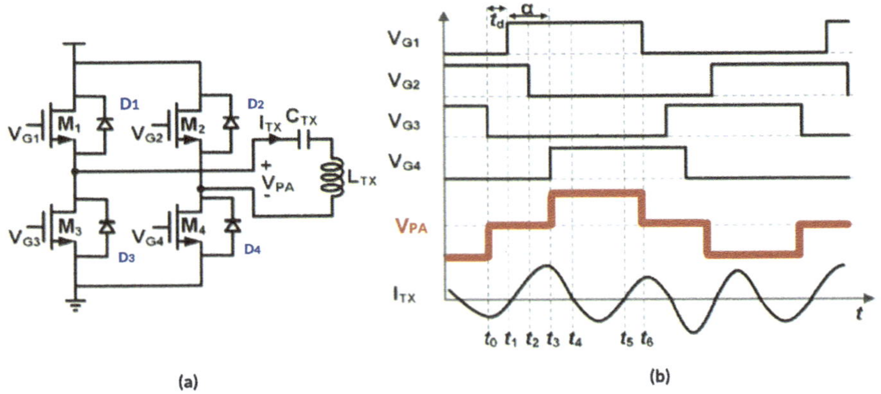

Fig. 12 **a** Differential power amplifier and **b** the waveforms in the harmonic-based phase-shifted modulation (HPSM) system with the 3rd harmonic

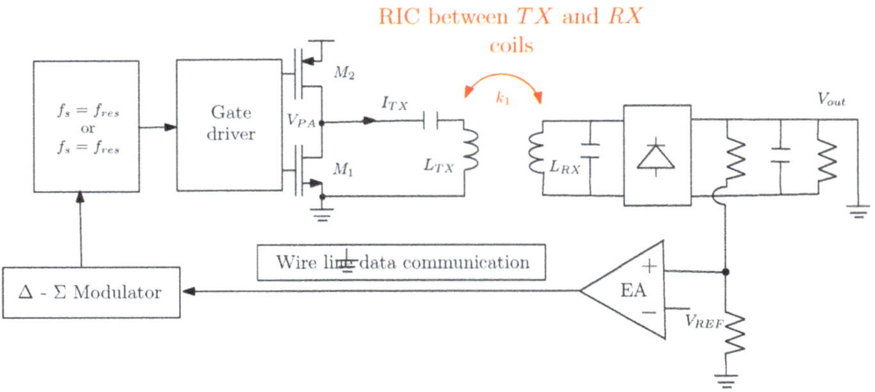

Fig. 13 Block diagram of the subharmonic resonant switching system using a RIC WPT approach to send power and a wireline to control the error of the amplifier (Shinoda et al. 2012)

MHz. In this system, the switching frequency f_s can be set to f_{res} as illustrates Fig. 13, respectively. V_{ref} can be regulated by switching f_s between f_{res} and $f_{\frac{res}{3}}$ using a PWM scheme while the resonant frequency is kept constant. In the experiments, the error amplifier (EA) output is fed back to the transmitter circuit through a wireline, and a $\Delta - \Sigma$ modulator is applied to switch the PA between the f_{res} and $f_{\frac{res}{3}}$ modes to reduce spurious emission (Shinoda et al. 2012). In the experiments, the feedback signal was sent to the transmitter by wireline. As a variable load, an electrically controlled load was used and the load was changed by a PC (Fig. 14) (Shinoda et al. 2012).

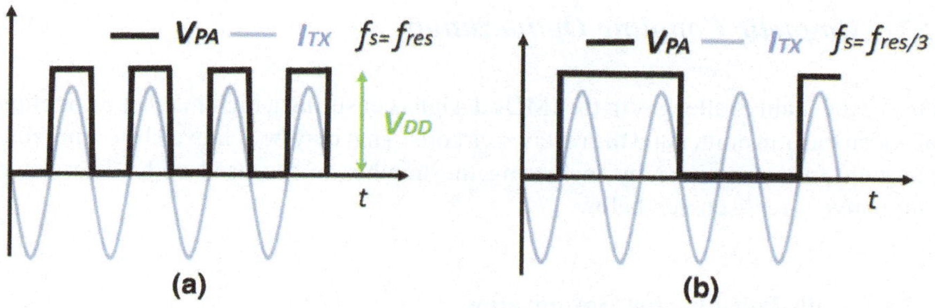

Fig. 14 The power amplifier V_{PA} and I_{TX} waveforms f_{sw} = **a** f_{res} and **b** $f_s = \frac{f_{res}}{3}$

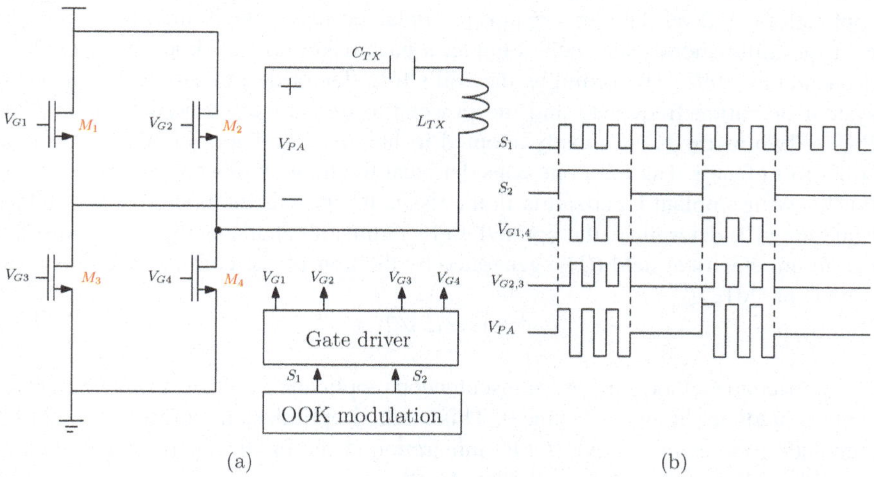

Fig. 15 **a** Schematic system of the TX regulation using PDM based on On-Off Keying (OOK) modulation, **b** Signal waveforms (S_1, S_2, V_{G1-4}, and V_{PA}) (Zhong and Hui 2018)

(c) Pulse density modulation

Zhong et al. proposed proposed a pulse density modulation (PDM) approach based on On-Off Keying (OOK) modulation (Zhong and Hui 2018). The system regulates the transmitted power by switching the PA between the working and the idle modes. By referring to Fig. 15, the deferential PA contains four power switches (from M_1 to M_4), that are driven by the gate control voltages V_{G1} and V_{G2}. Figure 15b shows that the switching signal S_1, modulated with an OOK signal envelope S_2, is applied to generate the control signals (from V_{G1} to V_{G4}). In the case where S_2 is logic high, the PA operates normally, and when S_2 is logic low, the PA operates in idle mode.

4.2 Magnetic Coupling Optimization

One of the main challenges in the IMDs design is ensuring a high magnetic coupling between the transmitter and the receiver circuits. That may be achieved by optimizing the coils form factor, or by increasing the number of the used coils. These two alternatives are discussed below.

4.2.1 Coils Form Factor Optimization

Coil's form factor miniaturization is one of the main challenges in transcutaneous applications. Indeed the receiver unit is implanted inside the human body, its size is at the millimeters scale. For example, a pacemaker size is about 49×46×6mm (Reach et al. 1992). Accordingly, the coil's form factor design should consider the system operating frequency and the power transfer range required by the active IMDs. Two methods are mostly adopted in literature to determine the transmitter coil's form factor. The First one considers that the transfer distance (d) is constant. For knowing implant localization in the tissue layers, this method is very suitable. Equation (10) determines the optimal coil's radius, it calculates (by derivation) the maximum electrical field ((H)) generated by the transmitter coil at a fixed distance (Sun et al. 2013).

$$r = d\sqrt{2} \tag{10}$$

The second method concerns transcutaneous applications with a moveable receiver (with a predicted transfer distance). This method optimizes the transmitter coil by calculating the electric field H (Ψ) integration at the predicted transfer range (d) using the following equitation (Sun et al. 2013):

$$\Psi = \int_{d_1}^{d_2} \vec{H} \times \vec{dd} = \int_{d_1}^{d_2} \frac{INr^2}{2\sqrt{(r^2+d^2)^3}} dd \tag{11}$$

where I is the current circling in the coil and N is the coil' turns number.

A commonly used method to optimize coils geometry during their design is the custom-made method (Xue et al. 2013). This method is useable when the operating frequency is already defined. It imposes four steps in the aim to optimize the transmitter and the receiver coils. (1) apply the design constraint of the receiver coil. (2) define the initial values for the coil dimension based on fabrication limitations and optimum transmitter coil dimension. (3) optimize the size of the transmitter coil. (4) step is to set the line width and full factor for the receiver coil side. Iterations are going until the required coils quality factor efficiency η value is achieved. Another method was proposed by Harrison et al. (2007) to optimize the planar spiral pancake-shape coils. Harrison and al. have concluded that the maximum η could be achieved by considering $dt = 0.18Dt$, and $dr = 0.75Dr$, where dt is the inner transmitter diameter, dr is the inner receiver diameter, and Dr is the outer receiver diameter.

The research results are based on derived equations and experimental study, which was conducting using multiples coils and their dimensions at maximum achieved η.

The coupling coefficient and the impedance matching are key parameters behind a high PTE. Hence, they allow the system to operate correctly with a small-received power. Moreover, they reduce tissue overheating and interference with other electronic devices. Skin and environment energy losses, device size, may limit the delivered power to the load. Therefore, matching and power management techniques should be present in the system design to increase the PTE.

4.2.2 Three Inductive Coils Coupling

Figure 16 illustrates the conventional WPT system where WPT is achieved via two resonant inductive coils (L_1, R_1), and (L_2, R_2), respectively. The two resonators (($L_1 R_1$,C_1) and ($L_2 R_2$, C_2)) are tuned to operate at a precise frequency. PTE efficiency depends on the mutual coupling between the two coils, k_{12} and the quality factors, Q_1 and Q_2; which all depends on the coils inductances, L_1 and L_2 , respectively. There is an optimal load (R_L) for a precise Q_1 and Q_2, which can maximize the system PTE. However, the optimal load value depends on the IMDs power consumption. Therefore, the two-coil WPT system may not achieve the highest PTE in certain transcutaneous applications.

Figure 17 shows the enhanced design that is a three-coil system. A new freedom degree (k_{23}) is added by adding the load coil L_3. This system was proposed to maximize the PTE for a precise load (R_L) by performing impedance matching.

PTE values reached in the conventional system (two coils), and the enhanced system (three coils) was compared in Kiani et al. (2011). The two investigated systems operate at 13.56 MHz. Results shows that in the 3-coils system, PTE is ensured for a full loads range from 10Ω to 1 kΩ. However, in the 2-coils system, optimal PTE is achieved only for a specific load value, R_{Load}= 200Ω.

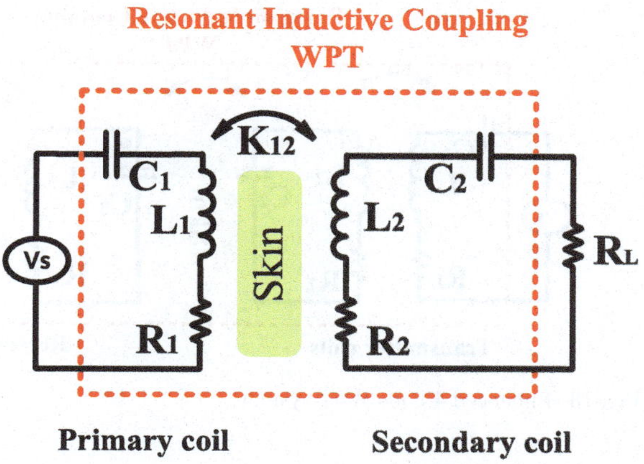

Fig. 16 2-coils WPT system for transcutaneous wireless power transfer applications

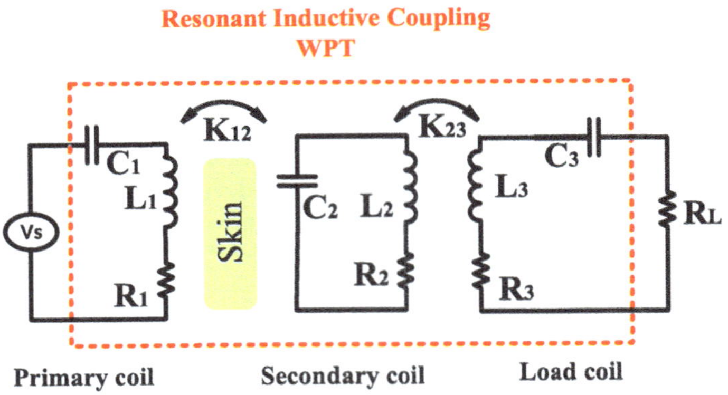

Fig. 17 3-coils enhanced WPT system: one coil in the transmitter circuit and two coils in the receiver circuit to improve the PTE by the load transformation

The provided results, showed that the conventional 2-coil system produces a high PTE for short coupling distance, which is the case of many transcutaneous applications. Thus, it is close to the optimal choice especially to reduce the system volume (Kiani et al. 2011).

4.2.3 Four Inductive Coils Coupling

Figure 18 presents a 4-coils WPT system which is an improved version of the 3-coils design (Mujeeb-U-Rahman et al. 2015). The system contains four resonant circuits magnetically coupled by the coefficients k_1, k_2, k_3 and k_4. Compared to the conventional system presented in Fig. 16, it adds two freedom degrees that are k_{23} and k_{34}. The coil quality factor is expressed as follows:

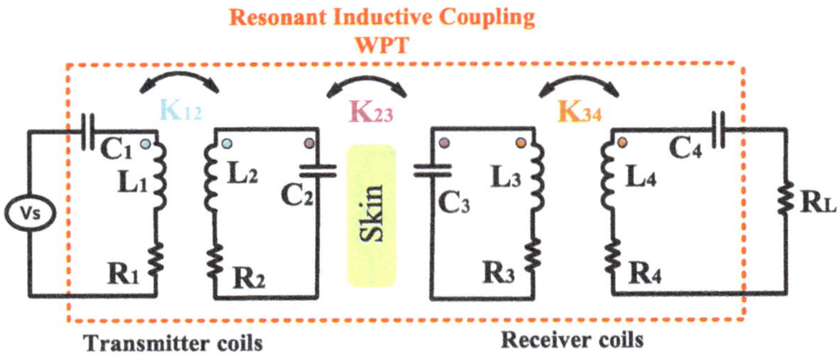

Fig. 18 Enhanced 4-coils WPT system

$$Q_i = \frac{1}{R_i} \times \sqrt{\frac{L_i}{C_i}} \tag{12}$$

where $i = 1, ..., 4$ is referred to the coil number. PTE is maximum if the coupling factor, k_c reaches a critical value given by the following formula:

$$k_c = \frac{1}{k_c} + k_{12}^2 \tag{13}$$

4-coils WPT system provides a more significant transfer distance and a higher transfer efficiency compared to the 2-coil systems. Moreover, it is less sensitive to the distance variations. Nevertheless, these values are relatively limited. A more detailed discussion can be found in Mujeeb-U-Rahman et al. (2015).

5 Receiver Circuit Optimization

The receiver conversion setup contains a circuit chain namely rectifiers, capacitors and regulators to convert the AC voltage to a DC voltage. Design specifications influence the power conversion structure choice like the frequency, the input voltage, the transferred power and the system volume. Power conversion efficiency in the receiver circuit is very important to improve the overall WPT system efficiency and delivers the required power to the load. Since the recovered voltage changes with the coil distance variations and misalignments, a DC–DC converter named also a voltage regulator often follows the AC–DC converter to provide a constant supply voltage to the implant. AC–DC and DC–DC converters are presented and discussed in the following sub-sections:

5.1 AC–DC Converters

Rectification is the first step in the receiver circuit conversion process, it is ensured by a rectifier that converts the received alternative voltage (AC) at the receiver coil to a direct voltage (DC). Rectifier circuits may be single-phase (half-wave rectification and full-wave rectification) or multi-phase (three-phase half-wave circuit, a three-phase full-wave circuit using a center-tapped transformer, or three-phase bridge rectifier controlled). Single-phase rectifiers are mostly used for low power consumption devices, but three-phase rectifiers are very used for high-power industrial applications. In general, the types of rectifiers are:

- Passive rectifiers: provide a fixed DC output voltage for a given AC supply where diodes are used only.

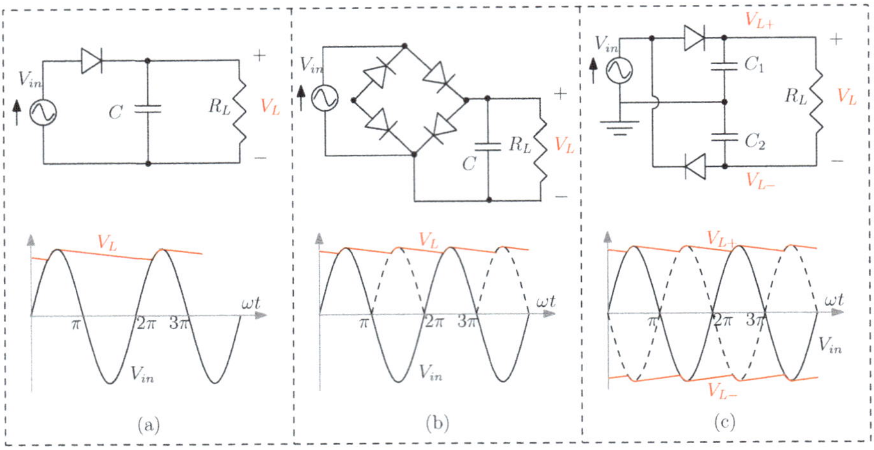

Fig. 19 Passive rectifiers: **a** half-wave rectifier, **b** full-wave rectifier and **c** voltage doubler

- Active rectifiers: is a technique for improving the efficiency of rectification by using active devices such as transistors and operational amplifiers..

The commonly used passive rectifiers for IMD are the half-wave rectifier, the full wave rectifier and the voltage doubler as shown in Fig. 19. In inductive coupling WPT systems, efficiency is always the major concern during system design. In general, the rectifier efficiency η_r is determined as follows:

$$\eta_r = \frac{V_L}{mV_{FW} + V_L} \tag{14}$$

Where V_L is the rectifier output voltage, m is the number of diodes and V_{FW} is the voltage drop.

The half-wave rectifier has the simplest design since it contains only one diode and a capacitor. In addition, it has the best efficiency compared with other types (Trigui 2013). Passive rectifiers have a small size which reduces the IMD system design complexity. Nevertheless, they create a lot of harmonics and operate at low power range. Moreover, they provide a fixed DC output voltage that can't satisfy the IMD load variations. The power conversion efficiency of the receiver unit in a RIC WPT is mainly limited to the efficiency of the rectifier circuit. Thus active rectifiers are usually used compared with the passive rectifiers to reduce the forward voltage across the employed diodes and consequently increase the efficiency of the circuit. The control of the power range and the DC voltage is an important advantage in active rectifiers. Therefore, these devices are widely used in IMDs (Xiang et al. 2018; Yogosawa et al. 2017). However, the disadvantages of this type of rectifier are the high switching losses and current harmonics that occur due to hard switching.

5.1.1 Reconfigurable AC–DC Converter

Active rectifiers require higher peak inputs compared to their outputs, which may temporarily unavailable at large coils distances due to the weak inductive coupling. In addition, the variations in the received voltage across the receiver coil resulting from coil misalignments or the load variations can reduce the supply voltage of the IMDs. In order to overcome such limitation, Lee et al. proposed a 13.56 MHz adaptive reconfigurable active voltage Doubler/Rectifier for extended-range inductive power transmission (Lee et al. 2011). The system contains a voltage doubler (VD) and a rectifier (REC) integrated into a single structure. Using an output voltage sensing circuit, the reconfigurable VD/REC can adaptively change its operating mode to either VD or REC depending on which one is the better choice for generating the desired output voltage with the highest PCE. This helps the VD/REC to accommodate with a wider range coil couplings. Figure 20 illustrates the schematic diagram of the proposed VD/REC. It contains active diodes (D_1, D_2, D_{N1}), in which comparators (CMP_{P1}, CMP_{P2}, CMP_{N1}) drives pass transistors (P_1, P_2, N_1) at the appropriate times due to their turn-on and turn-off offset functions, which lead to high PCE and low dropout voltage. Lee et al. provides more details about the circuit design and the measurements results in both air and muscles environments.

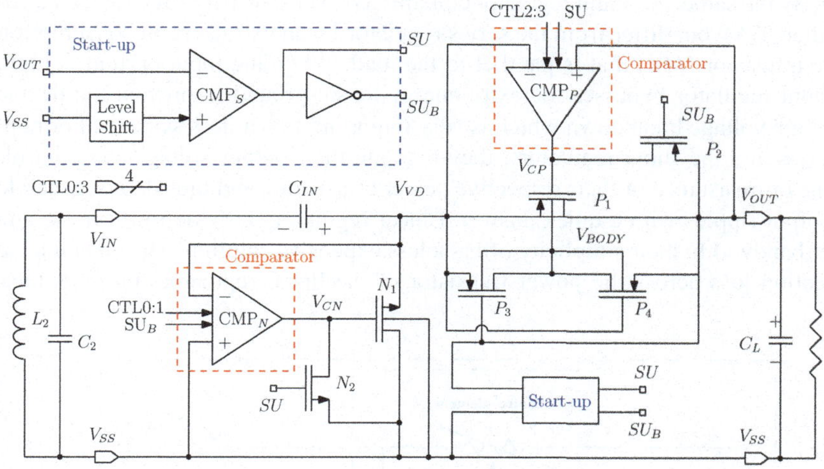

Fig. 20 Active voltage doubler employing high speed offset-controlled comparators (CMP_N and CMP_P) to achieve higher PCE for REC and VD modes

5.2 DC–DC Converters

The DC–DC voltage converter or regulator maintains the rectified output voltage constant regardless of load variations or changes in the input AC voltage. They are referred to as linear or switching regulators, depending on the method used for conversion process.

A. Linear regulators

The linear regulator uses linear components such as a resistive load with others components to regulate the voltage output. It maintains the output at a constant level, by dissipating excess power delivered by a voltage or current source. It is very suitable for low-power and miniature systems, such as wireless power receivers for portable and implantable applications. The simplest way to regulate the rectified voltage in the receiver circuit is to add a linear regulator as shown in Fig. 21 where V_{IN} is the regulator input voltage and V_{OUT} is the regulator output voltage.

As shown in Fig. 22 the linear regulator may be a series or shunt regulator. The series regulator is connected in series with the load. It contains a power transistor (MP) controlled by an error amplifier (EA) and a voltage reference. The series linear regulator adjusts the output voltage headroom $(V_{IN} - V_{OUT})$ by its transistor (M_P). By the same, the shunt regulator contains a power transistor (M_N) and an error amplifier (EA), but differ from the series regulator by an extra current consumption by the transistor (M_N) that is parallel to the load. When the input current is high, the shunt regulator bypasses the extra energy to the ground to prevent the device from over-voltage breakdown which is very important in wireless systems. For both the series and the shunt regulators, they regulate their output voltages by controlling the on-resistance of their respective power transistors, and thus do not generate any output ripples with a quick answer. Linear regulators provide lower noise and higher bandwidth; their simplicity offers a less expensive solution. The unavoidable conduction loss across the power transistor of the linear regulators presents their

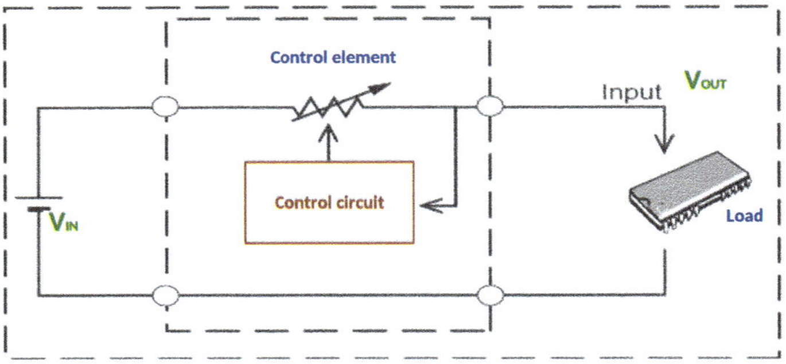

Fig. 21 Linear converter

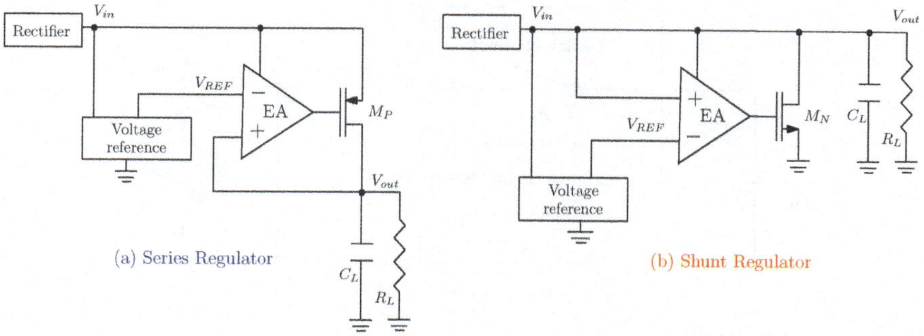

Fig. 22 Linear regulators: **a** series regulator, **b** shunt regulator

major drawback. These losses are caused by the load current, and they are proportional linear to $V_{IN} - V_{OUT}$. V_{DO} is defined as the difference between the minimum input voltage ($V_{IN,MIN}$) and the output voltage V_{OUT}. Therefore in order to get high efficiency, the linear regulator has to have a very low dropout voltage. In fact, the linear low-dropout (LDO) regulators are mostly used in portable applications. Series regulator is more suitable for WPT systems because the power transistor of an LDO regulator is commonly a P-type transistor as shown in Fig. 1.28 (a). Hence its gate can be driven by a low voltage and it can work in the active region even V_{IN} is very close to V_{OUT} (Lu et al. 2018).

B. Switching regulators

Figure 23 illustrates a simple schematic of a switching voltage regulator. The system uses a switching element to transform the incoming power supply into a pulsed voltage, which is then smoothed using capacitors, inductors, and other elements. Power is supplied from the input to the output by using non-dissipative elements, typically MOSFET and diodes to switch between an on-and-off state until the desired voltage is reached. The switching converter may be a buck-converter, a boost-converter, or a fly-converter, depending on the arrangement of its components. Switching voltage regulators offer three main advantages compared to linear regulators:

- Better switching efficiency.
- The power stored by the inductor can be transformed to output voltages that can be greater than the input (boost), negative (inverter), or can even be transferred through a transformer to provide electrical isolation with respect to the input.

Nevertheless, they also have disadvantages since they can be noisy, require more external parts (control loops) and the system design is complex. As we have illustrated in this section, each converter type has advantages and disadvantages, therefore the application requirements determine which one is more suitable for each case.

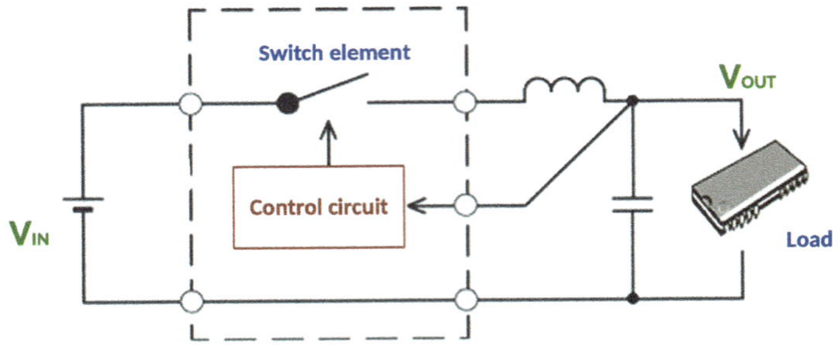

Fig. 23 Switching regulator

5.2.1 Reconfigurable DC–DC Converter

IMDs require precise power regulation and management. The ideal power converter for such applications must present high efficiency in order to avoid the human body tissues heating and must have a small form factor regarding the invasive nature of the implantable circuit. Power requirements of IMDs differ from a device to another. For example, a pacemaker needs 20 μW however a cochlear implant consumes 2400 μW. George et al. have proposed a reconfigurable smart DC–DC converter with wide voltage and power range output to be suitable for varieties of IMDs (George et al. 2016). The proposed system is able to provoid multiple levels of DC–DC voltage. Since it ensures both the buck and the boost voltage conversion. George et al. provide more technical details about the Buck-Boost DC–DC converter for biomedical implants.

6 Human Health Consideration for EM Field Exposure

Powering active IMDs is very delicate due to its in-vivo nature. Different parameters must be taken into account such as size miniaturization, biocompatibility, safe packaging and human body exposure limits to electromagnetic (EM) field. IMDs must operate under proper guidelines and regulations determined by the World Health Organization (WHO) (WHOQOL 1995), the Institute of Electrical and Electronics Engineers (IEEE) (IEEE 2006) and the International Commission for the Non-Ionizing Radiation Protection (ICNRIP) (ICNIRP 2010). Human body protection against the harmful effects of overexposure to the EM field is a critical subject in WPT systems. Indeed, by increasing the power level, the human body can be exposed to dangerous waves. The two main components that present a risk are the magnetic field frequency and intensity. For protection reasons, it is mandatory to respect standards and human safety limits during IMD designing and manufacturing.

Table 4 ISM Bands frequencies (ICNIRP 2010)

Frequency band	Values
Low frequencies (LF)	100–150 kHz
High frequencies (HF)	6.78 MHz; 13.56 MHz; 27.125 MHz and 40.68 MHz
Ultra high frequencies (UHF)	433.92 MHz; 869 MHz and 2.4 GHz

6.1 Safety Limits of Human Exposure to Magnetic, Electric and EM Field

Safety limits for human exposure to the EM field are established by analyzing various scientific evidence of the EM field effect on the human body. The WHO usually recommends two guidelines for safety levels to Radiofrequency (RF) and EM fields exposures, the ICNIRP and the IEEE. Both IEEE and ICNIRP precise in their reviews that human exposure to magnetic, electric, and EM fields may increase the human body temperature, stimulate muscle and nerve tissue, but there is no formal evidence that they cause cancer. Most of IMDs under research items are wirelessly powered by magnetic coupling, operating at frequencies that satisfy the ISM (Industrial, Scientific, and Medical) frequency bands. Hence, their use does not require any authorization from the authorities. Table 4 presents the ISM Bands values. Figures 24 and 25 shows the electric and magnetic field value corresponds to each frequency for both occupational and general public. The occupational population consists of adults who generally experience known electromagnetic field conditions. They are trained to be aware of potential risks and to take appropriate precautions. Oppositely, the general public consists of individuals of all ages and of varying health status. Mostly, these are unaware of their exposure to EM field. Moreover, individual members of the public cannot be expected to take precautions to minimize or avoid exposure to EM field. For the ultra-high frequencies (UHF) bands, if the frequency increases, the transmitted power also will be high, but the tissue warming effects will be severe (Mutashar et al. 2014). By respecting the EN 300 330 standards, it is preferable to work with frequencies lower than 30 MHz (Sparks et al. 2016). The 13.56 MHz frequency is the most adopted since it ensures a compromise between the authorized magnetic field strength the transmission range, and the biocompatibility (Trigui et al. 2015a).

6.2 Specific Absorption Rate

The Specific Absorption Rate (SAR) is the time derivative of the absorbed energy per mass of human body tissue when exposed to the (EM) field or radio frequency (RF) (Liu et al. 2019). It relates to the internal E-field by the following formula (Basaret al. 2018):

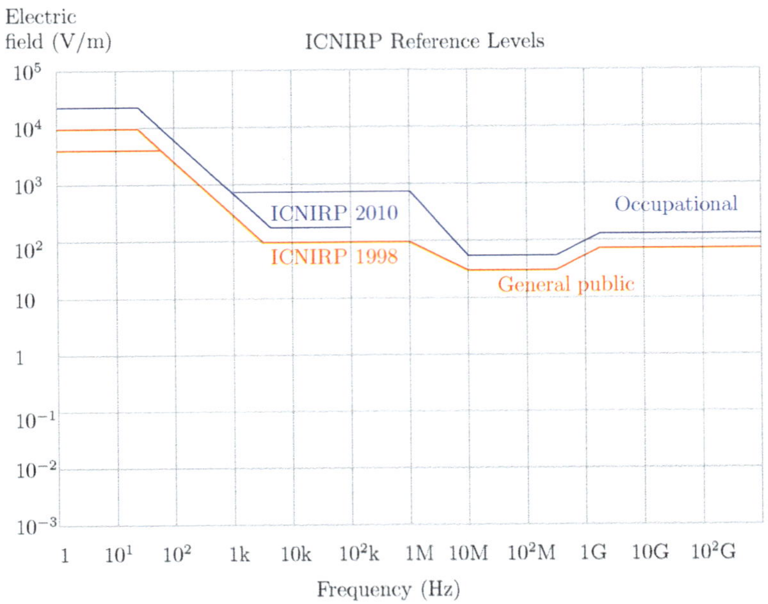

Fig. 24 Human exposure reference levels to time-varying electric field

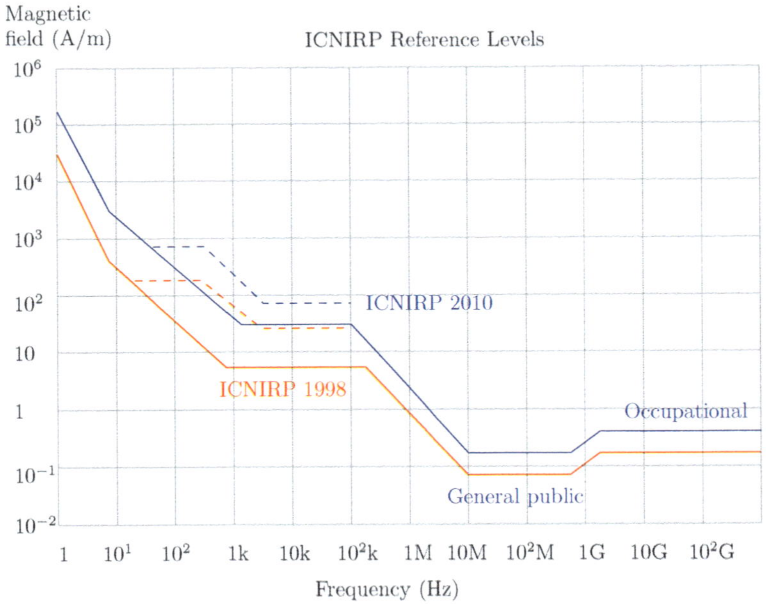

Fig. 25 Human exposure reference levels to time-varying magnetic field

Table 5 SAR limits for wireless applications for the frequency range 300 kHz to 300 GHz determined by ICNIRP and IEEE

Category	Organization	Head/trunk localized SAR (W/kg)	Limbs localized SAR (W/kg)
General public	ICNIRP	2	4
General public	IEEE	1.6	4
Occupational	ICNIRP	10	20
Occupational	IEEE	8	20

$$SAR = \frac{\sigma \mid E \mid^2}{\rho} \text{ (W/kg)} \tag{15}$$

with σ is the tissue conductivity, ρ denotes the mass density and E is the root mean square value (RMS) of the electric field strength. Very intense EM radiations can damage biological tissues. Therefore, harmful health effects may occur if exposure limits to the EM field are not taking into account. Hence respecting the SAR values is necessary for some exposure conditions, especially when the body is extremely near to the source. IEEE and ICNIRP limit the SAR for both occupational and general public exposure in the frequency range 300 kHz to 300 GHz for wireless applications as shown in Table 5.

7 Summary and Future Perspectives

Traditional IMDs will be replaced by a new generation of implantable smart, reconfigurable devices endowed with sensing, processing, communication, and actuation capabilities. Miniaturization, limited invasiveness, longevity, safety, and electromagnetic compatibility are some of the basic requirements that directly affect the design of appropriate powering methods for the next generation of smart IMDs. In this paper, firstly we have presented the general concept of implants with highlighting some of their medical applications and a brief recall of their history. Secondly, we have presented the batteries with their two categories: single-use and rechargeable. In addition, we have illustrated other types of embedded power sources like supercapacitors. As for independents systems, we have emphasized the near-field techniques. Hence, they promise specific advantages that can overcome the shortcoming of traditional methods such as electronic wires or batteries. Actually, the RIC WPT approach is the most suitable for wirelessly powering IMDs owing to its ex-tended transmission range, its relatively high efficiency, its ability to scale over a range of output power levels, from milliwatts to tens of watts, and its safety (less exposure to radio frequency radiation). In addition, it leads to more flexible range applications that were not otherwise possible. Nevertheless, recent IMDs design remains a challenge that may reduce the overall system efficiency. Accordingly, we have presented a survey that summarizes the main proposed designs in recent literature with the aim of increasing the wireless coupling efficiency, and providing a strong regulation in the transmitter and the receiver circuits. Due to the invasive criteria of IMDs, international guidelines should be well considered by designers/manufacturers of medical devices.

Ensuring high wireless transfer efficiency is still the most limiting factor in the development of futuristic IMDs. These smart devices will include processing, sensing, and wireless communication capabilities, which increase their power requirements that will be delivered to the load with more efficiency. Adding functionalities to IMDs increases not only its power requirement, but also its overall volume. Thus, considerable research should be conducted to power up cutting edge biomedical devices. Finally, all the proposed solutions for powering wireleeesly the IMDs with the RIC approach must be safe for the patient by using bio-compatible materials, and by respecting the human body exposure limits to the magnetic radiations.

References

Al Idrus, I. (2016). *Medtronic deep brain stimulation software earns FDA nod*. Fierce Biotech.

Al-Kalbani, A. I., Yuce, M. R., & Redoute, J. M. (2014). A biosafety comparison between capacitive and inductive coupling in biomedical implants. *IEEE Antennas Wireless Propagation Letters, 13,* 1169–1171.

Bazaka, K., & Jacob., M. V. (2012). Implantable devices: Issues and challenges. *Electronics, 2*(1), 3–33.

Basar, M. R., Ahmad, M. Y., Cho, J., & Ibrahim, F. (2018). An improved wearable resonant wireless power transfer system for biomedical capsule endoscope. *IEEE Transactions on Industrial Electronics, 65,* 7772–7781.

Ben Amor, A., Kouki, A., & Coo, H. (2015). Power approaches for implantable medical devices. *sensors*.

Ben Fadhel, Y., Katata, S., Rahmani, S., & Al-Haddad, K. (2016). Used methods to wirelessly powered implantable medical devices. *17th international conference on sciences and techniques of automatics and computer engineering (STA), Sousse,* (pp. 2–4).

Ben Fadhel, Y., Katata, S., Rahmani, S., & Al-Haddad, K. (2019). Near-field wireless power transfer is a promising approach to power-up active implants. In *20th international conference on sciences and techniques of automatics and computer engineering (STA 2019) Sousse, Tunisia*.

Ben Fadhel, Y., Katata, S., Sadraoui, K., Rahmani, S., & Al-Haddad, K. (2019). A modified wireless power transfer system for medical implant. *Energies*.

Biel, L., Pettersson, O., Philipson, L., & Wide, P. (2001). An electret-based angular electrostatic energy harvester for battery-less cardiac and neural implants. *IEEE Access, 5,* 19631–19643.

Bihr, U., Liu, T., & Ortmanns, M. (2014). Telemetry for implantable medical devices: Part 3-data telemetry. *IEEE Solid-State Circuit Magazine, 6*(4), 57–62.

Cai, H., Shi, L., & Li, Y. (2014). Harmonic-based phase-shifted control of inductively coupled power transfer. *IEEE Transactions on Power Electronics, 29,* 596–602.

Christ, A., Douglas, M., Nadakuduti, J., & Kuster, N. (2013). Assessing human exposure to electromagnetic fields from wireless power transmission systems. *Proceedings of the IEEE, 101,* 1482–1493.

Cruciani, S., Campi, T., Maradei, F., & Feliziani, M. (2014). Numerical Simulation of wireless power transfer system to recharge the battery of an implanted cardiac pacemaker. In *The international symposium on electromagnetic compatibility (EMC Europ 2014)*, Gothenburg, Sweden.

Council Directive of 20 June 1990 on the approximation of the laws of the member states relating to active implantable medical devices. *(90/385/EEC)*, 3.

Dai, J., & Ludois, D. C. (2015). A survey of wireless power transfer and a critical comparison of inductive and capacitive coupling for small gap applications. *IEEE Transactions on Power Electronics, 30*(3), 6018–6029.

Dai, J., & Ludois, D. C. (2015). Single active switch power electronics for kilowatt scale capacitive power transfer. *IEEE Journal of Emerging and Selected Topics in Power Electronics, 5*(1), 315–323.

Elizabeth Witherspoon(2019). Fighting hearing loss: What you need to know about cochlear implants. https://www.wraltechwire.com/2019/05/06/fighting-hearing-loss-what-you-need-to-know-about-cochlear-implants/

The Operating Principle in Electric Double-layer Capacitors.

George, L., Dario, G., Lehmann, T., & Hamilton, T. J. (2016). A 0.04 mm_2 Buck-Boost DC-DC converter for biomedical implants using adaptive gain and discrete frequency scaling control. *IEEE Transactions on Biomedical Circuits and Systems*, 3–7.

Harrison, R. R. (2007). Designing efficient inductive power links for implantable devices. In *IEEE international symposium on circuits and systems, USA* (pp. 2080–2083).

Hartfordhealthcare. (2020). Abnormal heart rhythms: Restore your heart to a normal beat. https://hartfordhealthcare.org/services/heart-vascular/departments/electrophysiology/pacemaker-defibrillator.

International Commission on Non-Ionizing Radiation Protection. (2010). Guidelines for limiting exposure to time-varying electric and magnetic fields for low frequencies (1Hz-100Khz). *Health Physics, 99,* 818–839.

Jiang, C., Chau, K., Liu, C., & Lee, C. H. (2014). An overview of resonant circuits for wireless power transfer. *Energies, 10,* 894.

Kanoun, O., Bradai, S., Khriji, S., Bouattour, G., El Houssaini, D., Ben Ammar, M., et al. (2021). Energy-aware system design for autonomous wireless sensor nodes: A comprehensive review. *Sensors, 2,* 548.

Kiani, M., Jow, U. M., & Ghovanloo, M. (2011). Design and optimization of a 3-coil inductive link for efficient wireless power transmission. *IEEE Transactions on Biomedical Circuits and Systems, 5*(6), 284–579.

Kiourti, A., & Nikita., K. S. (2017). A review of in-body biotelemetry devices: Implantable, ingestibles, and injectables. *IEEE Transactions on Biomedical Engineering, 464*(7), 1422.

Kosuge, A., Hashiba, J., Kawajiri, T., Hasegawa, S., Shidei, T., Ishikuro, H., et al. (2016). An inductively powered wireless solid-state drive system with merged error correction of high-speed wireless data links and NAND flash memories. *IEEE Solid-State Circuits, 51,* 1041–1050.

Lazzi, G. (2005). Thermal effects of bioimplants. *IEEE Engineering in Medicine and Biology Magazine, 24*(5), 76–81.

Lee, H., & Ghovanloo, M. (2011). Fully integrated power-efficient AC-to-DC converter design in inductively-powered biomedical applications. *IEEE, 1.*

Li, P., & Bashirullah, R. (2007). A wireless power interface for rechargeable battery operated medical implants. *IEEE Transactions on Circuits and Systems, 54*(10), 912–915.

Li, X., Tsui, C. Y., & Ki, W. H. (2015). A 13.56MHz wireless power transfer system with reconfigurable resonant regulating rectifier and wireless power control for implantable medical devices. *IEEE Solid-State Circuits, 50,* 978–989.

Lithium-Iodine battery - Flicker - the central intelligence Agency the central intelligence Agency.

Liu, L., Ye, D., Yu, Y., Liu, L., & Wu, Y. (2017) Carbon-based flexible micro-supercapacitor fabrication via mask-free ambient micro-plasma-jet etching. *2020 Chinese Academy of Sciences, 111,* 125.

Liu, C., Jiang, C., Song, J., & Chau, K. T. (2019). An effective sandwiched wireless power transfer system for charging implantable cardiac pacemaker. *IEEE Transactions on Industrial Electronics, 66,* 4108–4117.

Lu, Y., & Ki, W. H. (2018). Linear regulators for WPT. In *Book: CMOS integrated circuit design for wireless power transfer* (pp. 97–100).
Maleki, T., Cao, N., Song, S. H., Kao, C., Ko, S. C., & Ziaie, B. (2011). An ultrasonically powered implantable micro-oxygen generator (IMOG). *IEEE Transactions on Biomedical Engineering, 58*(11), 3104–3110.
Mashhadi, I. A., Pahlevani, M., Hor, S., Pahlevani, H., & Adib, E. (2019). A new wireless power-transfer circuit for retinal prosthesis. *IEEE Transactions on Power Electronics, 34,* 6425–6439.
Matthew, M., Valente, V., & Demosthenous, A. (2018). Practical inductive link design for biomedical wireless power transfer: A tutorial. *IEEE Transaction on Biomedical Circuits and Systems,* 15–16.
Mywood, I. (2017). Loyola first in Chicago to offer pill camera as colonoscopy alternative.
Mendoza-Ponce, P., John, B., & H. Krautschneider, W. (2018). Super-capacitors for implantable medical devices with wireless power transmission. In *2018 14th conference on Ph.D. research in microelectronics and electronics (PRIME)* (pp. 241–244).
Mohamad Jawad, A., Nordin, R., Kamel Gaghan, S., Mahmoud Jawad, H., & Ismail, M. (2017). Opportunities and challenges for Near-Field wireless transfer: A review. *Energies, 10*(1022), 5–14.
Mujeeb-U-Rahman, M., Adalian, O., Chang, C. F., & Scherer, A. (2015). Optical power transfer and communication methods for wireless implantable sensing platforms. *Journal of Biomedical Optics, 20*(9), 012–095.
Mutashar, S., Hannan, M., Samad, S., & Hussain, A. (2014). Analysis and optimization of spiral circular inductive coupling link for bio-implanted applications on air and within human tissue. *Sensors, 14*(7), 11522–11541.
Na, K., Jang, H., Ma, H., & Bien, F. (2015). Tracking optimal efficiency of magnetic resonance wireless power transfer system for biomedical capsule endoscopy. *IEEE Transactions on Microwave Theory and Techniques, 63*(1).
Orthoinfo. (2020). Fracture after total hip replacement. https://orthoinfo.aaos.org/en/diseases--conditions/fracture-after-total-hip-replacement/.
Osman, M. H., Shah, A. A., & Walsh., F. C. (2010). Recent progress and continuing challenges in bio-fuel cells. Part II: Microbial. *Biosensors and Bioelectronics*, 148.
Ozeri, S., & Shmilovitz, D. (2010). Ultrasonic transcutaneous energy transfer for powering implanted devices. *Ultrasonics, 50*(6), 556–565.
Pandey, A., Allos, F., Patrick Hu, A., Budgett, D. (2011). Integration of supercapacitors into wirelessly charged biomedical sensors. In *2016 6th IEEE conference on industrial electronics and applications electronics (PRIME)* (pp. 56–59).
Peña-Eguiluz, R., Flores-Fuentes., Antonio Mercado-Cabrera, A., López-Callejas, R., Rodriguez-Mendez, B. G., Valencia-Alvarado, R., Bethsabet Jaramillo-Sierra, B J., & Hernández-Arias., A. N. (2020). Stacked class-E amplifier. *IEEE Transactions on Industrial Electronics, 67*(10), 1.
Reach, G., & Wilson, G. S. (1992). Can continuous glucose monitoring be used for the treatment of diabetes?? *Analytical Chemistry, 64,* 382A-386A.
Schormans, M. Valente, V., & Demosthenous, A. (Oct, 2018). Practical inductive link design for biomedical wireless power transfer: A tutorial. **In IEEE Transactions on Biomedical Circuits and Systems, 12**(5), 1112–1130. https://doi.org/10.1109/TBCAS.2018.2846020.
Shinoda, R., Tomita, K., Hasegawa, Y., & Ishikuro, H. (2012). Voltage-boosting wireless power delivery system with fast load tracker by ds-modulated sub-harmonic resonant switching. In *Proceedings of the IEEE international solid-state circuits conference (ISSCC), Digest of technical papers*, San Francisco, CA, USA, (pp. 288–290).
Si, P., Hu, A. P., Malpas, S., & Budgett, D. (2008). A frequency control method for regulating wireless power to implantable devices. *IEEE Transactions on Biomedical Circuits and Systems, 2,* 22–29.
Sparks, C., Maughan, C., & Smith, L. (2016). Development of a glucose-powered biobattery for implantation and use in humans. *Research On Capitol Hill, 50*(3), 808–812.
Standard for safety levels with respect to human exposure to radio frequency electromagnetic fields, 3KHz to 300GHz (2006). IEEE Std C951-2005, 1–238.

Sun, T., Xie, X., & Wang, Z. (2013). *Wireless power transfer for medical microsystems*. New York, USA: Springer.
The World Health Organization quality of life assessment. *Social Science and Medicine IEEE Transaction on Antenna Propagation, 62*(12), 6492–6495.
Trigui, A. (2013). Asservissement de l'énergie inductive transmise aux implants électroniques. *Universitè de Montréal* (pp. 26–27).
Trigui, A., Hached, S., Mounaim, F., Ammari, A. CH., & Sawan, M. (2015) Inductive power transfer system with self-calibrated primary resonant frequency. *IEEE Transactions on Power Electronics*, 5–6.
Trigui, A., Mehri, S., Ammari, A., Ben Hadj Slama, J., & Sawan, M. (2001). Prosthetic power supplies. In: *Wiley encyclopedia of electrical and electronics engineering* (pp. 10–11).
Wang, J. (2001). 40 years of advances and challenges. *Electroanalysis, 13*, 984–988.
Wang, J., Leach, M., Lim, E. G., Wang, Z., & Huang, Y. (2019). Investigation of magnetic resonance coupling circuit topologies for wireless power transmission. *Microwave and Optical Technology Letters*, 5–9.
Wen, K. (2012). Early history and challenges of implantable electronics. *ACM Journal on Emerging Technologies in Computing Systems, 8*(2).
Xiang, Y., Wang, Y., & Richard Shi, C. J. (2018) A 13.56 MHz active rectifier with self-switching comparator for wireless power transfer systems. IEEE, 53–57.
Xue, R. F., Cheng, K. W., & Je, M. (2013). High-efficiency wireless power transfer for biomedical implants by optimal resonant load transformation. *IEEE Transactions on Circuits and Systems I: Regular Papers, 60*(4), 867–874.
Yogosawa, K.,Shinohara, H., & Kousuke Miyaji, K. (2017). A 13.56 MHz CMOS active diode full-wave rectifier achieving ZVS with voltage-time-conversion delay-locked loop for wireless power transmission. *IEEE*, 27–28.
ZeroHedge. (2020). Wisconsin Company is injecting RFID microchips into hands of employees. 32M company. https://www.mintpressnews.com/tag/rfid-microchip/
Zhong, W., & Hui, S. Y. R. (2018). Maximum energy efficiency operation of series-series resonant wireless power transfer systems using on-off keying modulation. In *Proceedings of the IEEE international solid-state circuits conference (ISSCC), Digest of technical papers*, San Francisco, CA, USA (pp. 288–290).

Fault Diagnosis for Nonlinear Biological Processes Based on Machine Learning Models

Radhia Fezai, Majdi Mansouri, Hazem Nounou, Mohamed Nounou, and Hassani Messaoud

Abstract Kernel-based learning techniques have been widely used to monitor and detect faults in biological systems. However, it is well known that the data used in the training phase must be stored and used for validation purposes. This results in a high computation cost when the training data set is very large. To address the above issue, we propose in this paper a novel approach to jointly enhance the detection accuracy and reduce the execution time required for fault detection. The developed approach, so-called, reduced kernel PLS (RKPLS)-based generalized likelihood ratio test (GLRT) aims to reduce the number of training samples to build a new KPLS model. Then, it consists to apply a GLRT to the evaluated residuals obtained from RKPLS model for fault detection purposes. A simulation using a Cad system in E.coli (CSEC) is performed to show how the reduction of the training data set affects the computation time and fault detection accuracy.

Keywords Partial least squares (PLS) · Kernel PLS (KPLS) · Reduced kernel PLS (RKPLS) · Fault detection · Generalized likelihood ratio test (GLRT) · Biological process

---

R. Fezai (✉) · M. Mansouri (✉) · H. Nounou · M. Nounou
Electrical and Computer Engineering Program, Texas A&M University at Qatar, Doha, Qatar
e-mail: majdi.mansouri@qatar.tamu.edu

R. Fezai
e-mail: radhia.fazai@gmail.com

H. Nounou
e-mail: hazem.nounou@qatar.tamu.edu

M. Nounou
e-mail: mohamed.nounou@qatar.tamu.edu

M. Nounou
Chemical Engineering Program, Texas A&M University at Qatar, Doha, Qatar

H. Messaoud
Laboratory of Automatic Signal and Image Processing, National Engineering School of Monastir, Monastir, Tunisia
e-mail: hassani.messaoud@enim.rnu.tn

O. Kanoun and N. Derbel (eds.), *Advanced Systems for Biomedical Applications*, Smart Sensors, Measurement and Instrumentation 39,
https://doi.org/10.1007/978-3-030-71221-1_4

1 Introduction

With the growing complexity of biological processes, the requirements for availability, security and reliability increase considerably. Failure detection is becoming a major problem in the biological applications. Thus, a large number of researchers, in the past years, are careful to develop an efficient process monitoring and fault detection approaches. In particular, with advances in instrumentation, computer and database technologies make it possible to collect and store large amounts of data from process variables. In many applications, process monitoring based on data-based methods is more appropriate. Multivariate statistical process control (MSPC) approaches have been widely used to solve the problem of detecting abnormal events in the given processes (Yi et al. 2017; Zhang and Zhang 2009; Lau et al. 2013; Gharahbagheri et al. 2017). PCA and PLS are the most popular MSPC approaches in the biological applications for process monitoring (Mansouri et al. 2017b, c). When the input-output information is available, PLS can provide more better modeling abilities through the use of input process data (X) and output data (Y), rather than focusing only on the variance of the input (X) as in an PCA. But, PLS is a linear model and can have unmodeled features when addressing a non-linear systems data. To address this issue, kernel PLS (KPLS) has been proposed (Rosipal and Trejo 2001; Yi et al. 2017; Zhang et al. 2015). KPLS models consist of transforming data into a higher dimensional space in which the data are linear. This makes the kernel-based machine learning models an attractive choice to modeling non-linear processes. KPLS provides an efficient detection abilities by obtaining latent variables that exhibit a non-linear correlation with the response variables and represents well the behavior of the model processes. KPLS is widely regarded as a data-driven tool for solving the problem of non-linear features since it has two advantages: (i) It has excellent capacity for non-linear mapping; (ii) It fully considers the non-linear relationships between variables of the process and the quality. For statistical process monitoring purposes, faults are detected with detection indices that trigger alarms when an index has violated its control limit; after a fault is detected. The most used indices are the Hotelling's T^2 index which is applied for monitoring principal component subspace in the feature space, and the squared prediction error (SPE) and generalized likelihood ratio test (GLRT) that are used for residual subspace monitoring. GLRT statistic is considered as the most efficient statistic in detecting faults in industrial processes (Botre et al. 2016; Mansouri et al. 2017b). However, the major drawback of KPLS-based fault detection technique is its computational complexity. Using KPLS for fault detection imposes a high computational cost when the training data set is large since the collected and stored data are used for both modeling and monitoring. To reduce the computational complexity of the KPLS model, a new reduced data set based on K-means clustering is extracted and then the new reduced KPLS model is built in the feature space using the new data set. Finally, GLRT statistic is used for fault detection. In monitoring biological processes, several approaches have been proposed. For instance, a real-time fault detection and localization scheme was developed for handwriting device on the plane (Chihi and Benrejeb 2018). The proposed approach allows connecting

IEMG signals/pen tip coordinates data as a parametric model for the multi-inputs multi-outputs system. This approach was proposed for detecting, in real-time, several types of faults in one or two inputs signals and in the same or different times.

The authors in Mansouri et al. (2018a) have proposed a new approaches based on kernel PCA for monitoring biological processes by detecting single and multiple faults in these processes. This technique was proposed to monitor the Cad System in E. coli (CSEC) systems through detecting faults in some variables like transport proteins, lysine, regulatory proteins, cadaverine and enzymes.

The authors in Lall et al. (2008) proposed an enhanced fault detection approach based on wavelet packet energy decomposition. This technique aims to identify the model, detect the fault and diagnose the biological electronic systems.

Other detection and monitoring approach was developed in Mansouri et al. (2018b). The developed approach uses the EWMA control chart for detecting aberrations in the genetic information of patients, which can help medical doctors to take a decision on diseases.

The objective of this paper is to propose a novel fault detection method that incorporates the GLRT statistic into reduced KPLS. The bases of the reduced KPLS-based GLRT is to extract useful information based on k-means clustering. Then, is to nonlinearly map the new data into a feature space to implement PLS in this space. In addition, monitoring statistic can be formulated from the reduced model for fault detection.

The remainder of this paper is organized as follows. In the Preliminaries section, PLS and KPLS are briefly introduced. The reduced KPLS-based GLRT for fault detection is given in the next section. In the Case studies section, a Cad system in E.coli (CSEC) process simulation study is given. Finally, some conclusions are drawn.

2 Preliminaries

2.1 *Kernel Partial Least Squares*

Kernel PLS consists of mapping the original non-linear data in a high-dimensional linear feature space, then performing a linear PLS (Rosipal and Trejo 2001; Li et al. 2010; MacGregor et al. 1994) in the feature space.

Figure 1 presents the main steps illustrating the kernel PLS algorithm.

ϕ is a nonlinear projection which maps the input vector from the original space into the feature space $\mathcal{F}$ in which it is approximately linearly linked as: Φ:

$$\Phi = \left[\phi(x_1)\ \phi(x_2)\cdots\phi(x_N)\right]^T \in \mathbb{R}^{N\times f} \tag{1}$$

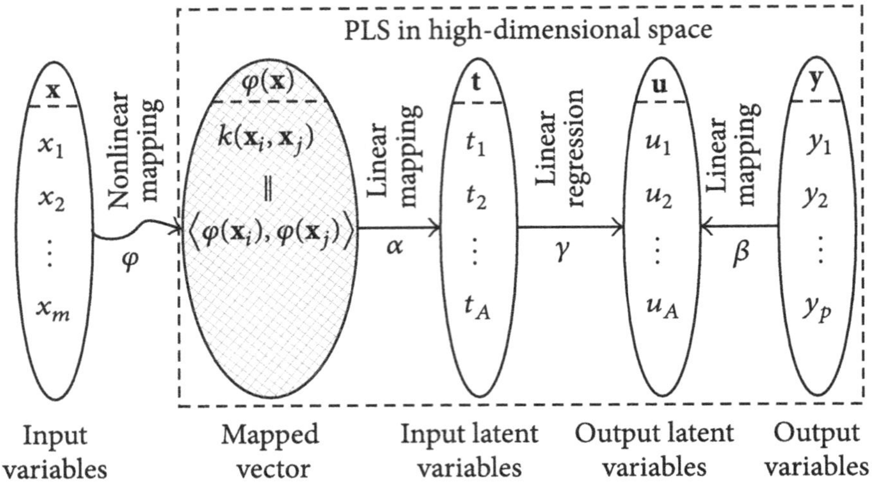

Fig. 1 Schematic diagram of kernel PLS

where, $I_N = \left[1 \cdots 1 \right]^T \in \mathbb{R}^N$ (Sheriff et al. 2017; Si et al. 2020; Botre et al. 2016; Fazai et al. 2019). The kernel PLS model of $(\overline{\Phi}, Y)$ can be identified as:

$$\begin{aligned} \Phi &= T P^T + F \\ Y &= U Q^T + G \end{aligned} \tag{2}$$

where, $P \in \mathbb{R}^{m \times A}$ and $Q \in \mathbb{R}^{N \times A}$ are the loading matrices of Φ and Y, respectively. $T \in \mathbb{R}^{N \times A}$ and $U \in \mathbb{R}^{N \times A}$ are the input and output score matrices, respectively. F and G are residuals. A is the number of latent variables (LVs) in the mapping space computed by the CPV metric (Sheriff et al. 2017). $\phi(.)$ is not explicitly represented and the dimension of Φ is arbitrarily large or even infinite. To deal with this issue, a kernel matrix K is presented as:

$$K = \Phi \Phi^T \tag{3}$$

where k_{ij} is the element of the i_{th} row and j_{th} column of K that is expressed as:

$$k_{ij} = \phi^T(x_i)\phi(x_j) = k(x_i, x_j) \tag{4}$$

The kernel matrix K is then computed as:

$$K = \Phi \Phi^T = \begin{bmatrix} k(x_1, x_1) & \cdots & k(x_1, x_N) \\ \vdots & \ddots & \vdots \\ k(x_N, x_1) & \cdots & k(x_N, x_N) \end{bmatrix} \tag{5}$$

Table 1 KPLS algorithm

1. Set $i = 1$, $K_1 = K$, and $Y_1 = Y$
2. Initialize the score-vector $\boldsymbol{u}_i (N \times 1)$ of the latent variable u_i of Y_i, as the maximum-variance column of Y_i
3. Compute the score-vector $\boldsymbol{t}_i (N \times 1)$ of the latent variable t_i of Φ_i, as $\boldsymbol{t}_i = K_i \boldsymbol{u}_i / \lVert K_i \boldsymbol{u}_i \rVert$, $\lVert \boldsymbol{t}_i \rVert = 1$
4. Regress the columns of Y_i on $\boldsymbol{t}_i$: $\boldsymbol{c}_i = Y_i \mathbf{t}_i$, where $\mathbf{c}_i$ is a weighting vector
5. Calculate the new score-vector: $\boldsymbol{u}_i = Y_i \boldsymbol{c} i / \lVert Y_i \boldsymbol{c}_i \rVert$, $\lVert \boldsymbol{u}_i \rVert = 1$
6. Repeat steps (3) to (5) until the convergence of $\boldsymbol{t}_i$
7. Deflate the matrices: $K_{i+1} = (I - \boldsymbol{t}_i \boldsymbol{t}_i^T) K_i (I - \boldsymbol{t}_i \boldsymbol{t}_i^T)$, $Y_{i+1} = Y_i - \boldsymbol{t}_i \boldsymbol{t}_i^T Y_i$
8. Save the data in the matrices: $T \leftarrow \boldsymbol{t}_i$, $U \leftarrow \boldsymbol{u}_i$
9. Set $i = i + 1$, and return to step (2). Stop when $i > \ell$, with ℓ being the selected number of latent variables

$k(.,.)$, so-called kernel function, can be expressed in different types (Cristianini et al. 2000; Nguyen and Golinval 2010). Gaussian kernel is the most applied functions and it is given by:

$$k(x_i, x_j) = exp\left(-\frac{(x_i -_j)^T (x_i - x_j)}{c}\right) \tag{6}$$

where c denotes the width of a Gaussian function. The regression coefficients matrix B can be computed from the K and Y matrices, as (Mansouri et al. 2017a; Godoy et al. 2014),

$$B = \Phi^T U (T^T K U)^{-1} T^T Y \tag{7}$$

The prediction of the response variables is expressed as:

$$\hat{Y} = \Phi B = K U (T^T K U)^{-1} T^T Y \tag{8}$$

From Eq. (8), we can show that the outputs can be obtained using the inner products of the mapped vectors. Thus, for a new sample x, the outputs are computed as follows:

$$\begin{aligned} \hat{y} = B^T \phi(x) &= Y^T T \left[U (T^T K U)^{-1}\right]^T \Phi \phi(x) \\ &= Y^T T \left[U (T^T K U)^{-1}\right]^T k(x) \end{aligned} \tag{9}$$

where $k(x)$ is expressed as

$$k(x) = \Phi \phi(x) = \left[k(x_1, x) \ldots k(x_N, x)\right]^T \tag{10}$$

Table 1 presents the main steps illustrating the KPLS model.

3 Fault Detection Approach: Reduced KPLS Based GLRT

3.1 Reduced KPLS Model

Building the KPLS model and using it for process monitoring can result in high computational cost if the training data set is large. One way to reduce the computation time consists in extracting a reduced number of samples from the training data using K-means and in mapping this reduced set in the feature space. After that, use the new mapped data to build the new KPLS model. K-means consists to classify the training dataset into N_r disjoint clusters, where the value of N_r is fixed prior to learning. Each cluster is defined by an adaptive change centroid (also called center of the cluster), based on certain initial values. K-Means calculates the squared distances between the centroids and the data then assigns the data to the closed centroid. Thus, N input-output data set $\{x_i, y_i\}$, $i = 1, \ldots, N$ are classified into N_r disjoint input-output subsets $\{C_1, C_2, \ldots, C_{N_r}\}$ and $\{S_1, S_2, \ldots, S_{N_r}\}$ each containing n_i observations, where $x(j) \in C_i and y(j) \in S_i \quad j = 1, \ldots, n_i, \; i = 1, \ldots, N_r$ by the minimization of the mean-square-error cost function

$$E1 = \sum_{i=1}^{N_r} \sum_{x(j) \in C_i} ||x_j - M_i||^2, \tag{11}$$

and

$$E2 = \sum_{i=1}^{N_r} \sum_{y(j) \in S_i} ||y_j - B_i||^2, \tag{12}$$

where $M_i = \sum_{x(j) \in; C_i} \frac{x_j}{n_i}$ is the centroid of cluster C_i, $B_i = \sum_{y(j) \in; S_i} \frac{y_j}{n_i}$ is the centroid of cluster S_i and n_i is the number of observations in C_i.

The resulting input data set obtained from the training data set can be defined as

$$X^r = \{ x^r(1) \; x^r(2) \cdots x^r(N_r) \} \tag{13}$$

where

$$x^r(i) = \frac{1}{n_i} \sum_{x(j) \in C_i} x(j), \quad i = 1, \ldots, N_r \tag{14}$$

with $x(j) \in \mathbf{R}^m, \quad j = 1, \ldots, n_i$ and $N_r = \ell + 1, \ldots, N$ is the number of reduced observations.

The resulting output data set obtained from the training data set can be given as

$$Y^r = \{ y^r(1) \; y^r(2) \cdots y^r(N_r) \} \tag{15}$$

where

$$y^r(i) = \frac{1}{n_i} \sum_{y(j) \in S_i} y(j), \quad i = 1, \ldots, N_r \tag{16}$$

In this case, the kernel matrix of the new reduced data set can be denoted as

$$K^r = \Phi^r \left(\Phi^r\right)^T \tag{17}$$

Similarly to Eqs. 7 and 8, we can see that the matrix of the regression coefficients B^r will have the following form

$$B^r = \left(\Phi^r\right)^T U^r (T^T K^r U^r)^{-1} \left(T^r\right)^T Y \tag{18}$$

where T^r and U^r are the input and output score matrices using the reduced data set.

3.2 Fault Detection Using Generalized Likelihood Ratio Test (GLRT)

In this section, the RKPLS modelling framework is integrated with a generalized likelihood ratio test (GLRT) to develop a fault detection scheme. The GLRT chart is used to measure the residuals of the responses variables obtained from the GPR model to detect anomalies. The GLR index is a commonly used hypothesis testing technique in model-based fault detection (Botre et al. 2016; Jiao et al. 2017; Zhang and Ma 2011; Mansouri et al. 2017c). Let us consider $E \in \mathbb{R}^N$ as a measured vector that follows one of the two Gaussian distributions (Harkat et al. 2019)

$$\begin{cases} H_0 = \{E \sim \mathcal{N}(0, \sigma^2 I_N)\} & \textit{(null hypothesis);} \\ H_1 = \{E \sim \mathcal{N}(\theta, \sigma^2 I_N)\} & \textit{(alternative hypothesis)} \end{cases} \tag{19}$$

where θ is the mean vector (the value of the fault in consideration) and σ^2 is the known variance.

The likelihood estimation of θ is determined by maximizing the GLRT (G(E)) as

$$G(E) = 2\log\left(\frac{\sup\limits_{\theta\in\mathbb{R}^N} exp\left(-\frac{\|E-\theta\|^2_2}{2\sigma^2}\right)}{exp\left(-\frac{\|E\|_2^2}{2\sigma^2}\right)}\right). \tag{20}$$

The probability density function f_θ can be defined as

$$f_\theta(E) = \frac{1}{(2\pi)^{\frac{N}{2}}\sigma^N} exp\left(-\frac{1}{2\sigma^2}\|E-\theta\|_2^2\right). \tag{21}$$

By inserting Eq. (21) into Eq. (20), we obtain

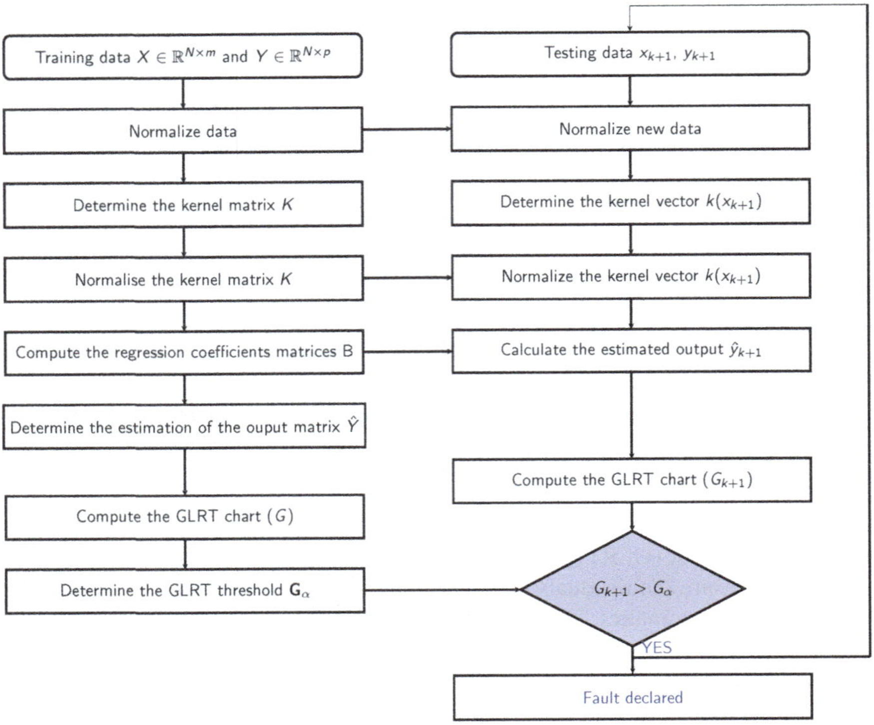

Fig. 2 Schematic illustration of RKPLS-based GLRT

$$
\begin{aligned}
G(E) &= \frac{1}{\sigma^2}\left(\underset{\theta}{min}\|E-\theta\|_2^2 + \|E\|_2^2\right) \\
&= \frac{1}{\sigma^2}\left(\|E-\hat{\theta}\|_2^2 + \|E\|_2^2\right) \\
&= \frac{1}{\sigma^2}\left(\|E\|_2^2\right),
\end{aligned}
\tag{22}
$$

where $\hat{\theta} = \underset{\theta}{argmin}\|E-\theta\|_2^2 = E$ and σ^2 is the variance of the residual.

The GLRT chart is distributed according to the χ^2 distribution (Harkat et al. 2019). The control limit G_α of GLRT G chart can be determined as

$$
G_\alpha = g\chi^2_{h,\alpha}, \tag{23}
$$

where $g = \frac{b}{2a}$ and $h = \frac{2a^2}{b}$, with a and b are the mean and variance of the GLRT chart, respectively. If the G is out of its respective threshold $G > G_\alpha$ then the fault is detected.

The RKPLS-based GLRT algorithm is illustrated schematically in Fig. 2.

4 Case Study Using the Cad System in E.coli (CSEC)

The evaluation studies are illustrated through a biological process representing a Cad System in E. coli (CSEC) model.

The detection performances are assessed in terms of: False alarms (a fault is detected when there does not occur a fault in the system. It is necessary a low rate of false alarms), good detection (a fault that occurs and it is detected), computation time (the required time for fault detection)

4.1 Cad System in E.coli (CSEC) Description

The Cad system in E.coli (CSEC) model is the conditional stress response module (Mansouri et al. 2017b, c, 2018a, b). It is induced only at low pH and a lysine-rich environment. The CSEC has been used in the context of parameter estimation of S-system models for biochemical networks. The three main parts of the CSEC system are the decarboxylase $CadA$, the transport protein $CadB$, and the regulatory protein $CadC$. The decarboxylase $CadA$ converts lysine Lys into cadaverine in a reaction that consumes H^+. The protein $CadB$ imports the product cadaverine and exports the substrate lysine. The decarboxylase $CadA$ and the protein $CadB$ decrease the intracellular H^+ concentration and hence participate in pH homeostasis. The cytoplasmic membrane protein $CadC$ is the positive regulator of $cadBA$ that detects the external conditions. The variables of the model are the decarboxylase $CadA$, the protein $CadBA$, the lysine Lys and the cadaverine $Cadav$ (see Fig. 3).

The qualitative model that shows the relationship between the variables decarboxylase $CadA$, protein $CadBA$, lysine Lys, and cadaverine $Cadav$ is defined as

$$\begin{aligned}
\frac{d[CadA]}{dt} &= \alpha_1[Cadav]^{g_{13}} - \beta_1[CadA]^{h_{11}} \\
\frac{d[cadBA]}{dt} &= \alpha_2[CadA]^{g_{21}} - \beta_2[cadBA]^{h_{22}} \\
\frac{d[Cadav]}{dt} &= \alpha_3[cadBA]^{g_{32}} - \beta_3[Cadav]^{h_{33}}[Lys]^{h_{34}} \\
\frac{d[Lys]}{dt} &= \alpha_4[CadA]^{g_{41}} - \beta_4[Lys]^{h_{44}} \qquad (24)
\end{aligned}$$

The parameters of the model are given in Table 2.

The data generated from the model (24) contains 4 variables and 2001 samples. Decarboxylase $CadA$, protein $CadBA$, and lysine are considered as input variables. The cadaverine Cadav is considered as output variable. Figure 4 shows time evolution of the 4 variables of the simulated CSEC model with total 2001 observations.

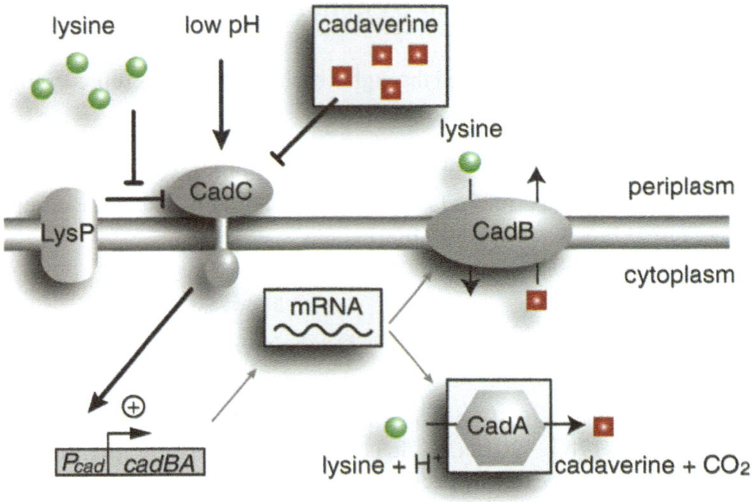

Fig. 3 Qualitative model of the CSEC system (simplified)

Table 2 Model parameters of CSEC system

Parameter	Value	Parameter	Value	Parameter	Value	Parameter	Value
α_1	12	β_1	10	g_{13}	−0.8	h_{11}	0.5
α_2	8	β_2	3	g_{21}	0.5	h_{22}	0.75
α_3	3	β_3	5	g_{32}	0.75	h_{33}& h_{34}	0.5 & 0.2
α_4	2	β_4	6	g_{41}	0.5	h_{41}	0.8

4.2 *Fault Detection Results*

KPLS and RKPLS based GLRT chart will be applied to the process for a comparison study. In this simulation, a training dataset of 2001 samples is collected under normal conditions and a testing data set of 2001 samples is generated under faulty operating conditions.

The RKPLS model is identified by only 1000 extracted samples thanks to the H-Kmeans clustering from which 10 kernel latent variables are maintained. The kernel latent variables is obtained by using cumulative percent variance (CPV) criterion. For KPLS and RKPLS, 99% confidence limit is utilized to detect faults. Figures 5 and 6, Table 3 show the time evolution of GLRT statistic based KPLS and RKPLS, respectively. If the GLRT statistic values are higher than confidence limit values this means that there are false alarms in the process. We can show from these figures that the GLRT chart present some false alarms in both case. The fault detection results are presented in terms of FAR and computation time under normal conditions in Table 3. The results show that RKPLS-based GLRT provides an improvement in terms of false alarm rate (0.59%) compared to KPLS-based GLRT. We can see

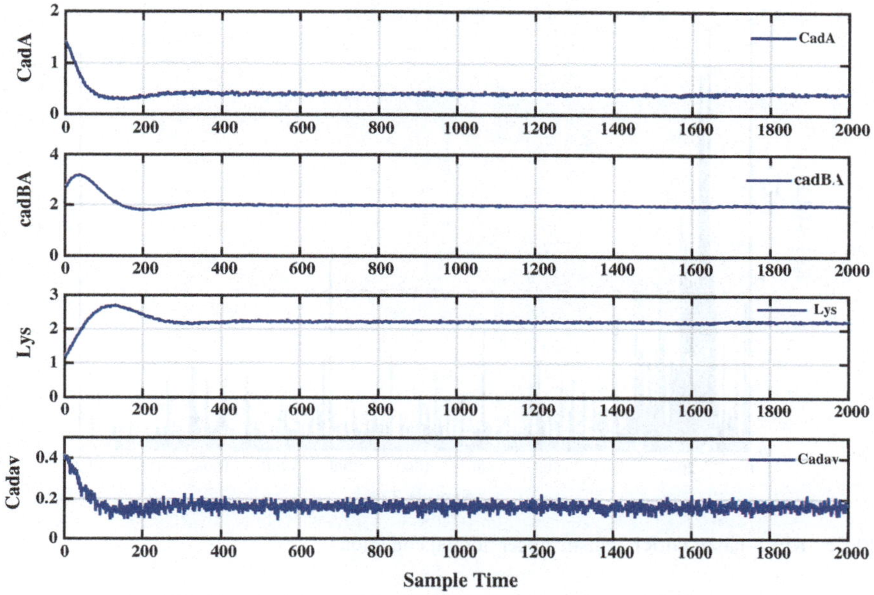

Fig. 4 Time evolution of $CadA$, $CadBA$, lysine and Cadav

Table 3 Summary of FAR, GDR and CT

Metrics	KPLS-GLRT	RKPLS-GLRT
FAR (%)	1.149	0.59
CT (s)	48.67	10.9688

also that using the developed method the time required for fault detection decreases significantly in comparison to the KPLS-based GLRT technique.

To illustrate the impact of the data reduction on the detection efficiency, we included a single fault in $cadBA$ in the interval [1400 2001]. The fault size is fixed to 3σ of this variable. Figures 7 and 8 present the results using the classical KPLS-based GLRT and the proposed RKPLS-based GLRT in the faulty case. From theses figures, it can be shown that the proposed technique is able to detect the fault in the given interval.

To further show the detection ability of the proposed approach, three indicators including FAR, MDR and computation time (CT) are calculated (see Table 4).

It can be seen from Table 4 that the proposed method provides less values of MDR, FAR and CT when compared to the KPLS approach. The proposed method reduce significantly the computation time when using using a number of samples equivalent to 50% of the number of samples of the training data. Thus, it can not only reduce the computation cost but also guarantee the fault detection performance(FAR and GDR). This fact is useful in many industrial applications where the updating procedure requires to be processed with other steps in a short time.

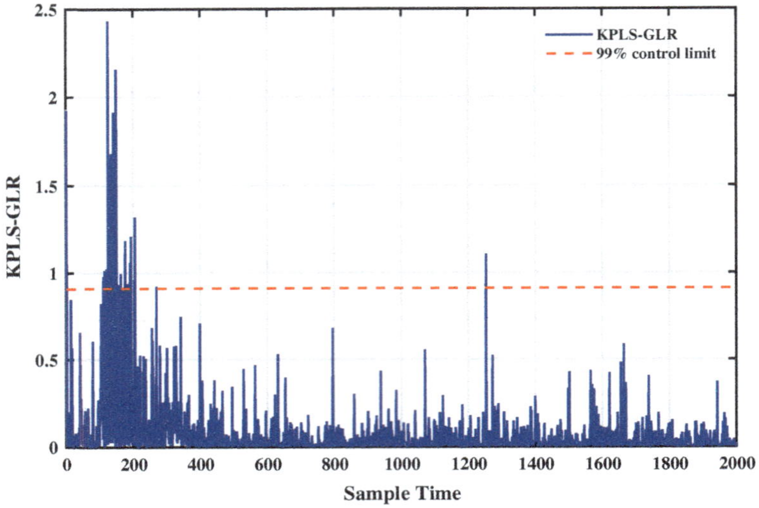

Fig. 5 KPLS-based GLRT statistic under normal conditions

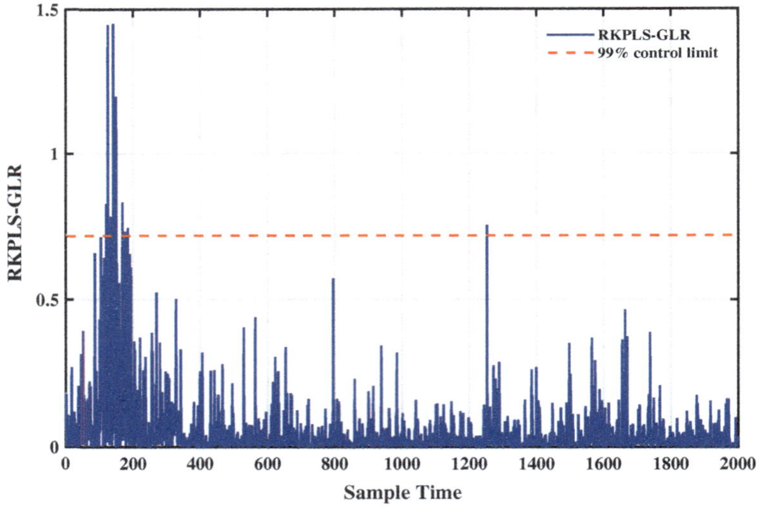

Fig. 6 RKPLS-based GLRT statistic under normal conditions

Table 4 Summary of FAR, GDR and CT

Metrics	KPLS-GLRT	RKPLS-GLRT
FAR (%)	1.71	0.00
MDR (%)	0.92	0.00
CT (s)	52.18	11.62

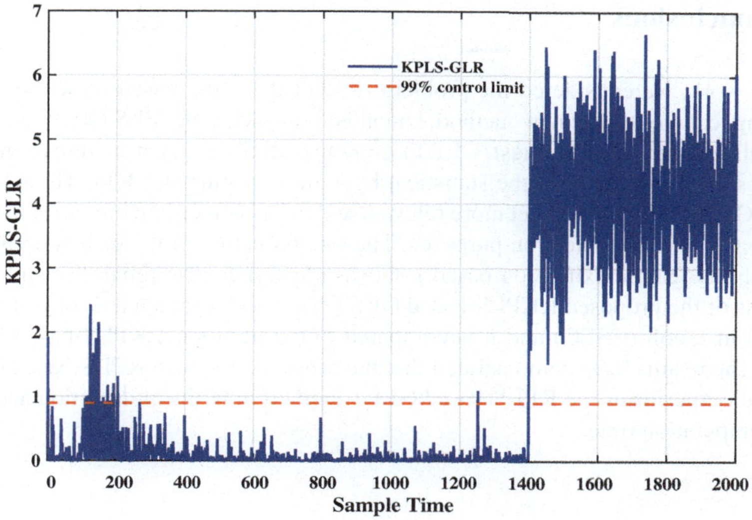

Fig. 7 Results using KPLS-based GLRT

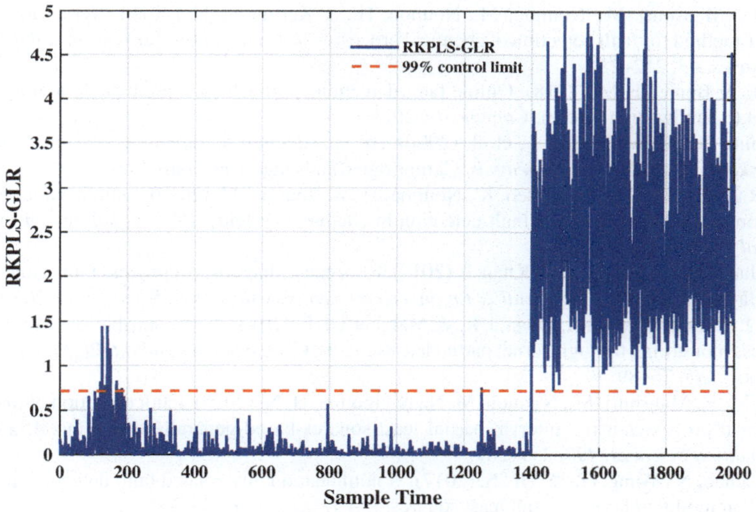

Fig. 8 Results using RKPLS-based GLRT

5 Conclusions

In this work, we considered the problem of fault detection based on kernel-based learning technique. The new method, so-called reduced kernel PLS (RKPLS)-based generalized likelihood ratio test (GLRT), combines RKPLS which captures nonlinear features from the data and the statistical hypothesis testing (GLRT). The RKPLS-based GLRT allows to extract more relevant and informative characteristics from the data to use later for detection purposes. The size reduction of the data is performed through the feature extraction-based k-means clustering. The reliability and effectiveness of the proposed RKPLS-based GLRT have been evaluated through the Cad system in E.coli (CSEC) and it is compared to the traditional KPLS-based GLRT chart. The results have demonstrated that the proposed reduced KPLS-based GLRT method outperforms the KPLS-based GLRT method in terms of detection accuracy and computation time.

References

Botre, C., Mansouri, M., Nounou, M., Nounou, H., & Karim, M. N. (2016). Kernel PLS-based GLRT method for fault detection of chemical processes. *Journal of Loss Prevention in the Process Industries*, *43*, 212–224.

Chihi, I., & Benrejeb, M. (2018). Online fault detection approach of unpredictable inputs: Application to handwriting system. *Complexity*, *2018*.

Cristianini, N., Shawe-Taylor, J., et al. (2000). *An introduction to support vector machines and other kernel-based learning methods*. Cambridge Cambridge University Press.

Fazai, R., Mansouri, M., Abodayeh, K., Nounou, H., & Nounou, M. (2019). Online reduced kernel pls combined with GLRT for fault detection in chemical systems. *Process Safety and Environmental Protection*.

Gharahbagheri, H., Imtiaz, S., & Khan, F. (2017). Root cause diagnosis of process fault using KPCA and Bayesian network. *Industrial & Engineering Chemistry Research*, *56*(8), 2054–2070.

Godoy, J. L., Zumoffen, D. A., Vega, J. R., & Marchetti, J. L. (2014). New contributions to non-linear process monitoring through kernel partial least squares. *Chemometrics and Intelligent Laboratory Systems*, *135*, 76–89.

Harkat, M.-F., Mansouri, M., Nounou, M. N., & Nounou, H. N. (2019). Fault detection of uncertain chemical processes using interval partial least squares-based generalized likelihood ratio test. *Information Sciences*, *490*, 265–284.

Jiao, J., Zhao, N., Wang, G., & Yin, S. (2017). A nonlinear quality-related fault detection approach based on modified kernel partial least squares. *ISA Transactions*, *66*, 275–283.

Lall, P., Gupta, P., Kulkarni, M., Panchagade, D., Suhling, J., & Hofmeister, J. (2008). Prognostication and health monitoring of electronics in implantable biological systems. In *ASME 2008 International Mechanical Engineering Congress and Exposition* (pp. 657–671). American Society of Mechanical Engineers Digital Collection.

Lau, C., Ghosh, K., Hussain, M., & Hassan, C. C. (2013). Fault diagnosis of Tennessee Eastman process with multi-scale PCA and ANFIS. *Chemometrics and Intelligent Laboratory Systems*, *120*, 1–14.

Li, G., Qin, S. J., & Zhou, D. (2010). Geometric properties of partial least squares for process monitoring. *Automatica*, *46*(1), 204–210.

MacGregor, J. F., Jaeckle, C., Kiparissides, C., & Koutoudi, M. (1994). Process monitoring and diagnosis by multiblock pls methods. *AIChE Journal*, *40*(5), 826–838.
Mansouri, M., Baklouti, R., Harkat, M. F., Nounou, M., Nounou, H., & Hamida, A. B. (2018a). Kernel generalized likelihood ratio test for fault detection of biological systems. *IEEE Transactions on Nanobioscience*, *17*(4), 498–506.
Mansouri, M., Harkat, M. F., Teh, S. Y., Al-khazraji, A., Nounou, H., & Nounou, M. (2018b). Model-based and data-driven with multiscale sum of squares double EWMA control chart for fault detection in biological systems. *Journal of Chemometrics*, *32*(12), e3068.
Mansouri, M., Nounou, H., Harkat, M. F., & Nounou, M. (2017a). Fault detection of chemical processes using improved generalized likelihood ratio test. In *22nd International Conference on Digital Signal Processing (DSP)* (pp. 1–5). IEEE.
Mansouri, M., Nounou, M. N., & Nounou, H. N. (2017b). Improved statistical fault detection technique and application to biological phenomena modeled by s-systems. *IEEE Transactions on Nanobioscience*, *16*(6), 504–512.
Mansouri, M., Nounou, M. N., & Nounou, H. N. (2017c). Multiscale kernel pls-based exponentially weighted-GLRT and its application to fault detection. *IEEE Transactions on Emerging Topics in Computational Intelligence*, *99*, 1–11.
Nguyen, V. H., & Golinval, J.-C. (2010). Fault detection based on kernel principal component analysis. *Engineering Structures*, *32*(11), 3683–3691.
Rosipal, R., & Trejo, L. J. (2001). Kernel partial least squares regression in reproducing kernel Hilbert space. *Journal of Machine Learning Research*, *2*, 97–123.
Sheriff, M. Z., Botre, C., Mansouri, M., Nounou, H., Nounou, M., & Karim, M. N. (2017). *Process monitoring using data-based fault detection techniques: Comparative studies*. In *Fault diagnosis and detection*. InTech.
Si, Y., Wang, Y., & Zhou, D. (2020). Key-performance-indicator-related process monitoring based on improved kernel partial least squares. *IEEE Transactions on Industrial Electronics*.
Yi, J., Huang, D., He, H., Zhou, W., Han, Q., & Li, T. (2017). A novel framework for fault diagnosis using kernel partial least squares based on an optimal preference matrix. *IEEE Transactions on Industrial Electronics*, *64*(5), 4315–4324.
Zhang, Y., Du, W., Fan, Y., & Zhang, L. (2015). Process fault detection using directional kernel partial least squares. *Industrial & Engineering Chemistry Research*, *54*(9), 2509–2518.
Zhang, Y., & Ma, C. (2011). Fault diagnosis of nonlinear processes using multiscale KPCA and multiscale KPLS. *Chemical Engineering Science*, *66*(1), 64–72.
Zhang, Y., & Zhang, Y. (2009). Complex process monitoring using modified partial least squares method of independent component regression. *Chemometrics and Intelligent Laboratory Systems*, *98*(2), 143–148.

Prospects of Internet of Things (IoT) and Machine Learning to Fight Against COVID-19

Khandaker Foysal Haque and Ahmed Abdelgawad

Abstract IoT and Machine Learning has improved multi-fold in recent years and they have been playing a great role in healthcare systems which includes detecting, screening and monitoring of the patients. IoT has been successfully detecting different heart diseases, Alzheimer disease, helping autism patients and monitoring patients' health condition with much lesser cost but providing better efficiency, reliability and accuracy. IoT also has a great prospect in fighting against COVID-19. This chapter discusses different aspects of IoT in aiding healthcare systems for detecting and monitoring Coronavirus patients. Two such IoT based models are also designed for automatic thermal monitoring and for measuring and real-time monitoring of heart rate with wearable IoT devices. Convolutional Neural Networks (CNN) is a Machine Learning algorithm that has been performing well in detecting many diseases including Coronary Artery Disease, Malaria, Alzheimer's disease, different dental diseases, and Parkinson's disease. Like other cases, CNN has a substantial prospect in detecting COVID-19 patients with medical images like chest X-rays and CTs. Detecting Corona positive patients is very important in preventing the spread of this virus. On this conquest, a CNN model is proposed to detect COVID-19 patients from chest X-ray images. Two CNN models with different number of convolution layers and three other models based on ResNet50, VGG-16 and VGG-19 are evaluated with comparative analytical analysis. The proposed model performs with an accuracy of 97.5% and a precision of 97.5%. This model gives the Receiver Operating Characteristic (ROC) curve area of 0.975 and F1-score of 97.5. It can be improved further by increasing the dataset for training the model.

Keywords Internet of Things (IoT) · Sensors · COVID-19 · Coronavirus · Detection of COVID-19 · Deep learning · Convolutional Neural Networks (CNN)

K. F. Haque (✉) · A. Abdelgawad
Central Michigan University, Mount Pleasant, MI, USA
e-mail: haque1k@cmich.edu

A. Abdelgawad
e-mail: abdel1a@cmich.edu

O. Kanoun and N. Derbel (eds.), *Advanced Systems for Biomedical Applications*, Smart Sensors, Measurement and Instrumentation 39,
https://doi.org/10.1007/978-3-030-71221-1_5

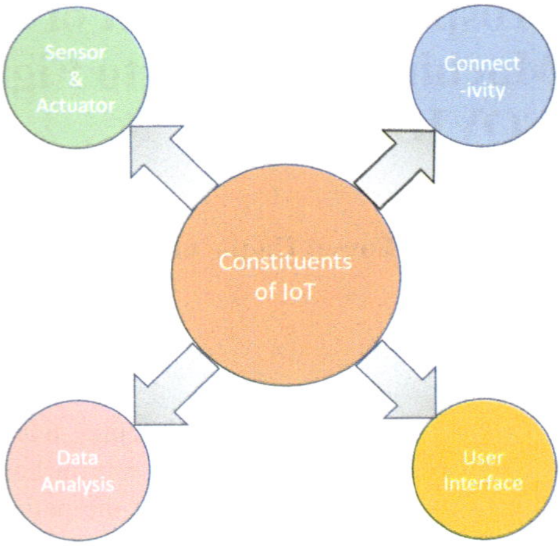

Fig. 1 Basic constituents of IoT

1 Introduction

Internet of Things (IoT) is the network of inter-connected objects that can transfer data among themselves. This is networking of the objects which usually involves sensors, actuators, communication transceivers, micro-controller or processing unit (Haque et al. 2020a). Almost every IoT architecture has four basic constituents:

- sensors and actuators,
- connectivity,
- data analysis,
- and user interface as depicted in Fig. 1.

In general, IoT architectures are based on the communicating sensor nodes. There are one or more processing units in the nodes which are equipped with different sensors and actuators depending on the field of applications. These IoT nodes collect necessary data from the deployed field which is further analyzed for monitoring or real-time decision tacking. Various kinds of sensors and actuators are used in nodes depending on the task of the deployment. Connectivity is another important part of IoT as it keeps things connected. Most IoT nodes are energy constrained as they are battery powered. So, they have to operate and communicate with very low power to last longer. Internet Engineering Task Force (IETF) has standardize Routing Protocol for Low Power and Lossy Networks (RPL) for IoT. RPL is an IPV6 routing protocol which is based on IEEE 802.15.4 with the adaptation layer of 6LoWPAN. But the communication standard for IoT is not limited to RPL rather things can be connected with any suitable wireless technology. Wi-Fi, GSM, 4G, 5G, radio communication technologies like Zigbee, LoRa, Sigfox are few popular choices for IoT connectivity. Data analysis is another component of basic IoT. This section monitors and analyzes the data which is collected from the deployment of IoT node. These analyzed data and decisions are presented to the end user through the designated websites, clouds and cell phone applications.

With the recent development of low power networks and sensors, IoT has grown forth and created numerous application fields (Haque et al. 2020b). Its application field includes industry, medical science, health monitoring, economy, weather forecasting, deep sea exploration underwater mining and daily life (Haque et al. 2020c; Yelamarthi et al. 2017; Abdelgawad and Yelamarthi 2017. The medical and health sector has always been one of the major application area of IoT. It is saving a lot of lives by detecting diseases early and remotely monitoring patients' vitals in real-time. It has made different disease detection more cheaper and a lot faster with higher accuracy and reliability.

In recent years, IoT has played an important role in healthcare by increasing the efficiency, accuracy, reliability, remote accessibility and availability of the medical devices. As IoT in the medical sector gains the popularity, it is sometimes termed as Internet of Medical Things (IoMT). IoMT architectures are used in Medical Nursing System (MNS) with Wireless Sensor Networks (WSN), Sensors, Wi-fi, Zigbee and Bluetooth for patients' data transferring (Joyia et al. 2017; Huang and Cheng 2014). An extensive research is also going on to monitor the patients' physiological condition with low cost medical sensing devices (Istepanian et al. 2011). Researchers are also interested in monitoring autism patients which can keep track of the data collected from the brain signals of the individual (Kumar and Bairavi 2016). Krishna et al. propose an IoT based algorithm to detect the abnormality in the kidney (Krishna et al. 2016). In depth research is also going to make the hospital systems more efficient by making them smart hospitals with IoT (Zhang et al. 2018; Yu et al. 2012; Catarinucci et al. 2015). IoT has a great prospect also in disease detection. Kumar and Gandhi propose an IoT based architecture with machine learning for early detection of heart diseases (Kumar and Gandhi 2018). Varatharajan et al. propose wearable sensor devices for early detection of Alzheimer's disease. Yang et al. propose an IoT-cloud based wearable ECG monitoring system which enables smart healthcare. IoT can also play a vital role in fighting against the most recent COVID-19 pandemic.

The world has been suffering the formidable outbreak of the novel Coronavirus which was first detected in December 2019. This creates a respiratory infectious disease which has been emerging and spreading very fast causing a real threat to the public health. Transmission rate and mode of transmission are very important factors for any contagious disease like COVID-19. According to the World Health Organization (WHO), respiratory droplets of size greater than 5–10 μm acts as a mode of transmission which potentially involve airborne transmission (WHO Organization 2020). This creates an alarming threat to the public health as interaction without necessary safety measures can be highly contagious. So, this disease poses a high growth factor with an estimated fatality rate of 2–5% (Wu et al. 2020). According to WHO, to the date of 6 July 2020, the worldwide total confirmed COVID-19 cases are 11.327 M with the fatality of 0.532 M (WHO coronavirus disease 2020a). Figure 2, Shows the time-series graph of total confirmed cases and deaths over the time to the date.

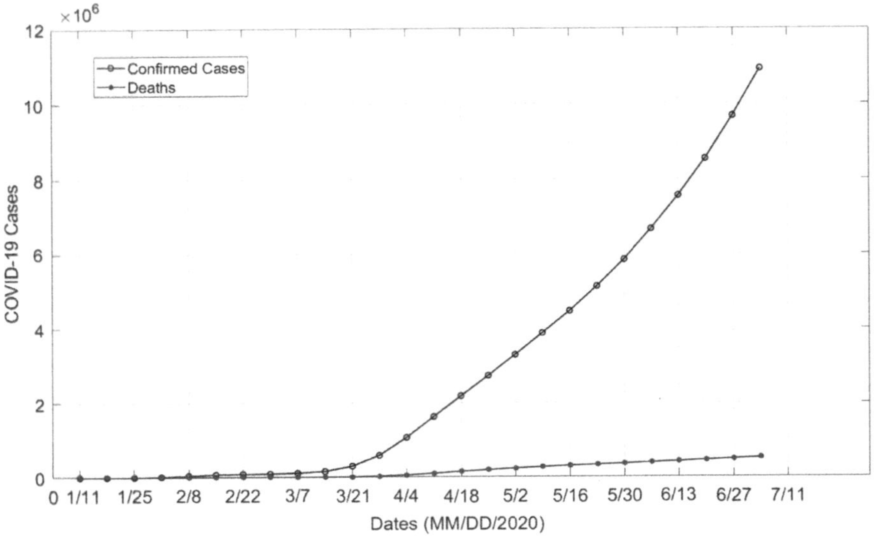

Fig. 2 Time-series graph of world-wide total confirmed cases and deaths due to COVID-19

It shows that this spreads very fast with geometric progression. Early detection of COVID-19 patients is one of the most important aspects to limit the spread of this virus. WHO listed a few rapid and detailed diagnostic tests for COVID-19 detection including genesig Real-Time PCR Coronavirus (COVID-19) testing and cobas SARS-CoV-2 for use on the cobas 6800/8800 systems (WHO coronavirus disease 2020b). These tests take a lot of time and money whereas IoT and Machine Learning can play a major role in automatic positive patient detection and patients' health monitoring. This can save both time and money which will eventually save lives. To find the cure and vaccine for this disease is another important aspect of this critical time to save lives and bring the world back to normal. The combined approach of IoT and Machine learning is vastly helping the humanity to fight against this pandemic. The next part of the chapter focuses on the different aspects of IoT and Machine Learning in fighting against COVID-19.

2 Fight Against COVID-19: An IoT Perspective

IoT is one of the most creative platforms to fight against this pandemic. IoT provides higher reliability and accuracy in screening, detecting and treatment of the COVID-19 patients. It can be used for thermal screening, measuring blood pressure, measuring heart rate and SpO_2 level, measuring glucose level and many other aspects of the detection and treatment of the disease (Sing et al. 2020; Mohammed et al. 2020;

Vaishya et al. 2020). The next part of this section discusses designing two such model to aid COVID-19 detection and treatment.

2.1 Thermal Monitoring with IoT

High temperature is one of the symptoms in COVID-19 affected people. Thermal screening using infrared thermometers is one of the ways to sort out suspected patients for further testing. But this involves the possibility for healthy people to be exposed to affected patients as they have to report to testing center for screening. A continuous real-time monitoring is needed to sort out the suspected patients for further testing. IoT can offer this much needed thermal monitoring accurately in real-time for 24/7. AMG8833 8x8 Infra-Red (IR) thermal camera is an excellent choice for distant monitoring of the thermal profile. It can measure the heat profile from a distance of 7 m (23 feet) and has the sensitivity of $\pm 2.5\,^{\circ}$C. It has the versatility of working with different micro controller platforms. It works really well with Arduino and Raspberry Pi. Here, an IoT node is designed to monitor the real-time thermal temperature from a distance and transfer the data to the cloud for monitoring 24/7. This can be mounted in the front door of the house or testing center for remote monitoring and early suspect detection. Feather Huzzah ESP8266 WI-Fi micro-chip is used as the processing unit as it provides full Wi-Fi stack. A TFT LCD display is also used to monitor the thermal activity. The system has some important components:

- thermal monitoring node,
- wireless connectivity,
- IoT cloud,
- user interface,
- and notification system.

Thermal monitoring node is equipped with ESP8266, AMG8833 IR thermal camera and TFT LCD display. It reads the subjects thermal profile with the IR thermal camera and the thermal image is simultaneously displayed in the LCD display. The node is connected to the Wi-Fi and it transfer the collected data to the cloud. ThingSpeak, Adafruit IO or Google Drive are a few common cloud storage that can be used for such implementation. The data is displayed in any desired interface such as: personal websites or Android/iOS applications. The working principle of the system is depicted the Fig. 3

The thermal node keeps monitoring the temperature and the thermal profile of the subjects put in front of it. ESP8266 is connected to the cloud over Wi-Fi, and it sends the collected data to the designated cloud. The node sends the temperature and thermal image of the subject to the cloud which can be analyzed further for screening out the abnormalities. User can access the collected data and thermal profile from the designed website or phone applications from anywhere in the world.

The capability of the system is improved further by enabling the auto facial recognition with machine learning and automatic notification system. For this, Raspberry

Fig. 3 Architecture of the thermal monitoring system with IoT

Pi is used as the node's processing unit which is equipped with AMG8833 IR thermal camera, Raspberry Pi camera and TFT LCD display. Here, IR thermal camera is used to measure the temperature and to get the thermal image whereas the Raspberry Pi camera takes the corresponding face images for detecting the person. Eigenface algorithm is used for facial recognition of the subjects (Gunawan et al. 2017). The system measures each of the subject's thermal profiles and puts the profile corresponding to the identified person's name. Upon detecting any abnormalities in the thermal profile, the system sends an email notification to the corresponding person. This makes the system fully autonomous and it does not require any human interaction. But the system needs to be trained with all the subjects' images to detect them later while testing.

2.2 *Heart Rate and SpO_2 Monitoring for Primary COVID-19 Screening*

SpO_2 is the measurement of the oxygen carrying hemoglobin in the blood compared to the hemoglobin that is not carrying oxygen. The normal SpO_2 percentage of a healthy human is 96–99%. SpO_2 or oxygen saturation play an important role in early detection of the disease. COVID-19 causes severe respiratory distress which eventually decrease the oxygen saturation in the blood which can cause severe damage to the patient's health and might cause death. One of the early symptoms of the COVID-19 affected patient is lower SpO_2 rate, according to WHO, SpO_2 rate goes down to less than 90% causing severe respiratory problem (WHO Organization Coronavirus disease 2019). Moreover, hospitals treating COVID-19 patients need to monitor this rate to support the patients with respiratory aid. So, the remote real-time monitoring of SpO_2 would help the medical sectors to detect and treat the patients better with more distancing and safety.

A real time heart-rate and SpO_2 monitoring system is designed in this section of the chapter. Arduino pro mini along with ESP8266 is used due to its smaller dimension such that it can be used as wearable device. Arduino pro mini is based on ATmega328 micro-controller and ESP8266 micro chip provides Wi-Fi connectivity

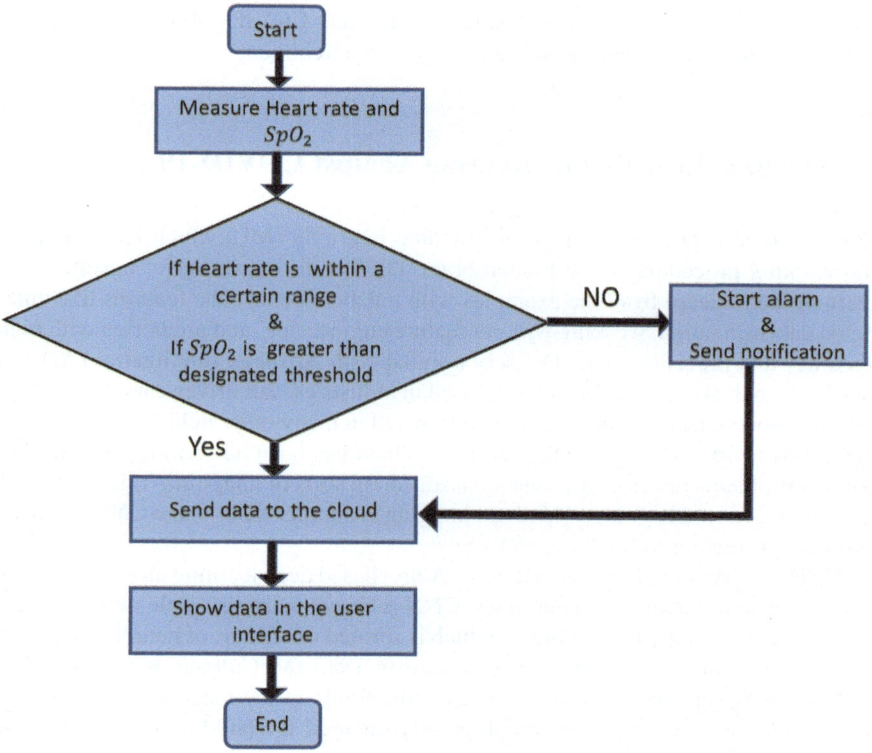

Fig. 4 Flow chart for real-time heart rate and SpO_2 measuring system

to the processing unit. MAX30102 pulse oximeter and heart rate monitor sensor is integrated to the node for measuring the heart rate and SpO_2 rate. It is connected to the Arduino pro mini board with I2C bus. A buzzer is also added to the node for alarming when heart rate or SpO_2 rate go beyond the specified threshold values. This can be used as a wearable device to monitor the heart rate and oxygen saturation at a regular interval. The working principle of the whole system is depicted in Fig. 4.

At first, the wearable node measures the heart rate and SpO_2 with MAX30102 sensor. This sensor passes a beam of light through the finger tip and receives the reflected light which depends on the absorption of the light by blood. It can measure the oxygen saturation rate of blood as absorption rate of the oxygenated blood varies from the deoxygenated one. This sensor can also measure the heart rate of the patient. After measuring, data is sent to the micro-controller where it compares if the heart rate is within a designated range and if SpO_2 value is greater than a designated value. If the obtained values are as expected it sends the data to the cloud and publishes the data in desired user interface. If the measured values are not within the designated range it creates an alarm with a buzzer and sends an email notification to the assigned authority. This system helps in screening and treating COVID-19 patients to a great extent and increase the safety and reliability of the treatment.

3 Machine Learning in Fighting Against COVID-19

Deep Learning (DL) is a branch of Machine Learning (ML) which is inspired by the working procedure of the human brain. DL has the capability of unsupervised learning i.e. to learn from the examples with unlabeled data. The features like unlabeled data utilization, working without feature engineering, and prediction with high accuracy and precision make DL very popular with Artificial Intelligence (AI) and Big Data analysis. DL has been vastly used in industries, self driven cars, face recognition, object detection, image classification and in many other fields. Convolutional Neural Network (CNN) is a DL algorithm which has been performing very well in solving problems like document analysis, different sorts of image classification, pose detection, action recognition. Medical imaging is another field where CNN has been showing promising results in recent years.

CNN is an Artificial Neural Network (ANN) based deep learning algorithm which has grown significantly in recent times. CNN is based on the principle of human nervous system specially human brains which is formed of billions of neurons. The idea of artificial neuron was first conceptualized in 1943 (McCulloch and Pitts 1943). Hubel and Wisel first found that for detecting lights in the receptive fields, visual cortex cells have a major role which greatly inspired the building of a model like Neocognitron (Gu et al. 2018). This model is considered to be the base and predecessor of CNN. CNN is formed of artificial neurons which has the property of self optimization with learning like the brain neurons. Due to this self optimizing property, it can extract and classify the features extracted from images more precisely than any other algorithm. Moreover, it needs very limited preprocessing of the input data though it yields highly accurate and precise results. CNN is vastly used in object detection and image classification including medical imaging.

CNN has been performing really well with medical imaging. For recent years, it has been used vastly for a different disease or anomaly detection. CNN does the diagnosis of Coronary Artery Disease (CAD), recognition of stages from bright-field microscopy images of malaria-infected blood, detection of Parkinson's disease from electroencephalogram (EEG) signals, the study of Alzheimer's disease and classification of many other diseases. CNN can also play a great role in COVID-19 detection from CT or X-ray images.

In the rest part of the section, a CNN model is proposed to detect COVID-19 positive patients from chest X-ray images as in authors' earlier work (Haque et al. 2020d). With very little time and resources, this model successfully detects Coronavirus patients with high accuracy. This work also includes the comparative analytical analysis of the CNN models in detecting COVID-19. This can help to implement testing of COVID-19 on a much greater scale which would save both money and time.

3.1 Related Works

Extensive research work is going on for classifying COVID-19 patient image data. Few researchers have proposed different DL models for classifying chest x-ray images whereas some others have taken CT images into consideration. Narin et al. proposed three CNN models based on pretrained ResNet50, InceptionV3, Inception-ResNetV2 for detecting COVID-19 patient from chest X-ray radiographs (Narin et al. 2020). These models are pretrained on ImageNet database and use 2×2 average pooling layers and two fully connected layers on the top of the pre-trained layers. It is mentionable that ImageNet provides a huge number of a generalized dataset for image classifications. The Softmax activation function is used for each of the models to finally classify the images. It is found that ResNet50 gives the classifying accuracy of 98% whereas InceptionV3 and Inception-ResNetV2 perform with the accuracy of 97% and 87% respectively. But these models have taken only 100 images (50 COVID-19 and 50 normal Chest X-rays) into consideration for training which might result in declined accuracy for a higher number of training images. Zhang et al. propose a CNN model for Coronavirus patient screening using their chest X-ray images (Zhang et al. 2020). This research group has used 100 chest X-ray images of 70 COVID-19 patients and 1431 X-ray images of other pneumonia patients where they are classified as COVID-19 and non-COVID-19 respectively. This model is formed of three main parts: backbone networks, classification head, and anomaly detection head. The backbone network is a 18 residual CNN layer pre-trained on ImageNet dataset. This model can diagnose COVID-19 and non-COVID-19 patients with an accuracy of 96% and 70.65% respectively. But this model also have few drawbacks like– using smaller dataset and its false positive rate is almost 30%. Hall et al. also worked on finding COVID-19 patients from a small set of chest X-ray images with DL (Hall et al. 2020). They have used pre-trained ResNet50 and VGG 16 along with their own CNN and this model generates the overall accuracy of 94.4% and false positive rate of 6%. The model shows the true positive rate of 0.969 whereas the true negative rate is 0.94. Sethy and Behea have also utilized deep features for Coronavirus disease detection (Sethy and Behera 2020). Their model is based on ResNet50 plus SVM where the features extracted from each CNN layers are utilized by SVM for classification. This model achieved the accuracy and F1-score of 95.38% and 91.41% respectively. The false positive rate and F-1 scores are 95.52 % and 91.41% which is better than other discussed models. But this model is trained with a very small dataset of 25 images for each of the class: COVID-19 and Normal.

3.2 *Proposed CNN Model for COVID-19 Detection*

3.2.1 Dataset Collection and Modeling

For training the proposed model, 161 chest X-ray images of COVID-19 patients are used which are obtained from open Github repository by Cohen et al. (2020). This repository contains patients' chest X-ray images of COVID-19, SARS, ARDS, Pneumocystis, Streptococcus, Chlamydophila, E.Coli, Legionella, Varicella, Lipoid, Bacterial, Pneumonia, Mycoplasma Bacterial Pneumonia, Klebsiella and Influenza. For training, only the COVID-19 positive X-rays have been taken into consideration and the patient age ranges from 12–93 years. The training also needs the normal or non-COVID-19 chest X-rays, which is obtained from Kaggle dataset naming "Chest X-ray Images (pneumonia)" (Mooney 2020). This repository contains 5863 images in two categories- normal and pneumonia. But we have taken 161 (same number as the COVID-19 chest X-ray images) normal chest X-ray images for the training purpose. The whole dataset is primarily split into two categories: training and validation maintaining the ratio of 80% and 20% respectively. Each group of training and validation dataset contains two subcategories: 'Normal' and 'COVID-19', containing the respective types of X-ray images. So, for the training, both the categories- 'Normal' and 'COVID-19' contain 161 chest X-ray images each whereas, the validation dataset contains 40 images for each of the 'Normal' and 'COVID-19' sub-categories. For maintaining unanimity and the image quality at the same time, all the images are converted to 224×224 pixels. Moreover, all the X-ray images that are used for training and validation of the model are in Posteroanterior (PA) chest view.

3.2.2 CNN Modeling

CNN has been playing a great role in classifying images, in particular medical images. This has opened new windows of opportunities and made the disease detection much more convenient. It also successfully detects recent novel Coronavirus with higher accuracy. One of the constraints that researchers encounter is a limited dataset for training their model. Being a novel disease, the chest X-ray dataset of COVID-19 positive patients is also limited. Therefore, to avoid overfitting, a sequential CNN model is proposed for classifying X-ray images. Figure 5 depicts the proposed CNN model for COVID-19 detection. This model has 4 main components:

- input layers,
- convolutional layers,
- fully connected layers,
- and (iv) output layers.

The tuned data set is fed into the input layers of the model. This model is trained on 165 X-ray images of each category: normal and COVID-19. It has four convolutional

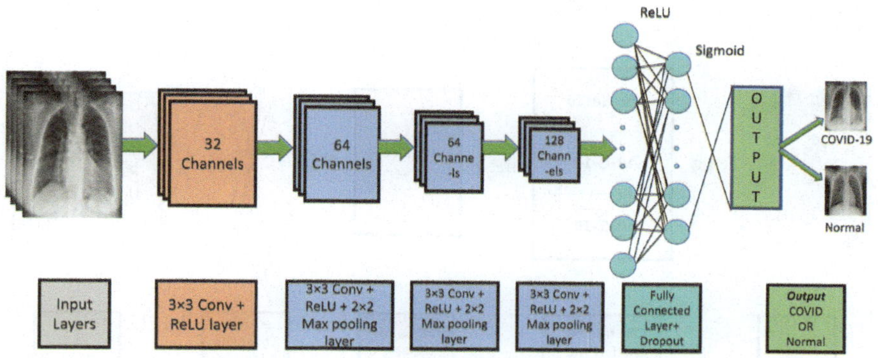

Fig. 5 Workflow diagram of the proposed CNN model for COVID-19 detection

layers, first one is a 2D convolutional layer with 3×3 kernels and Rectified Linear Unit (ReLU) activation function. ReLU is one of the most popular and effective activation functions that are being widely used in DL. ReLU does not activate all the neurons at the same time making it computationally efficient in comparison to other activation functions like tanh.

The next three layers are 2D convolutional layer along with the ReLU activation function and Max pooling. Max pooling accumulates the features of the convolutional layer by convolving filters over it. It reduces the computational cost as it minimizes the number of parameters thus it helps avoid overfitting. In each of three layers a 2×2 Max pooling layer is added after the convolutional layer to avoid overfitting and to make the model computationally efficient. In the next step of the model, the output of the convolutional layers is converted to a long 1D feature vector by a flatten layer. This output from the flatten layer is fed to the fully connected layer with dropout. In a fully connected layer, every input neuron is connected to every activation unit of the next layer. All the input features are passed through the ReLU activation function and this layer categorizes the images to the assigned labels. The Sigmoid activation function makes the classification decision depending on the classification label of the neurons. Finally, in the output layer, it is declared if the input X-ray image is COVID-19 positive or normal. This model is termed as 'Model 1'.

For comparative analysis, two more CNN models are also developed with 3 and 5 conv layers respectively instead of the 4 conv layers of the Model 1. These models with 3 and 5 conv layers are termed as 'Model 2' and 'Model 3' respectively. Model 2 has one 3 3 conv layer with ReLU having 32 channels and two more layers with 3 conv layers with ReLU and 2 Max Pooling layers having 64 channels each.

This work also takes few pretrained models into consideration in terms of their performance with COVID-19 image classification. Three pretrained models based on ResNet50, VGG-16 and VGG-19 are also developed and tuned to detect the COVID-19 cases from the same chest x-ray datasets (He et al. 2016; Simonyan and Zisserman 2014). ResNet is based on ImageNet and it has achieved excellent results with only 3.57% error. It has five stages each having one convolution and one

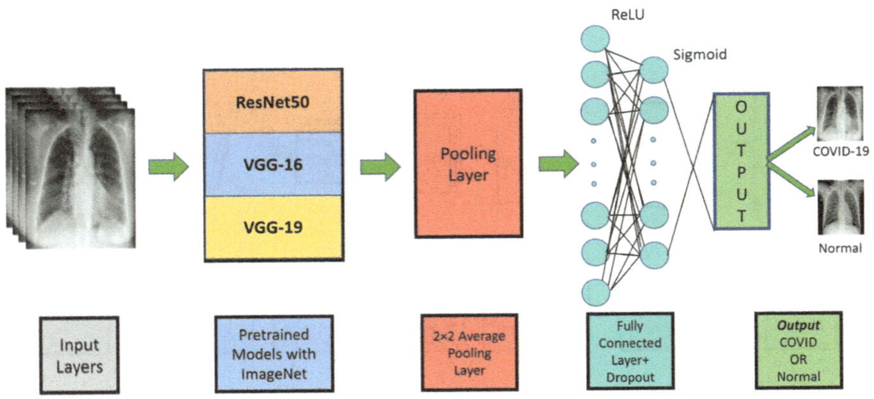

Fig. 6 Workflow diagram of the pretrained models for COVID-19 detection

identity block. Each of the convolution and identity blocks have 3 convolution layers. VGG-16 is CNN model which is 16 layers deep as its name suggests. This is one of the most excellent CNN architecture for image classification. This model doesn't have a large number of hyper -parameters rather it use 3 × 3 convolution layers and 2 max pooling layer with stride of 1 and 2 respectively. The whole architecture is based on this consistent convolution and max pooling layer. VGG-19 is of the same architecture as VGG-16 except for VGG-19 has 19 deep layers instead of 16. The pretrained model of these three CNN architecture are used to extract features and outputs are fed to 2 × 2 pooling layer. A flatten layer converts the outputs to a 1D feature vector. The output from the flatten layer is fed to the fully connected layer with dropout which has the same architecture as Model 1, Model 2 and Model 3. Figure 6 depicts the workflow diagram of the pretrained models.

3.3 Results and Analysis

The proposed model is trained for 25 epochs with 10 steps per epoch. In this section, the proposed model (Model 1) is analyzed along with Model 2 and Model 3 and pretrained ResNet50, VGG-16 and VGG-19. These six models are trained and validated with the same dataset and machine. The validation accuracy and corresponding epochs for all the six models are plotted in Fig. 7.

The overall accuracy is 97.5%, 93.75%, and 95% for Model 1, Model 2, and Model 3 respectively whereas the pretrained model achieved the accuracy of 88.5%, 78.75% and 60% respectively by ResNet50, VGG-16 and VGG-19. It clearly shows that the proposed model (Model 1) performs better than the other models in terms of accuracy. The performance of the models is more evident from the metrics like precision, recall, and F-1 score. These performance metrics are calculated from the

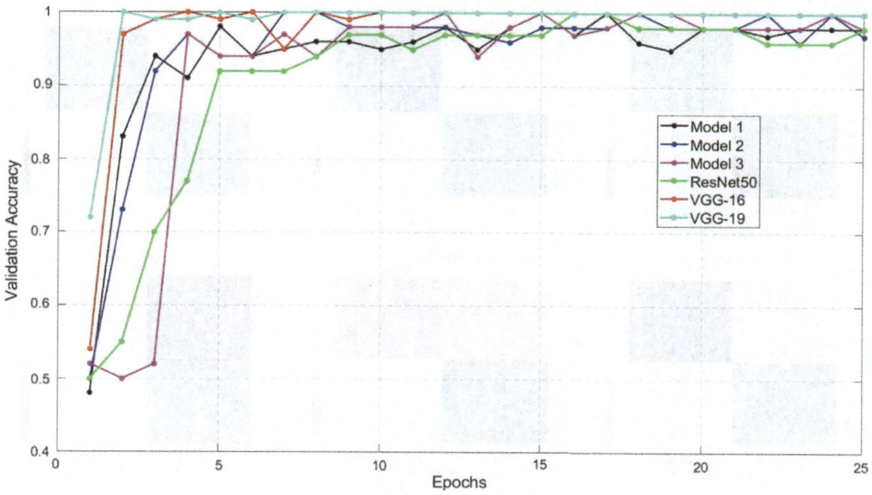

Fig. 7 Validation accuracy and corresponding epochs for all the six models

possible outcomes of the validation dataset which is obtained by the confusion matrix. A confusion matrix has four different outcomes: True Positive (TP), True Negative (TN), False Positive (FP), False Negative (FN). In this case, TP denotes the number of Corona positive patients detected as positive, TN denotes the number of Corona negative cases detected as negative, FP presents the number of cases which are actually negative but detected as Corona positive and FN gives the cases which are actually Corona positive but detected as negative. Receiver Operating Characteristics (ROC) curve represents the performance of the classifier at different threshold values which plot the TP rates vs FP rates. Confusion Matrices and corresponding ROC curves for all the six models is depicted in Figs. 8 and 9 respectively for analytical analysis.

Model 1 detects 39 TP and 39 TN cases, Model 2 finds 35 TP and 40 TN cases whereas, Model 3 detects 40 TP and 36 TN cases. On the other hand, the pretrained models perform very well in detecting the TN cases which is 40 for each models whereas the TP cases detected are 31, 23 and 8 by ResNet50, VGG-16 and VGG-19 respectively. The ROC curve area of the Model 1 is 0.975 which outperforms the other discussed models. On the contrary, VGG-19 achieved the lowest ROC curve area of 0.60 in compared to others. It is evident from the confusion matrix that Model 1 performs better in terms of case detection.

Accuracy defines how close the generated result is to the actual value whereas precision measures the percentage of the relevant results. Recall or sensitivity is another important factor for evaluating a CNN model. It is defined by the percentage of the total relevant results that a model can correctly classify. F1-score combines both precision and recall and it is designated as the weighted average of these two. Equations 1–4 represents accuracy, precision, recall, and F-1 sore respectively.

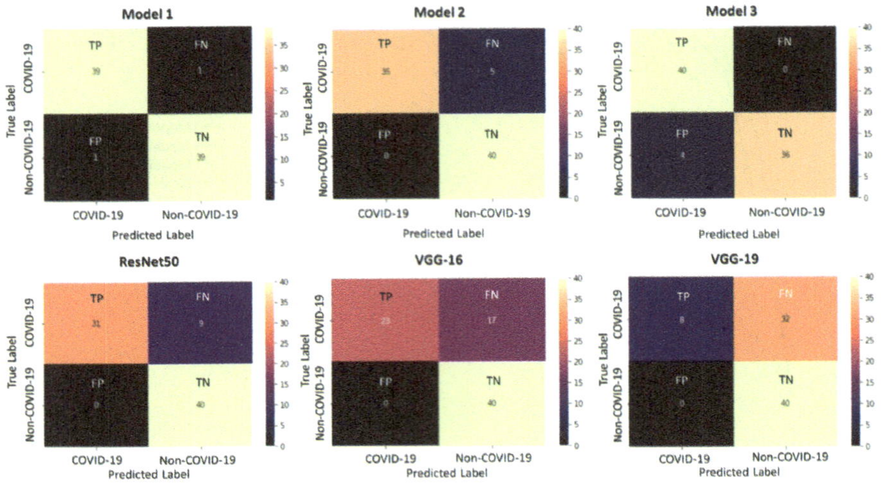

Fig. 8 Confusion matrices of all the six models on validation data set

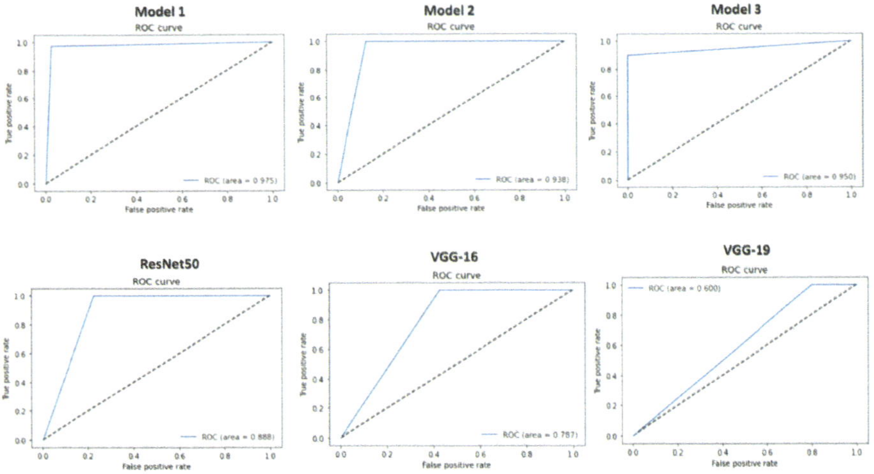

Fig. 9 ROC curves and curve area of all the six models

$$Accuracy = \frac{TP + TN}{TP + TN + FP + FN} \tag{1}$$

$$Precision = \frac{TP}{TP + FP} \tag{2}$$

$$Recall = \frac{TP}{TP + FN} \tag{3}$$

$$F1\ Score = 2 \times \left(\frac{Precision \times Recall}{Precision + Recall} \right) \tag{4}$$

Table 1 Confusion matrix parameters and performance metrics of the models

Model	TP	TN	FP	FN	TP (%)	TN (%)	Accuracy (%)	Precision (%)	Recall (%)	ROC Area	F1-Score
Model 1	39	39	1	1	97.5	97.5	97.5	97.5	97.5	0.975	97.5
Model 2	35	40	0	5	87.5	100	93.75	100	87.5	0.938	93.34
Model 3	40	36	4	0	100	90	95	90.9	100	0.950	95.23
ResNet50	31	40	0	9	77.5	100	88.75	100	77.5	0.888	87.32
VGG-16	23	40	0	17	57.5	100	78.75	100	57.5	0.787	73.01
VGG-19	8	40	0	32	20	100	60	100	20	0.60	33.33

Table 1 shows the confusion matrix parameters, accuracy, precision, recall, ROC curve area and F1-score of the mentioned six models.

Model 1 achieves the highest F1-score of 97.5, contrarily, VGG-19 performs with the lowest F1-score of 33.33. The overall performance and also the F1-score of the proposed model (Model 1) show better performance than that of the other models. The accuracy of the proposed model is 97.5% with the precision and recall value of 97.5% for both the parameters. Though this model lacks accuracy, the overall performance including accuracy and F1-score can be improved further by training the model with a larger dataset.

4 Conclusion

Mass testing and early detection of COVID-19 play an important role in preventing the spread of this recent global pandemic. Proper monitoring and treatment also play an important part in fighting against this. Time, cost, accuracy and reliability are the few major factors in any disease detection and treatment process, especially COVID-19. IoT and Machine Learning do a great job in addressing these issues. This chapter discusses different prospects of IoT in fighting against COVID-19. It also includes designing two different IoT model for thermal monitoring which would do the preliminary screening for detecting COVID-19 and heart rate monitoring system for affected patients. A CNN based model is proposed in the later part of the chapter for detecting COVID-19 cases from patients' chest X-rays. A set of 322 chest X-ray images which are equally divided into two classes: 'COVID-19' and 'Normal', are used for training the model. Similarly, an equally divided image set of 80 chest X-rays are used for validation of the model. This model performs with an accuracy and precision of 97.5% and 97.5% respectively. Moreover, this model is compared to pretrained ResNet50, VGG-16, VGG-19 and other two CNN models with a different number of convolutional layers. The comparative studies show better F1-score and overall performance of the proposed model (Model 1) than that of other models. This model can be improved further with the availability of the larger dataset. So, CNN has great prospects in detecting COVID-19 with very limited time, resources, and costs. Though the proposed model shows promising results, it is in no way clinically tested. This model needs further improvements and clinical testing for it to work in clinical diagnosis.

Acknowledgements The authors would like to thank Dr. Lisa Gandy for her suggestions to improve the manuscript.

References

Abdelgawad, A., & Yelamarthi, K. (2017). Internet of things (IoT) platform for structure health monitoring. *Wireless Communications and Mobile Computing*.

Catarinucci, L., De Donno, D., Mainetti, L., Palano, L., Patrono, Stefanizzi, M. L., & Tarricone, L. (2015). An IoT-aware architecture for smart healthcare systems. *IEEE Internet of Things Journal*, *2*(6), 515–526.

Cohen, J. P., Morrison, P., & Dao, L. (2020). Covid-19 image data collection. arXiv:2003.11597. https://github.com/ieee8023/covid-chestxray-dataset

Gu, J., Wang, Z., Kuen, J., Ma, L., Shahroudy, A., Shuai, B., et al. (2018). Recent advances in convolutional neural networks. *Pattern Recognition*, *77*, 354–377.

Gunawan, T. S., Gani, M. H. H., Rahman, F. D. A., & Kartiwi, M. (2017). Development of face recognition on raspberry pi for security enhancement of smart home system. *Indonesian Journal of Electrical Engineering and Informatics (IJEEI)*, *5*(4), 317–325.

Hall, L. O., Paul, R., Goldgof, D. B., & Goldgof, G. M. (2020). Finding covid-19 from chest x-rays using deep learning on a small dataset. arXiv preprint arXiv:2004.02060.

Haque, K. F., Zabin, R., Yelamarthi, K., Yanambaka, P., & Abdelgawad, A. (2020). An IoT based efficient waste collection system with smart bins. In *IEEE 6th World Forum on Internet of Things (WF-IoT)*. IEEE.

Haque, K. F., Abdelgawad, A., Yanambaka, P., & Yelamarthi, K. (2020). An energy-efficient and reliable RPL for IoT. In *2020 IEEE 6th World Forum on Internet of Things (WF-IoT)*. IEEE.

Haque, K. F., Kabir, K. H., & Abdelgawad, A. (2020). Advancement of routing protocols and applications of underwater wireless sensor network (UWSN)-a survey. *Journal of Sensor and Actuator Networks*, *9*(2), 19 (2020).

Haque, K.F., Haque, F.F., Gandy, L., & Abdelgawad, A. (2020). Automatic detection of COVID-19 from chest X-ray images with convolutional neural networks. In *3rd IEEE International Conference on Computing, Electronics & Communications Engineering (IEEE iCCECE'20)*.

He, K., Zhang, X., Ren, S., & Sun, J. (2016). Deep residual learning for image recognition. In *Proceedings of the IEEE Conference on Computer Vision and Pattern Recognition* (pp. 770–778).

Huang, C.-H., & Cheng, K.-W. (2014). RFID technology combined with IoT application in medical nursing system. *Bulletin of Networking, Computing, Systems, and Software*, *3*(1), 20–24 (2014).

Istepanian, R. S., Sungoor, A., Faisal, A., & Philip, N. (2011). Internet of m-health things 'm-IoT'. In *IET Seminar on Assisted Living*.

Joyia, G. J., Liaqat, R. M., Farooq, A., & Rehman, S. (2017). Internet of medical things (IOMT): Applications, benefits and future challenges in healthcare domain. *Journal of Communication*, *12*(4), 240–247.

Kumar, K. S., & Bairavi, K. (2016). IoT based health monitoring system for autistic patients. In *Proceedings of the 3rd International Symposium on Big Data and Cloud Computing Challenges (ISBCC–16')* (pp. 371–376). Springer.

Kumar, P. M., & Gandhi, U. D. (2018). A novel three-tier internet of things architecture with machine learning algorithm for early detection of heart diseases. *Computers & Electrical Engineering*, *65*, 222–235.

Krishna, K. D., Akkala, V., Bharath, R., Rajalakshmi, P., Mohammed, A., Merchant, S., et al. (2016). Computer aided abnormality detection for kidney on FPGA based IoT enabled portable ultrasound imaging system. *IRBM*, *37*(4), 189–197.

McCulloch, W. S., & Pitts, W. (1943). A logical calculus of the ideas immanent in nervous activity. *The Bulletin of Mathematical Biophysics*, *5*(4), 115–133.

Mohammed, M., Syamsudin, H., Al-Zubaidi, S., AKS, R. R., & Yusuf, E. (2020). Novel covid-19 detection and diagnosis system using IoT based smart helmet. *International Journal of Psychosocial Rehabilitation*, *24*(7).

Mooney, P. (2020). Chest x-ray images (pneumonia). https://www.kaggle.com/paultimothymooney/chest-xray-pneumonia

Narin, A., Kaya, C., & Pamuk, Z. (2020). Automatic detection of coronavirus disease (covid-19) using x-ray images and deep convolutional neural networks. arXiv preprint arXiv:2003.10849.

Sethy, P. K., & Behera, S. K. (2020). Detection of coronavirus disease (covid-19) based on deep features. Preprints *2020–030300*.

Singh, R. P., Javaid, M., Haleem, A., & Suman, R. (2020). Internet of things (IoT) applications to fight against covid-19 pandemic. *Diabetes & Metabolic Syndrome: Clinical Research & Reviews*.

Simonyan, K., & Zisserman, A. (2014). Very deep convolutional networks for large-scale image recognition. arXiv preprint arXiv:1409.1556.

Vaishya, R., Javaid, M., Khan, I., & Haleem, A. (2020). Artificial intelligence (AI) applications for covid-19 pandemic. *Diabetes & Metabolic Syndrome. Clinical Research & Review*.

WHO Organization Coronavirus Disease (covid-19). (2019). https://www.who.int/emergencies/diseases/novel-coronavirus-2019.

WHO Organization. (2020). *Modes of transmission of virus causing covid-19: Implications for IPC precaution recommendations: Scientific brief, 27 March 2020*. Technical report, World Health Organization.

WHO Coronavirus Disease (COVID-19) Dashboard. (2020). Geneva world health organization. https://covid19.who.int/.

WHO Coronavirus Disease (COVID-19) Newsroom. (2020). Geneva world health organization. https://www.who.int/news-room/detail/07-04-2020-who-lists-two-covid-19-tests-for-emergency-use.

Wu, Y.-C., Chen, C.-S., & Chan, Y.-J. (2020). The outbreak of covid-19: An overview. *Journal of the Chinese Medical Association*, *83*(3), 217.

Yelamarthi, K., Aman, M. S., & Abdelgawad, A. (2017). An application-driven modular IoT architecture. *Wireless Communications and Mobile Computing*, *2017*, 2017.

Yu, L., Lu, Y., & Zhu, X. (2012). Smart hospital based on internet of things. *Journal of Networks*, *7*(10), 1654.

Zhang, H., Li, J., Wen, B., Xun, Y., & Liu, J. (2018). Connecting intelligent things in smart hospitals using NB-IoT. *IEEE Internet of Things Journal*, *5*(3), 1550–1560.

Zhang, J., Xie, Y., Li, Y., Shen, C., & Xia, Y. (2020). Covid-19 screening on chest x-ray images using deep learning based anomaly detection. arXiv preprint arXiv:2003.12338.

Development of an IoT-Based System for Training in Cardiopulmonary Resuscitation

Yassine Bouteraa, Hisham M. Alzuhair, and Naif M. Alotaibi

Abstract With the increase in heart disease and frequent heart attacks, the death rate has increased in a terrifying and appalling manner worldwide due to the lack of first aid practices such as cardiopulmonary resuscitation (CPR). Cardiopulmonary resuscitation is a dual process in which the paramedic resuscitates the lung and heart. As for resuscitation of the lung, it is through the delivery of air and oxygen to it through artificial respiration, as for reviving the heart, it is accomplished through manual pressure on the area of the injured heart "in the area between the thoracic bone and the spine" so that blood is pumped to human vital organs of the injured person's body, especially the brain. This paper involves designing a manikin that interacts with humans to facilitate the application of CPR in a correct and scientific form by measuring the strength and depth of pressure and calculating the number of compression per minute (CPM). The system control architecture is based on IoT concept where data that is read from integrated sensors and sent to IoT hubs such as Amazon Web Services. The purpose of this work is to provide trainees with a training tool on resuscitation approaches.

Keywords Cardiopulmonary resuscitation · E-Health · IoT

Y. Bouteraa (✉)
Prince Sattam bin Abdulaziz University, Al-Kharj, Saudi Arabia

CEM-Lab, Digital Research Center of Sfax, University of Sfax, Sfax, Tunisia
e-mail: yassine.bouteraa@isbs.usf.tn

H. M. Alzuhair · N. M. Alotaibi
Computer Engineering Department, Prince Sattam bin Abdulaziz University, Al-Kharj, Saudi Arabia
e-mail: Halzuhair111@gmail.com

N. M. Alotaibi
e-mail: Nalotaibi640@gmail.com

O. Kanoun and N. Derbel (eds.), *Advanced Systems for Biomedical Applications*, Smart Sensors, Measurement and Instrumentation 39,
https://doi.org/10.1007/978-3-030-71221-1_6

1 Introduction

Cardiopulmonary resuscitation (CPR) is an emergency procedure that saves lives when the heart stops beating. This procedure or technique is considered as a form of first aid by which victims of cardiac arrest can be saved by applying the appropriate method. Immediate cardiopulmonary resuscitation can double or triple the chance of survival. With the community's lack of knowledge of cardiopulmonary resuscitation and the often-late arrival of ambulances at the scene of the incident, the percentage of victims has been increased. A cardiac arrest can kill in few minutes, hence the need for emergency resuscitation. Education and awareness of society are essential to introduce this culture of practice of emergency medical assistance to save as many lives as possible. After several studies applied in the United States by the American Heart Association in 2016, the statistics show that over the 2013–2016 period, cardiovascular diseases cause more victims every year than all forms of cancer and respiratory diseases combined. In fact, in the United States, every year, 610,000 people die from heart disease, or 1 in 4 deaths. These scary statistics were the main motivation for this project. The goal is to simulate cardiopulmonary resuscitation to train people to cope with this type of emergency. The interactive CPR manikin will contain various hardware components, such as a microcontroller, sensors, displays, and other integrated components. The sensors are designed to provide feedback such as correct depth detection, hand compression, and the number of compressions per minute applied to the patient's chest. An RGB led gives feedback on the quality of the rescue function. The LED is set to the red color by default. If the cardiopulmonary resuscitation (CPR) process is done at the appropriate depth for an adult, as well as the compression per minute (CPM) is correct, the RGB LED turns green. Otherwise, the indicator stays red. An incorrect method of cardiopulmonary resuscitation has two serious effects. One is to break the rib cage, which means a stronger application of force than recommended. Second, the number of compressions per minute (CPM) is not enough for the heart to pump again. This can decrease the percentage of survival of the patient. CPR is an emergency procedure applied to patients with cardiac arrest. The goal of CPR is to provide a flow of oxygenated blood to the brain. Research has shown that chest compressions are often unsuccessful because the person practicing them usually does not use enough force. Some people may be reluctant to use more force and risk breaking ribs. Therefore, educating people to appropriate cardiopulmonary resuscitation procedures with the proper strength can increase the percentage of survival.

According to American studies, the time required for an ambulance to reach the scene of the incident would be 10–14 min. This is considered long, as it takes only a few minutes for the patient who has suffered a cardiac arrest to die. For example, people need to know which procedure can help victims of cardiac arrest and distribute them to the community. Cardiopulmonary resuscitation (CPR) is the first step to saving these types of victims. Therefore, we focus on the development of a prototype allowing people to master these acts of rescue. When a patient suffers a cardiac arrest, every minute that passes decreases his rate of life by 10%, which means that after 10 min, his chances of living are almost non-existent. A study was conducted to measure the community's knowledge of the CPR procedure. The result of the study showed that only 41.5% of people knew the approach correctly. This percentage is low and could make us lose more lives. The percentage of Arab societies is worse.

Sudden Cardiac Arrest (SCA) is the third leading cause of death in society. The use of the correct cardiopulmonary resuscitation method has a significant impact on society in many ways. One of the most important aspects is to save as many lives as possible by providing CPR to the patient. The cardiopulmonary resuscitation manikin can be used in many places, such as schools, universities, social centers, to train people on the risks of cardiac arrest and rescue intervention techniques. This training aims to raise awareness and provide knowledge about effective intervention. The cardiopulmonary resuscitation technique will help reduce the number of deaths of victims with heart attacks.

Various works have focused on the design and development of medical electronic devices (Leung et al. 2019; Chen et al. 2019) and different studies dealing with the cardiopulmonary resuscitation have been presented in Bray et al. (2017), Nishiyama et al. (2019), Higashi et al. (2017). The perceived barriers to providing cardiopulmonary resuscitation (CPR) education are discussed in Hoyme and Atkins (2017). Cardiopulmonary Resuscitation Training Disparities in the United States is presented in Blewer et al. (2017). An automated closed-loop resuscitation system that uses a fuzzy logic controller to adjust infusion rate based on arterial pressure has been developed in Ying et al. (2002). Similar works dealing with dummy-based systems for cardiopulmonary resuscitation are presented in Pastrick et al. (1999), Pastrick and Charlton (2004), Voss et al. (2017), Barash et al. (1999), Shaw and Marvin (2019), Yang et al. (2016). Robotics has recently contributed massively in the development of the medical sector (Benabdallah et al. 2015; Bouteraa and Abdallah 2017; Bouteraa et al. 2018, 2019, 2020). A manipulator robot has been developed for cardiopulmonary resuscitation (CPR) (Jung et al. 2016). A parallel robot is designed to perform the cardiopulmonary resuscitation (CPR) (Li and Xu 2007). In Li and Xu (2005), authors designed a parallel mechanisms-based medical robot for chest compressions. A 2-DOF parallel manipulator was developed for performing chest compressions (Yedukondalu et al. 2018).

In this paper, the American Heart Association guidelines is applied to build a CPR dummy-based system. The major differences between Cardiac Arrest and Heart Attack are discussed. The developed system is used to train people on performing Cardiopulmonary Resuscitation (CPR) method as well as to educate them about the risks of Cardiac Arrest.

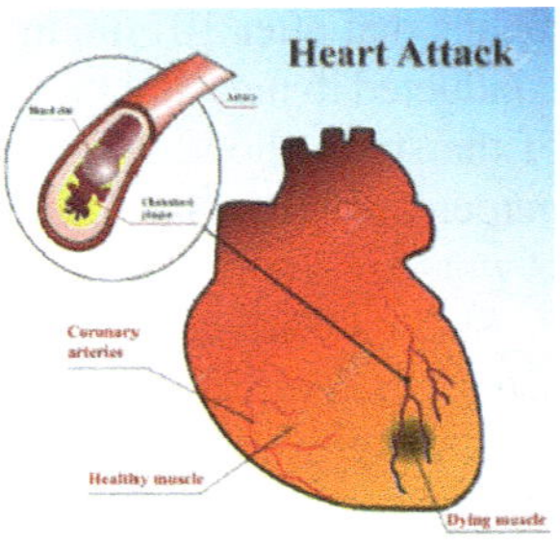

Fig. 1 Heart attack process

2 Background

2.1 Heart Attack

The Heart Attack is a circulation issue as shown in Fig. 1 and it occurs when blood flow to the heart is either blocked or reduced. Heart Attacks can be classified as mild or very serious accidents. During the attack, the heart might continue beating normally, however, if the blockage is still not resolved speedily, parts of the cardiac muscle begin to die as a result of the lack of oxygen. The faster the treatment applies, the least the damage occurs. We can prevent a heart attack from happening if we know the signs (symptoms) of a heart attack such as pain or discomfort in the chest, shortness of breath, back pain, and vomiting since the symptoms can occur hours, days, or weeks.

Figure 1 indicates that when the artery is blocked or reduced a blood clot occurs in which the blood is not flowing as normally does. Thus, the muscles of the heart start dying.

In case of a human who does not have a heart attack, blood flows normally. Otherwise, arteries are either completely blocked or partially blocked by the accumulation of cholesterol and fatty deposits (called plaques) on the inner walls of the arteries where blood and oxygen flow abnormally.

2.2 Cardiac Arrest or Sudden Cardiac Arrest

The Cardiac Arrest is an electrical disturbance issue and occurs when the heart malfunctions and stops beating unexpectedly. In which the blood plus the oxygen cannot be flowing to human vital organs such as the brain, heart, kidneys, liver, and lungs. The electrical disturbance deactivates the heart's electrical signal so that the heart is either beating very fast or very slow. Dissimilar to Heart Attack, Sudden Cardiac Arrest has been always categorized as a serious matter. Thus, an assistant must be provided immediately by either a cardiopulmonary resuscitation (CPR) or

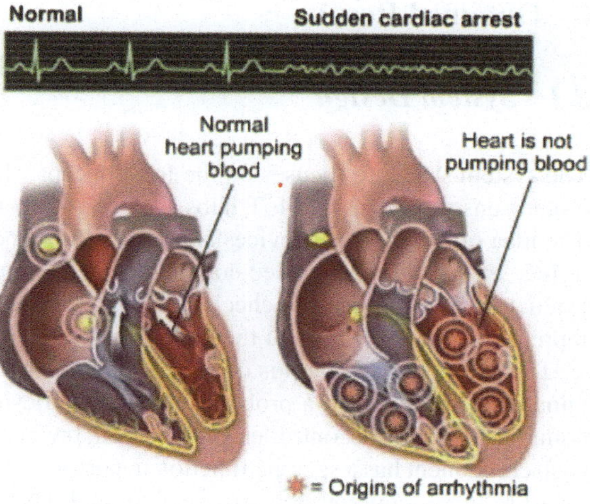

Fig. 2 Rhythm of a normal and abnormal heart

an automated external defibrillator (AED) to restore the normal rhythm of the heart, otherwise, death can occur within minutes. Moreover, Sudden Cardiac Arrest can occur with little or even no warning because symptoms are instant. However, there are some signs to identify whether a person is having an SCA. For instance, a sudden loss of consciousness, lack of breath, and no pulse. SCA can be caused by multiple events such as a sudden blow to the chest, drug abuse, heart attack, drowning, and low body temperature which is known as Hypothermia.

Figure 2 shows what happens when the heart pumping normally and when it stops.

2.3 Adult CPR

Cardiopulmonary resuscitation is a fundamental skill that a person must learn to save someone's life. CPR is a combination of two key components, the first of which is chest compression and the second of which provides breathing, also called artificial ventilation. To provide high quality CPR to an adult, three essential factors must be considered: the number of compressions per minute (CPM), airway management and artificial respiration. Chest compressions have the highest impact for survival; however, many rescuers fail because they do not push hard or fast. As for the CPM, it should be administered at a compression rate of 100–120 hands at a depth of 2–2.4 in. when the person is unconscious and not breathing or out of breath.

3 Obtained Results

3.1 System Design

The system structure focuses on the IoT concept where data that is read from integrated sensors and sent to IoT hubs such as Amazon AWS and stored as raw data. The idea of using cloud services allows real-time monitoring as well as identifying trainee errors. When a trainee completes the CPR process, the developed software evaluate the performance to check if he passes the test or if he fails. The architecture representing the operation of the system is shown in Fig. 3.

The architecture system is designed such way to be efficient, modern and maintainable in the event of a problem. The circuit design is composed of a variety of components as microcontroller, LCD screen, bread board, sensors and LEDs, where each component has a specific function to perform. The sensors built into the dummy measure force, compressions per second, and the number of breaths delivered per second. Moreover, the values of these sensors are sent via the Arduino MKR 1000 microcontroller to an IoT hub to be stored and displayed. We worked on a solution design by analyzing the main problem in which we identified the inputs, the outputs and the processing operations. We design an algorithm by writing a pseudocode, then transferring it to a complete solution design using a flowchart. The organization chart illustrates the complete resolution. To go deeper, the organization chart is divided into two sections in which there will be two separate modes from which the trainee can choose. The first part, as shown in Fig. 4, is the tutorial mode where the trainees will review all the cardiopulmonary resuscitation procedures by displaying

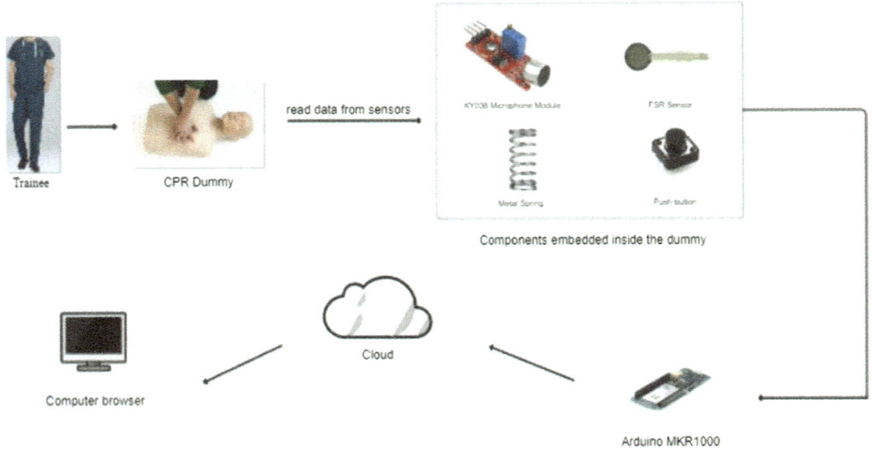

Fig. 3 System architecture

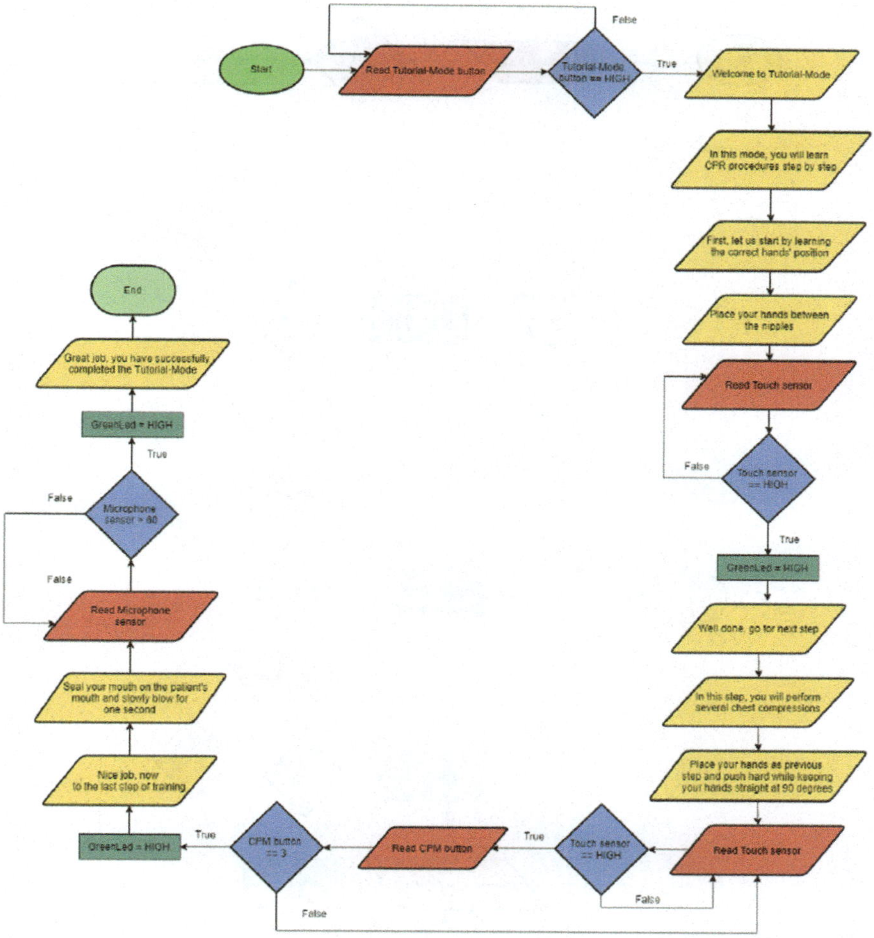

Fig. 4 Tutorial-mode flowchart

each AHA instruction to be executed by the trainees to prepare them for the real test that will take place next.

The second part, as shown in Fig. 5, is the real mode, this mode is the cardiopulmonary resuscitation test that a trainee can pass (certified) or fail (not certified). The test simulates a cardiac arrest disease in which the patient is having sudden heart failure, unconscious, and not breathing. Moreover, this test is linked with a timer such that when the trainees begin cardiopulmonary resuscitation, each stage has a minimal and maximum time. For example, the first step of CPR which is the chest compressions has a minimal time of 14 s and a maximum of 16 s. Once the trainees complete this stage with the ranged time, they can move to the next step which is providing breaths. The reason for the restricted procedures with timer is the fact

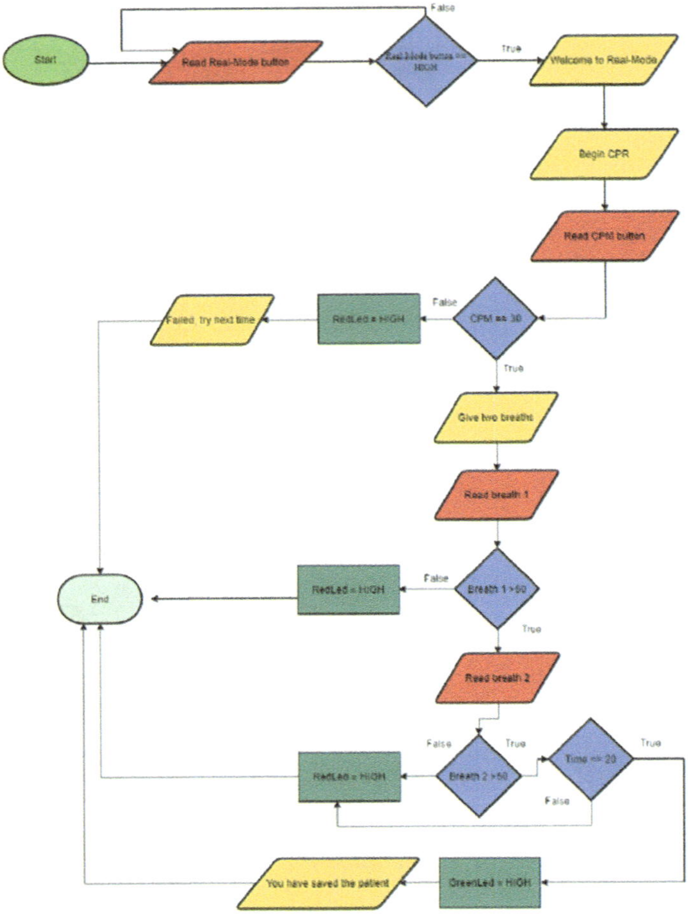

Fig. 5 Real-mode flowchart

that to save a patient, we do not have enough time because it takes a maximum of 4 min to recover a patient otherwise the patient could die. Therefore, it is extremely important to work with a ratio of 30: 2 for trainees or 120 compressions per minute for untrained people.

The first objective is to measure the force applied to the manikin's chest by a force-sensitive resistance sensor (FSR) in order to check if it meets the required force, i.e. around 60 ponds. To detect the depth of the compressions, a push button is attached to the inside of the manikin at a depth of 2–2.4 in. To meet the American Heart Association recommendation that the CPM be between 100 and 120 per minute, an additional push button is used as a counter to record the number of compressions per minute (CPM). First, the lung bag that simulates the human lungs was assembled

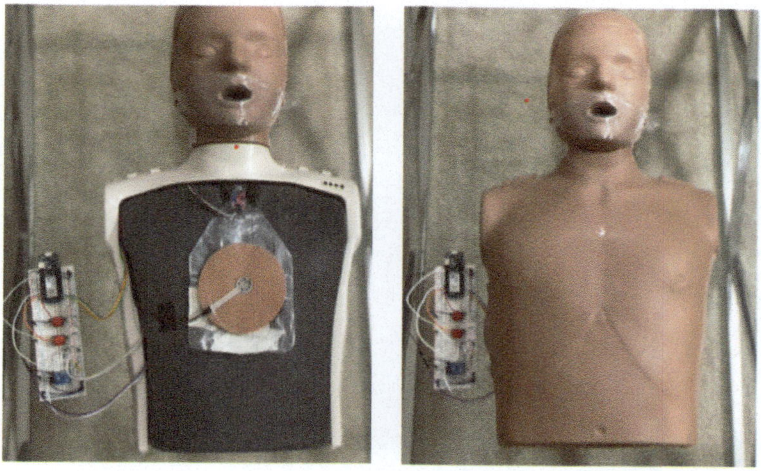

(a) (b)

Fig. 6 Developed dummy system

inside the mannequin by lifting the chin and positioning the lung bag through the throat to be under the chest. Second, the microphone sensor is placed to measure the breaths provided by the trainee. The correct position of the heart between the nipples is correctly identified to set up the push button for chest compressions. Then, the Force-sensitive sensor is placed to measure the force applied by the trainee. The FSR has to be above the push-button which has been lay previously. The reason behind placing it as shown is we wish to use the FSR to sense the hand's placements when a trainee performs CPR. Figure 6 shows the developed dummy.

The screenshot shown in Fig. 7a, is the result of completing the test successfully of the Real-Mode in which we simulate a cardiac arrest event where the heart is stopped beating, the patient is unconscious, and the lungs have stopped providing the body with oxygen. We started by working on a 30 : 2 ratio per minute such that the trainee performed 30 compressions without interruption with a range of time 14–16 s. After finishing the chest compressions, the trainee had to provide a total of two breaths to the dummy. The amount of time in this process should not be more than 4 s in which the trainee places the left hand in the forehead and the right hand on the chin then gently tilt the head to open the airway and provide breath for one second. After that, the trainee waited for the chest to recoil then provided another breath. In Fig. 7b, the result shown is for the trainee that has failed in the real mode such that the trainee did not finish the rescue operation in time such as taking a long time in the stage of chest compressions, blowing too hard in the lungs of the patient, not waiting for a second to let the chest to recoil.

For these reasons, the trainee has led to the death of the patient. Moreover, the mistakes that the trainee has done can occur in real life, hence we set up a test that

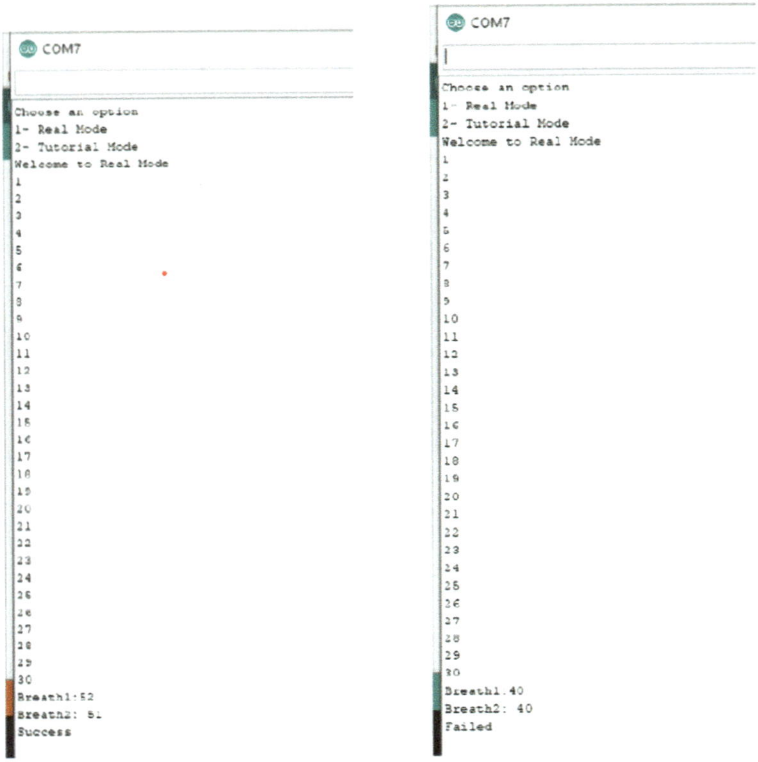

(a) (b)

Fig. 7 Cardiopulmonary resuscitation test: **a** Real-mode success, **b** Real-mode failure

simulates real situations and by considering most of the mistakes that could a rescuer do to be in our test as a measurement of failure.

The result, shown in Fig. 8, demonstrates the second mode where a trainee learns the basics of cardiopulmonary resuscitation by displaying the procedures and explaining some of the essential concepts such as heart attack, cardiac arrest, and brain stroke so the trainee can differentiate each disease and its symptoms. Moreover, in this mode trainees perform cardiopulmonary resuscitation procedures step by step in which they learn and become competent in the hand's placement, how much of depth needed for the first shock of the heart, number of compressions per minute, and how to provide breaths.

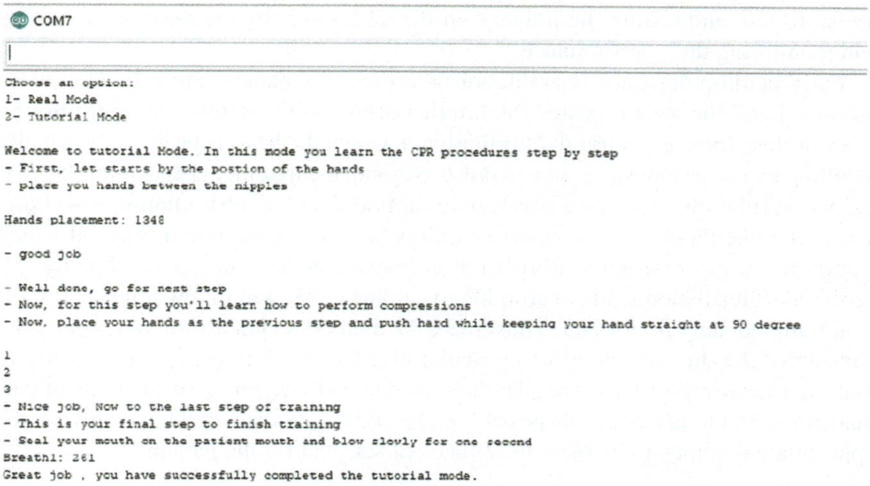

Fig. 8 Tutorial-mode

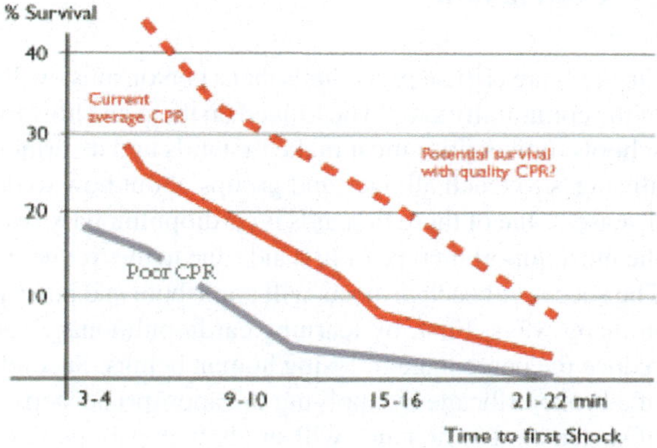

Fig. 9 Percentage of survival

4 Discussion

As can be shown from the graph in Fig. 9, the quality of CPR indeed matters. In fact, the survival of the patients who suffer from out of hospital cardiac arrest raises by 40%. Moreover, research has shown that by delivering a professional CPR it impacts the victim's chance of living directly. Moreover, when a professional delivers a high-quality CPR as a comparison to non-professional, the chances of survival of the victim varies.

For these reasons, we have concentrated on building and designing a system that trains and prepare trainees to be ready for the real world by setting a test of their knowledge, awareness of the First-Aids in general as well as focusing on the cardiac arrest disease and testing the trainees on the CPR skills by the developed dummy and monitoring their performance.

Early cardiopulmonary resuscitation is critical for patient survival for various reasons. First, the most repeated rhythm that occurs with victims who are suffering from cardiac arrest is ventricular fibrillation in which the heart is beating abnormally resulting in unconsciousness and no pulse. Second, the treatment needed for the ventricular fibrillation is an early and effective cardiopulmonary resuscitation procedure. Third, the likelihood of a successful cardiopulmonary resuscitation method which can diminish the ventricular fibrillation is reduced as time moves on. Finally, the ventricular fibrillation tends to crumble to cardiac arrest within few minutes if CPR is not immediately performed. Thus, the CPR method and defibrillation rapidly are considered the standard care for ventricular fibrillation. Moreover, we had to ensure that the trainees must have complete knowledge and awareness of the risks of cardiac arrest and to act as fast as possible to rescue the cardiac arrest patient with the appropriate methods to increase the chance of survival for the patient.

5 Conclusion

First aids are critical procedures that a person must well know. Therefore, all members of the community should be trained on them and how to perform them correctly. Also, schools, universities must make first aids and its principle within all curricula and in all stages to teach all ages and groups about how to deal with various incidents and diseases. One of these first aids is cardiopulmonary resuscitation and is considered as the most important type of first aids due to the frequent occurrence of these accidents. The success of such projects will contribute and have an impact on the whole society in many ways. First, by learning cardiopulmonary resuscitation will save lives and reduce the percentage of losing human beings. Second, to prevent the common issue of a broken ribcage by applying the appropriate steps. Finally, having the awareness of such a kind of first aids will surely help paramedics before they arrive at the scene of accidents.

In this chapter, guidelines from the American Heart Association have been applied to build and design a Dummy-based system for Cardiopulmonary Resuscitation (CPR). The objective of the system must be used to train people in the implementation of the method of cardiopulmonary resuscitation (CPR) as well as to make them aware of the risks of cardiac arrest. The results obtained showed the effectiveness of the developed system.

References

Barash, D., Totman, M., Freeman, G. A., & McGough, T. S. (1999). CPR competition system. U.S. Patent, 5,885,084.

Benabdallah, I., Bouteraa, Y., Boucetta, R., & Rekik, C. (2015). Kinect-based computed torque control for lynxmotion robotic arm. In *Proceedings of the International Conference on Modelling, Identification and Control (ICMIC)*.

Blewer, A. L., Ibrahim, S. A., Leary, M., Dutwin, D., McNally, B., Anderson, M. L., et al. (2017). Cardiopulmonary resuscitation training disparities in the united states. *Journal of the American Heart Association*, *6*, e006124.

Bouteraa, Y., Ben Abdallah, I., & Elmogy, A. (2020). Design and control of an exoskeleton robot with EMG-driven electrical stimulation for upper limb rehabilitation. *Industrial Robot: The International Journal of Robotics Research and Application*.

Bouteraa, Y., Ben Abdallah, I., & Elmogy, A. M. (2019). Training of hand rehabilitation using low cost exoskeleton and vision-based game interface. *Journal of Intelligent & Robotic Systems*, *96*(1), 31–47.

Bouteraa, Y., Ben Abdallah, I., & Ghommam, J. (2018). Task-space region-reaching control for medical robot manipulator. *Computers & Electrical Engineering*, *67*, 629–645.

Bouteraa, Y., & Ben Abdallah, I. (2017). A gesture-based telemanipulation control for a robotic arm with biofeedback-based grasp. *Industrial Robot: An International Journal*.

Bray, J. E., Smith, K., Case, R., Cartledge, S., Straney, L., & Finn, J. (2017). Public cardiopulmonary resuscitation training rates and awareness of hands-only cardiopulmonary resuscitation: A cross-sectional survey of victorians. *Emergency Medicine Australasia*, *29*, 158–164.

Chen, Y., Ding, S., Xu, Z., Zheng, H., & Yang, S. (2019). Blockchain-based medical records secure storage and medical service framework. *Journal of Medical Systems*, *43*(1), 5.

Higashi, E., Fukagawa, K., Kasimura, R., Kanamori, Y., Minazuki, A., & Hayashi, H. (2017). Development and evaluation of a corrective feedback system using augmented reality for the high-quality cardiopulmonary resuscitation training. In *Proceedings of the IEEE International Conference on Systems, Man, and Cybernetics (SMC)*.

Hoyme, D. B., & Atkins, D. L. (2017). Implementing cardiopulmonary resuscitation training programs in high schools: Iowa's experience. *The Journal of Pediatrics*, *181*, 172–176.

Jung, J., Kim, J., Kim, S., Kwon, W. Y., Na, S. H., Kim, K. S., etc. (2016). Application of robot manipulator for cardiopulmonary resuscitation. In *Proceedings of the International Symposium on Experimental Robotics*.

Leung, L. W. S., Roudsari, A., Kuo, A., & Courtney., K. L. (2019). System dynamics in remote monitoring service for cardiovascular implantable electronic devices. In *ITCH*.

Li, Y., & Xu, Q. (2005). Dynamic analysis of a modified DELTA parallel robot for cardiopulmonary resuscitation. In *Proceedings of the IEEE/RSJ International Conference on Intelligent Robots and Systems*.

Li, Y., & Xu, Q. (2007). Design and development of a medical parallel robot for cardiopulmonary resuscitation. *IEEE/ASME Transactions on Mechatronics*, *12*, 265–273.

Nishiyama, C., Kitamura, T., Sakai, T., Murakami, Y., Shimamoto, T., Kawamura, T., et al. (2019). Community-wide dissemination of bystander cardiopulmonary resuscitation and automated external defibrillator use using a 45-minute chest compression-only cardiopulmonary resuscitation training. *Journal of the American Heart Association*, *8*, e009436.

Pastrick, J. J., Lindseth S. W., Kopco, J. J., Nottingham, J. R., & Plantz J. S. (1999). Cardiopulmonary resuscitation manikin. U.S. Patent, 5,885,084.

Pastrick, J. J., & Charlton, C. M. (2004). Cardiopulmonary resuscitation manikin with replaceable lung bag and installation tool. U.S. Patent, 6,780,017.

Shaw, R. W., & Marvin, J. D. (2019). Single use mannequin having chest compression feedback mechanism for use in simulated cardiopulmonary resuscitation (CPR) training. U.S. Patent, 10,482,791.

Voss, G., Metzger, A., Lurie, K., & Yannopoulos, D. (2017). Enhanced guided active compression decompression cardiopulmonary resuscitation systems and methods. U.S. Patent Application, 15/638,452.

Yang, S.-J., Kim, D.-Y., & Chang, I. B. (2016). Cardiopulmonary resuscitation (CPR) simulator enabling repeated defibrillation training. U.S. Patent, 9,286,809.

Yedukondalu, G., Patnaik, S., Lakshmi Venkatesh, P., & Siva Jagadeesh, S. (2018). Chest compression with 2-DOF parallel manipulator for cardiopulmonary resuscitation. *International Journal of Engineering & Technology*, *7*, 211–215.

Ying, H., Bonnerup, C. A., Kirschner, R. A., Deyo, D. J., Michell, M. W., & Kramer, G. C. (2002). Closed-loop fuzzy control of resuscitation of hemorrhagic shock in sheep. In *Proceedings of the Second Joint 24th Annual Conference and the Annual Fall Meeting of the Biomedical Engineering Society, Engineering in Medicine and Biology*.

Portable Cardiopulmonary Resuscitation and Ventilator Device: Design and Implementation

Abdullah W. Al-Mutairi and Kasim M. Al-Aubidy

Abstract The demand for large numbers of ventilators and cardiopulmonary resuscitation (CPR) has increased recently after the global spread of the Corona epidemic (known as COVID-19), and the equipment available in hospitals and health centers has become insufficient. Providing a large number of these expensive devices in a short period of time is difficult, and providing suitable places for them is not easy. Hence the need to apply the concepts of reverse engineering in the design and manufacture of low-cost portable devices for use by patients, wherever they are. This chapter provides the main components of the mechanical ventilator and how it works as a first stage to apply the concepts of reverse engineering to the design and construction of a low-cost portable device as well as a cardiopulmonary resuscitation system. Wireless sensor network technologies were applied in order to access the device and adjust its main parameters by the specialist according to the patient's condition. This device can be used at home, ambulances, health centers and remote hard-to-reach places. The main parameters of the device can be adjusted directly by the operator or remotely by a specialist according to the patient's condition. Experimental results on the first prototype, when compared to available portable devices, indicate its ability to assist a patient with difficulty breathing. The prototype efficiently responds to the input settings provided by the paramedic or specialist, depending on the patient's condition. The respirator prototype is small and weighs 4.7 kg, equipped with a wireless guidance system for CPR, and its cost does not exceed 800 USD.

Keywords Ventilation · Bag volume mask · Low-cost ventilator · Portable ventilator · CPR · Real-time monitoring

A. W. Al-Mutairi (✉) · K. M. Al-Aubidy (✉)
Faculty of Engineering & Technology, Philadelphia University, Amman, Jordan
e-mail: kma@philadelphia.edu.jo

A. W. Al-Mutairi
e-mail: awm_mutairi@hotmail.com

O. Kanoun and N. Derbel (eds.), *Advanced Systems for Biomedical Applications*, Smart Sensors, Measurement and Instrumentation 39,
https://doi.org/10.1007/978-3-030-71221-1_7

1 Introduction

Medical information indicates that the effect of viruses on the performance of the lungs is very large, and may lead to a loss of the patient's life (Villaverde and Banga 2014). In many cases, the patient's respiratory system collapses due to the presence of viscous mucus in the lungs and makes it difficult for the patient to breathe on his own. Hence the importance of providing ventilation and recovery equipment to assist patients. In the beginning of 2020, the world was exposed to the Corona virus, known as COVID-19 pandemic, and ventilation and resuscitation equipment became necessary to help patients breathe. The ventilation and cardiac resuscitation devices available in hospitals are complex, expensive and limited. Moreover, providing a large number of these devices to deal with emergency situations (as is the case with the global epidemic of Corona) in a short period is difficult. Hence the importance of manufacturing ventilators and resuscitation to help patients so that they are easy to use and affordable.

The mechanical ventilators used in the intensive care units are complex in design and need compressed air to provide optimal lung ventilation. When transporting patients receiving ventilation support, there is a need to equip the ambulance with a mechanical ventilation device that is appropriate in size, external power requirements and average gas consumption (Fludger and Klein 2008). Portable ventilators are small in size, light weight, and are designed to provide mechanical ventilation using parameters settings. Advances in computer technology have played a major role in the development of medical devices used in diagnosis and treatment. It is now possible to design portable ventilators that come close to the traditional intensive care ventilator. Also, this technology in the ventilators made a qualitative shift in the interaction between the ventilator and the patient in order to provide the best advanced ventilation modes (Chatburn 2004). The graphical user interface gave an advanced advantage to interact between the operator and the patient's need.

Mechanical ventilation devices are very important aid to improving patient breathing. These devices need to set the parameters of the ventilator to suit the patient's exact condition, as it directly affects pulmonary gas exchange. A low-cost portable ventilator has been developed with wooden pieces and 6 kg weight (Ghafoor et al. 2017). It delivers breaths through the compression of a bag valve mask (BVM). Two parameters (pressure and required number of breathes per minute) are managed by the operator through a keyboard. Another similar prototyping of a low-cost portable mechanical ventilator was built out of acrylic and weighing 4.1 Kg. (Al Husseini et al. 2010). The ventilator delivers breaths by compressing a conventional bag-valve mask with a pivoting cam arm actuated by a DC motor. Tidal volume and number of breaths per minute are set by the operator. The control units of these prototypes are simple where the actuating signals are generated based on given set-points of pressure and breathing rate.

The overall ventilation system is a multivariable system with nonlinear interactions, and its response to changes are often non-linear and unpredictable (Sun et al. 1994). Therefore, it is very difficult to obtain an accurate mathematical model. Most of the ventilators used closed-loop control systems including PID controllers, adaptive and optimal controllers, these controllers are based on mathematical model to update the set-points or controller parameters. Other controllers, such as knowledge-based controllers and fuzzy logic controllers did not depend on the mathematical model of the ventilator.

The mathematical model can simulate the exchange of pulmonary gases for an individual patient to help operators know the patient's exact condition. They concluded that this approach would improve gas exchange and provide adequate ventilation for the patient. A fuzzy approach has been applied to the closed-loop feedback control of mechanical ventilation during general anesthesia (Schaublin et al. 1996). This control system automatically adjusts ventilator frequency (f) and tidal volume (VT) in order to maintain the end-tidal carbon dioxide fraction (CO_2) at set-point level. They concluded that the fuzzy-based control of mechanical ventilation is reliable and safe to maintain a desired CO_2 during anaesthesia. A fuzzy logic controller was used independently for both the simulation and experimental setup of a mechanical ventilator (Guler and Ata 2014). In the simulation, three linear lung models have been used to represent the lung. A fuzzy-LabVIEW based mechanical ventilator prototype was considered. It is concluded that "when lung resistance increased in the simulation and the experiments, the inspiration time and pressure in the lung increased just like in practice".

This chapter contains extensive information on the design and testing of a portable ventilator and CPR that was presented at an international conference (Al-Mutairi and Al-Aubidy 2020). The main objective of this chapter is to apply the concepts of reverse engineering in designing and building a ventilator with a CPR supervisory control and real-time monitoring of the patient's heart rate, in addition to enabling the specialist or health center to access the device wirelessly. The proposed device offers the following features:

- Portable and can be used easily in ambulances, home or hard-to-reach areas.
- Easy use and reliable with minimal risks to the patient's life.
- Low cost.
- It can be considered as a node in a wireless sensor network so that it can be easily accessed by a specialist or a medical center to update its parameters and provide medical advice.

The importance of the proposed device lies in assisting viral pulmonary patients or those who need artificial respiration as a result of sports injuries or the elderly, or any injuries that require rapid intervention, since the device is portable and can be used in hard-to-reach places for first aid. The next section explains the concepts of mechanical ventilators. The rest of the chapter is organized as follows; Sect. 3 covers concepts of reverse engineering and how they can be used in designing and manufacturing the proposed respiratory and cardiac resuscitation. Section 4 presents the operation of the proposed system. Elements of the mechanical ventilator and CPR unit are given in Sect. 5. Embedded controller design including hardware and software is presented in Sect. 6. Computer control of the mechanical ventilation device is discussed in Sect. 7. Results of experiments and simulations are discussed in Sect. 8. Finally, Sect. 9 covers conclusion and some suggested proposals for future work.

2 Mechanical Ventilators

Ventilation provides healthy air to breathe by moving the outside air into the room and distributing it inside the room. A mechanical ventilator is a device that takes care of breathing when a person is unable to breathe sufficiently alone. There are many reasons why a patient needs a ventilator, including low levels of oxygen or severe shortness of breath due to an infection.

Air enters the lungs through a pressure gradient between the airways and the alveoli. This can be accomplished by either raising the pressure in the airway (positive pressure ventilation) or by using negative pressure ventilation by reducing the pressure at the level of the alveoli (Kaynar and Sharma 2020).

2.1 Negative Pressure Ventilation

Negative pressure ventilators were the dominant devices to provide ventilation assistance until the middle of the last century to provide ventilation assistance. Mechanical ventilators were used to diversify the continuous negative pressure of ventilation around the entire patient's body or only around the chest and abdomen. At the beginning of the nineteenth century, a negative pressure operating room was developed that was used by surgeons to perform surgery (Woollam 1976). The patient's body, except the head, was maintained inside the chamber. The patient's lower body was encased in a flexible sack so that positive pressure could be applied to this part of the body, preventing blood from accumulating in the abdomen and lower extremities. The first intensive care units were established to manage dozens of patients requiring negative pressure ventilation (Kaynar and Sharma 2020).

2.2 Negative Pressure Ventilation

Positive pressure ventilators are devices used to help patients who are unable to breathe on their own. It provides air under pressure by using a mechanical respirator to improve air exchange between the lungs and the air. Positive pressure ventilation is only around the chest and abdomen. World war II contributed to the development of a positive pressure ventilation device to deliver oxygen and gas volume to high-altitude combat pilots. In the early 1950s, mechanical ventilation devices of positive pressure were used extensively during the polio epidemic in Scandinavia and the United States (Jackson and Muthiah 2020). Over the past few decades, mechanical ventilators have developed and become of the utmost importance in medical centers. These devices require supervision by medical experts to increase the effectiveness of ventilation and ensure patient safety.

Four main parameters need to be carefully defined and monitored while providing mechanical ventilation to the patient, these include (Mandal 2019):

- The ventilated air pressure that flows to and from the lungs.
- The volume of breath taken from the lungs.
- Air flow rate to the lungs.
- The time of inhale and exhale.

At positive end expiratory pressure (PEEP), a constant amount of positive pressure is applied during the mechanical ventilation cycle. As for continuous positive airway pressure, a steady amount of positive airway pressure is added to the breaths that the patient automatically takes (Mandal 2019).

Despite the importance of using mechanical ventilation for patients with respiratory problems, there are a number of potential risks, including; lung injury, lung infections caused by ventilation, lowering blood pressure, reducing cardiac output, and fluid retention in the body.

3 Ventilator Design Based on Reverse Engineering Concepts

Reverse engineering is a process that examines an existing product to determine detailed information and specifications in order to learn how it was made and how it works (Thayer 2018). Reverse engineering is the process by which a certain device is deconstructed to reveal its design or to extract knowledge regarding its principle of operation. There are many reasons for performing reverse engineering in various fields, these include:

- Products that are no longer manufactured and have no apparent designs for reproduction.
- Competitive product analysis to make design improvements.
- Carry out maintenance or repair of products for which documents are not available.

Reverse engineering process is an analysis in order to deduce design features from products with little or no additional knowledge about the procedures involved in their original production. The reverse engineering process involved analyzing a device to extract the design and how each part worked. In some cases, the goal of a reverse engineering process can be simply to perform competitor analysis to re-document devices and systems, not to copy them. Reverse engineering can also be used to create new products with additional specifications, features, and at a reasonable cost.

Recently, after the global spread of the new coronavirus (known as COVID-19), the demand for large numbers of ventilators and CPRs has increased in hospitals and health centers. In fact, providing a large number of these devices is difficult,

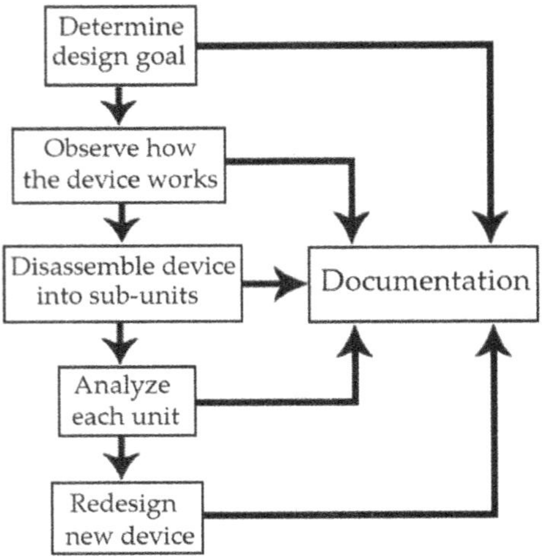

Fig. 1 General steps for reverse engineering process

in addition to their high cost and providing a suitable place for them. Therefore, there is an urgent need to use the concepts of reverse engineering in the design and manufacture of low-cost portable devices for use by patients on site.

The reverse engineering process of the ventilating and cardiopulmonary resuscitation devices included observing how the device works and performing an analysis of each part to determine its functional and organizational mechanisms and finally extracting the necessary design with careful documentation of each stage, as illustrated in Fig. 1.

After observing how the mechanical ventilator works and analyzing the functions of each part, five main units can be considered according to the reverse engineering processes, as shown in Fig. 2. The simple way to achieve the concept of the ventilator is by using a bag valve mask (BVM). A valve can be used to adjust the input Oxygen to the BVM. The necessary Oxygen is delivered to the lungs, according to the percentages set by the operator. This setting will be transferred to mechanical movement using an actuator controlled through a special control algorithm. The work of the human hand can be simulated using a straight lever arm made of PTFE connected to a DC motor. The computer control unit generates required signals to the motor drive unit according to feedback signals including angular position, pressure and time.

Early ventilators used simple mechanical control units, while recently advanced control units were used that made the ventilator more accurate and flexible in respiratory variables to suit the patient's need. Open-loop control is simplest type of ventilation control, in which the operator can adjust the pressure, and the control unit regulates a valve on and off at a set frequency and inspiratory-expiratory ratio. Such a controller is not recommended in medical applications, since the actual pressure and volume delivered were dependent on the real-time changes in the patient's respiratory system impedance (Chatburn 2004). Modern ventilators use a closed-loop control to maintain pressure and flow waves depending on the actual need of the patient

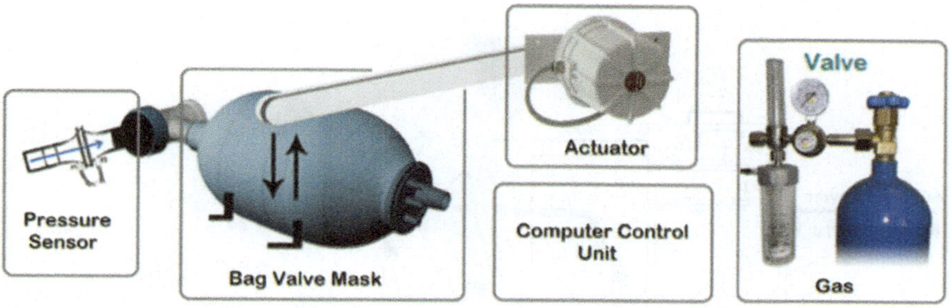

Fig. 2 Main parts of a mechanical ventilator

as well as to face any changing effects that may occur. This is accomplished using digital algorithms and feedback signals from the measured variables. Several optimal and adaptive controllers can be implemented using microcontrollers to improve the performance of ventilators. These controls are based on a fixed mathematical model, which is difficult to obtain accurately in real time applications. On the other hand, knowledge-based control does not depend on a stable mathematical model, it captures the expertise of human experts and extends control to all ventilation mode variables.

In this chapter, the concepts of reverse engineering will be applied to the design and construction of a low cost portable mechanical ventilator with CPR unit controlled by an embedded microcontroller. A real-time control algorithm will be chosen that relies on the feedback signals generated by the sensors, as well as the ability to wirelessly adjust the set points by the specialist and according to the patient's condition.

4 Overall System

To design a low-cost portable ventilator to meet CPR needs, a combination of sensors and actuators with an embedded microcontroller and a real-time monitoring unit will be used as shown in Fig. 3.

4.1 Ventilator

The ventilation device in this project can be used for less sick patients, and it can be helped by using the CPAP ventilation concept to provide constant air at the same flow rate and pressure. It can also be used for more serious patients who need a breathing machine that changes pressure and flow according to inhale and exhale and relies on the concept of PEEP to keep the lungs working continuously.

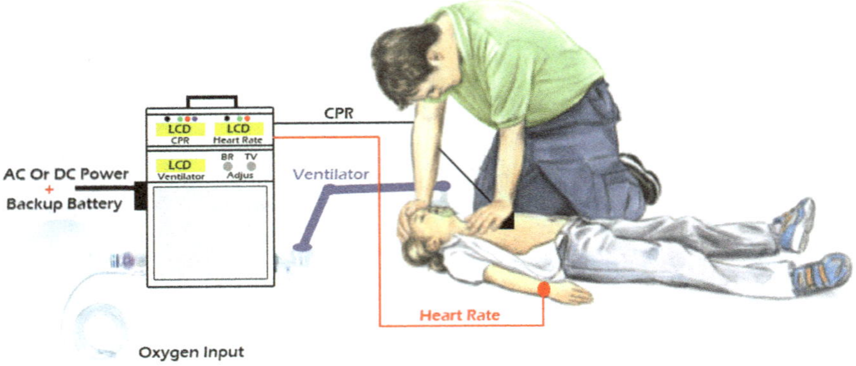

Fig. 3 Portable cardiopulmonary resuscitation and ventilator device

The implemented ventilator will simulate the working mechanism of the respiratory system by transporting oxygen through tubes, into our lungs and then diffusing it into the bloodstream, while carbon dioxide makes the opposite journey, by controlling the lung volumes and required vital capacity. This ventilator can be used for adults, children and infants, just by changing the mask size and the setting of the breathing parameters without any change for the internal components of the device.

The operator or specialist via the internet can monitor the device's operation and modify the following variables:

- Vital capacity: is the maximum amount of air that can be breathed out after breathing in as much air as possible.
- Breathing rate (BR): is the number of breaths in a minute.
- Tidal volume (TV): is the amount of air breathed in with each normal breath.
- Minute ventilation (VE): is the total volume of air entering the lungs in a minute, and can be obtained directly by:

$$VE = BR \times TV \tag{1}$$

By adjusting the above variables, the operator can select the required values as illustrated in Table 1.

Table 1 Parameters of the respiratory system (Barrett et al. 2012; Beardsell 2009)

Age	Breathing rate (/min)	Tidal volume (mL)	Minute ventilation (L/min)	Weight (kg)
Child	20–30	200–270	$20 \times 0.2 = 4$	33
Teenager	15–25	330–440	$15 \times 0.33 = 4.9$	55
Adult	12–20	440–580	$12 \times 0.5 = 6$	73

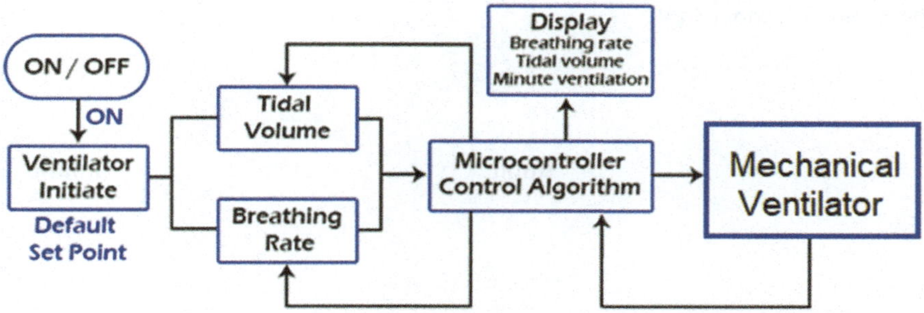

Fig. 4 Block diagram of the proposed ventilator

Tidal volume and respiratory rate parameters will be used to control the operation of the mechanical ventilator, as shown in Fig. 4. These two parameters can be determined according to the patient's medical data including gender, weight, and medical condition.

4.2 *Cardiopulmonary Resuscitation*

The cardiopulmonary resuscitation is an emergency procedure that combines chest compressions with mechanical ventilation in an effort to manually maintain healthy brain function until more measures are taken to restore automatic blood circulation and breathing to a person with a cardiac arrest (Atkins et al. 2015).

An emergency may occur with a sudden cardiac arrest of a person and there is an urgent need to work the CPR procedures correctly, but not everyone has knowledge of how to deal with this emergency. Therefore, a real-time monitoring unit was added to instruct the paramedic to perform the procedures for CPR as necessary. This system allows everyone (professional or trained) to perform CPR according to the patient's need as well as with the help of a specialist via the Internet. Operation of the proposed CPR unit depends on real-time measurement of variables and indicators, including heart rate, depth index, chest index, and CPR cycle counter.

Medical studies indicate that immediate CPR followed by defibrillation within 3–5 min of sudden (caused by abnormal heart rhythm) cardiac arrest dramatically improves survival (Speedy 2015). By using a sensitive measurement of the pressing force on the patient's chest, see Fig. 5, the CPR unit will help the operator by monitoring the CPR cycles procedures (5 cycles × 30 push) with the heart pulse and rate and give artificial respiration (Fig. 6).

The proposed system is an effective solution for resuscitation and artificial respiration due to lack of oxygen in the lungs as a result of infection. The proposed device is low-cost, lightweight, and easy to carry. It works by connecting it directly to a power source or via a battery. An embedded controller is used to manage and control all tasks of the CPR process.

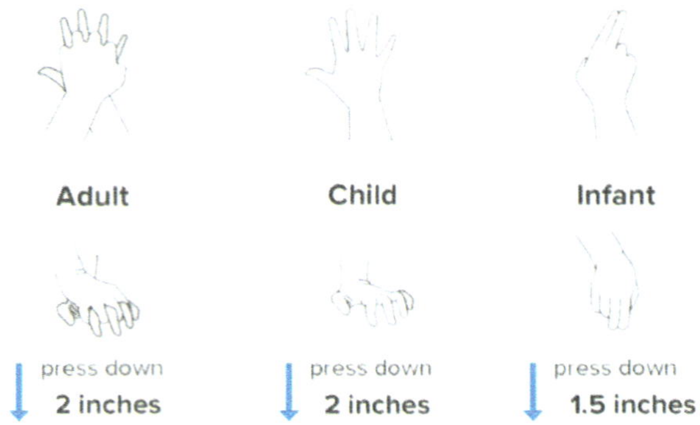

Fig. 5 How to perform CPR

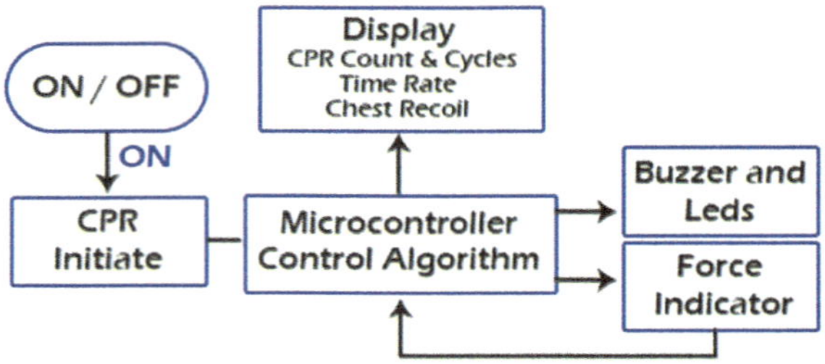

Fig. 6 General layout of the proposed CPR system

5 System Elements

The most important goal of this chapter is to design a portable low-cost ventilator with respiratory unit that meets the needs of patients. The implemented device is a node in a wireless sensor network and can be accessed by the health center or the specialist to update some specific set points according to the patient's condition. To achieve all these goals, the reverse engineering methodologies were used to design and implement every part of the device.

5.1 Ventilator Elements

Continuous positive airway pressure (CPAP) devices apply continuous positive airway pressure throughout the breathing cycle. The ventilator itself does not cycle during CPAP, no additional pressure above the level of CPAP is provided, and patients must initiate all of their breaths (Al-Mutairi and Al-Aubidy 2020). To achieve this concept a bag valve mask (BVM) has been used. The work of the human hand can be simulated using a straight lever arm made of PTFE connected to a DC motor, as shown in Fig. 7. The necessary oxygen is delivered to the lungs, according to the

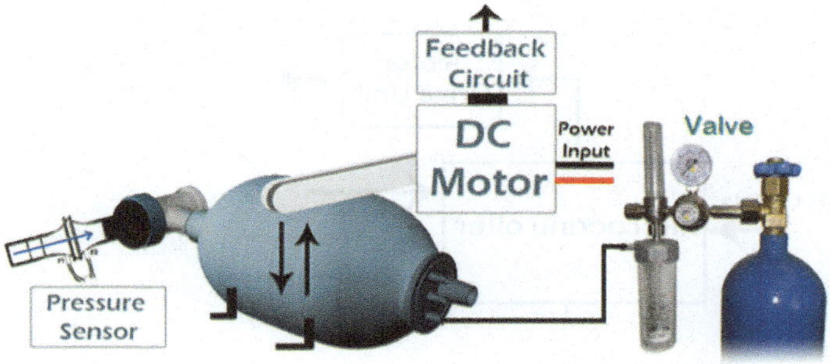

Fig. 7 General layout of the proposed ventilator

percentages set by the operator. An embedded microcontroller generates the required control signals for the motor based on set points and feedback signals, including angular position, speed, load, and time.

5.1.1 Bag Valve Mask

The adult bag valve mask has a specific volume and gives 2000 mL of air volume (Oxygen). If the patient needs 750 mL, for example, the BVM will be compressed to reach the required volume based on a high accuracy pressure sensor. A built-in valve is used to regulate the supply of Oxygen to the bag valve mask. The BVM will control the input oxygen by using a manual valve to adjust the Oxygen input from the bottle.

5.1.2 DC Motor

In the implemented project a car wiper motor has been used for its highly efficient worm drive motor specifications. The motor-gearbox unit has extremely zero backlash, and smooth operation. It has a rating of 5 Nm, and will overdrive to a maximum (29 Nm). It is a powerful motor for its small size, and operates from 12 V and has 2 operational speeds (either 45 rpm or 65 rpm). In this project, an embedded microcontroller has been used to regulate the operation of the DC motor based on a feedback signal generated from a rotary potentiometer. The motor will work up to 295° for both directions. A rotary potentiometer has been used as a sensor to provide the embedded microcontroller with position feedback signal, as illustrated in Fig. 8. The position detection sensitivity becomes 3.47 for each degree, since the real value of the position is given by;

$$RealValue = \frac{AnalogInput - MinimalPosition}{MaxPosition - MinimalPosition} \times MaxInput \tag{2}$$

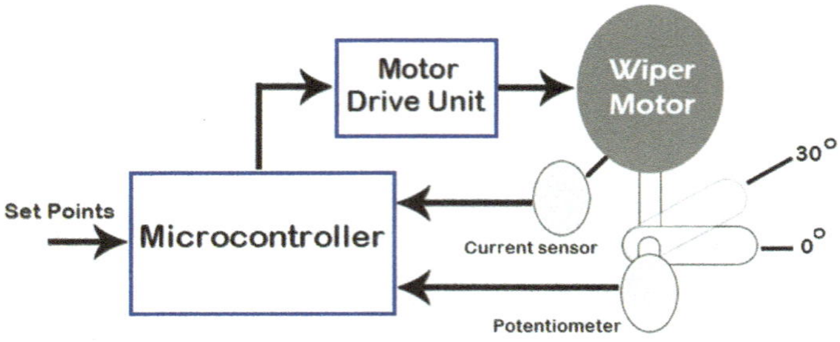

Fig. 8 Microcontroller based DC motor control

To move the actuator for the required position, clockwise and counter clockwise, the real value of the actuator position can be achieved by:

If (Real Value ≤ Required position) and (Real Value – Required position) > Error of potentiometer), Then Motor CW (clockwise)
If (Real Value > Required position) and (Real Value – Required position) > Error of potentiometer), Then Motor CCW (clockwise)

After installing the ventilator parts, several tests were conducted to ensure that the prototype works according to the given data and feedback signals. The motor used in this prototype consumes up to 8 amps at startup, and after starting it consumes 3–5 amps without any load.

5.1.3 Spirometer

Ventilation is one of the respiratory variables to be measured, and it has occupied an important position in pulmonary physiology and clinical medicine. Ventilation is the volume of gas that is transported to or out of the lungs and airways during a specific time period and is usually expressed in liters per minute. The measurement of ventilation requires only a clock and a flow meter, as it can be measured directly with a spirometer.

A spirometer is an apparatus for measuring the volume of air inspired and expired by the lungs. A spirometer measures the amount of air you breathe in and out and the speed of your breath. The spirometer principle of operation has been used for calculating the flow rate (liters/sec), by connecting a pressure sensor into the output of the bag valve mask, as illustrated in Fig. 9. This will be used as a feedback for the tidal volume (the amount of air breathed in with each normal breathe). The flow rate is proportional to the difference in pressure on each side of the constriction (for laminar flow). So by measuring the differential pressure on both sides of the constriction, the flow rate can be calculated.

In this project, a pressure sensor type (MPX5010) has been used since it is more suitable compared with others, as in Table 2. The analog generated signal from this

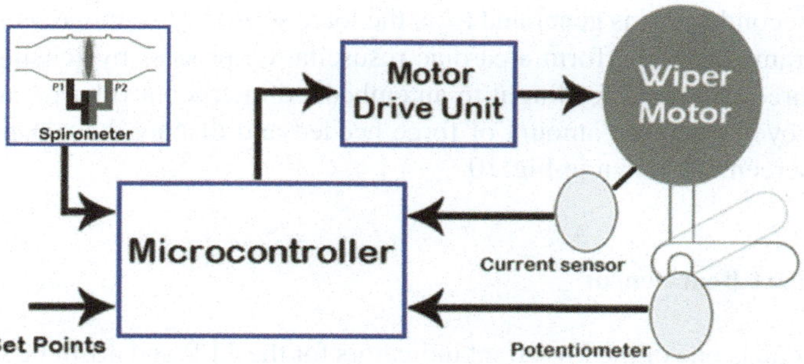

Fig. 9 Spirometer feedback

Table 2 Pressure sensor comparison

Feature	Honeywell SCXL004DN	Freescale MPX5010DP	Omron D6F-PH0505AD3
Pressure range (mm H_2O)	0–101.6	0–1019.78	±50.8
Output type, range	Analog, (0–18) V	Analog, (0–5) V	Digital, 12-bit
Sensitivity (mV per mm H_2O)	0.394	4.413	017 mm H_2O
Accuracy	±1%	±5%	±3%
Supply voltage and current	18V max, 2 mA	5 V, 10 mA	3.3 V, 6 mA
Response time	0.5 ms	1 ms	50 ms
Price (Qty 100)	$ 75	$ 7	$ 20

sensor is less than 50 mV in response to blowing into the spirometer, which is less than 1% of the full range. In this case, a 16-bit ADC with programmed reference voltage has been used.

Beside the position feedback signal from the DC motor, the spirometer will be used to insure the output tidal volume (the amount of air breathed in with each normal breath) value for the critical cases, as shown in Fig. 9.

5.2 *Cardiopulmonary Resuscitation Element*

This part is designed to support the mechanical ventilator, as there are cases of severe shortness of breath or heart failure as a result of emergency injuries occur continuously in addition to viruses such as Covid-19, and such cases require immediate and urgent emergency procedures to revive the heart using the technique of CPR. Two feedback signals are required; the first signal is generated from the heart sensor, while the second signal is generated from the force sensor. This unit assists the medic (or the paramedic) to perform a cardiac resuscitation process by sensing the force on the injured chest and feeding it to an embedded microcontroller to calculate the necessary cycles and the amount of force needed and display these parameters on the LCD screen, as shown in Fig. 10.

5.2.1 Heart Beat Sensor

The heart rate is one of the important indicators for the CPR and artificial respiration. The real-time heart rate monitoring will help the operator to start recovery without using the traditional manual methods of sensing the pulse of the heart specifically in places that are difficult for ambulances to reach or difficult to enter medical devices or that do not contain any electricity such as forests.

The pulse Oximeter waveform is one of the most commonly displayed clinical waveforms. As illustrated in Fig. 11, it is a simple optical technique that can be used to detect blood volume changing in the micro vascular bed of tissues. It is a low cost and non-invasive method that makes measurements at the surface of the skin. It is relatively easy to detect the pulsatile component of the cardiac cycle according to this theory.

The detected waveform is an amplified and highly filtered measurement of light absorption by the local tissue over time. All modern pulse Oximeter extract and

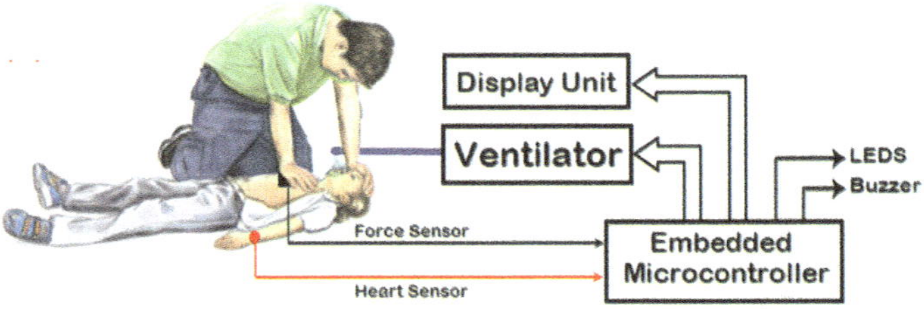

Fig. 10 Layout of the CPR unit

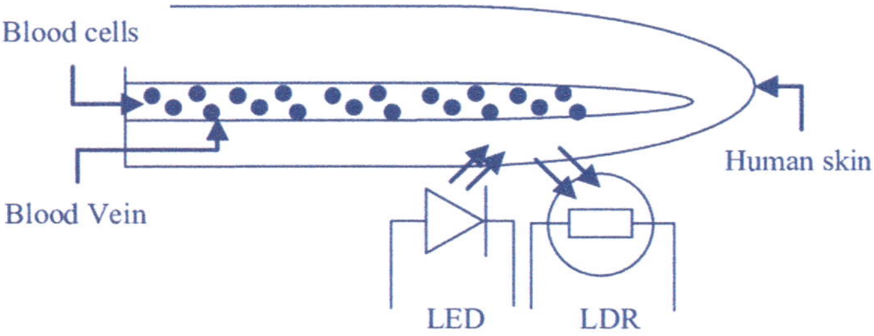

Fig. 11 Heart rate measurement using optical technique

display the heart rate and Oxygen saturation derived from the optical measurements at multiple wavelengths (Alian and Shelley 2014). The user can monitor the heart rate up to date on the screen, allowing him to focus on performing CPR, or monitoring the patient's vital condition.

5.2.2 Force Sensitive Sensor

Force sensor resistor (FSR) sensor is basically a low-cost piezoresistive type, where the resistance value changes when applying force or pressure. The sensor has a high resistance when unloaded, and drops as force is applied. The FSR sensor is used in this prototype to find the compressive strength of the medic on the patient's chest. It is about 18.28 mm diameter and 0.2–1.25 mm in thickness, and its measurement range is between 0.10–100 N. Figure 12 illustrates general characteristics of the FSR sensor.

5.3 Operator Interface

The operator can select the required values suitable for the patient by adjusting the breathing rate and tidal volume for each breath directly from the device. The device is considered as a note in a wireless sensor network and it can be accessed by the specialist or a medical center to monitor the case of the patient and also to modify the required parameters.

5.3.1 Tidal Volume (TV)

It represents the amount of air breathed in with each normal breath. The maximum tidal volume required for the patient is usually between 200 mL to 600 mL. Therefore, the device allow the user to control the TV output from 100 mL to 1500 mL maximum.

5.3.2 Breathing Rate (BR)

It is the number of breaths in a minute, It depends on the age, where the adult needs 12–20 breath/min with 350–500 mL of tidal volume, while the child needs more breathing rate (up to 50 breath/min) with less tidal volume. For example if the BR is set to 25 breath/min, then the microcontroller will move the actuator 25 times each minute for the required position depending on the tidal volume input. The microcontroller counts the operating time to calculate required number of cycles each minute as follows:

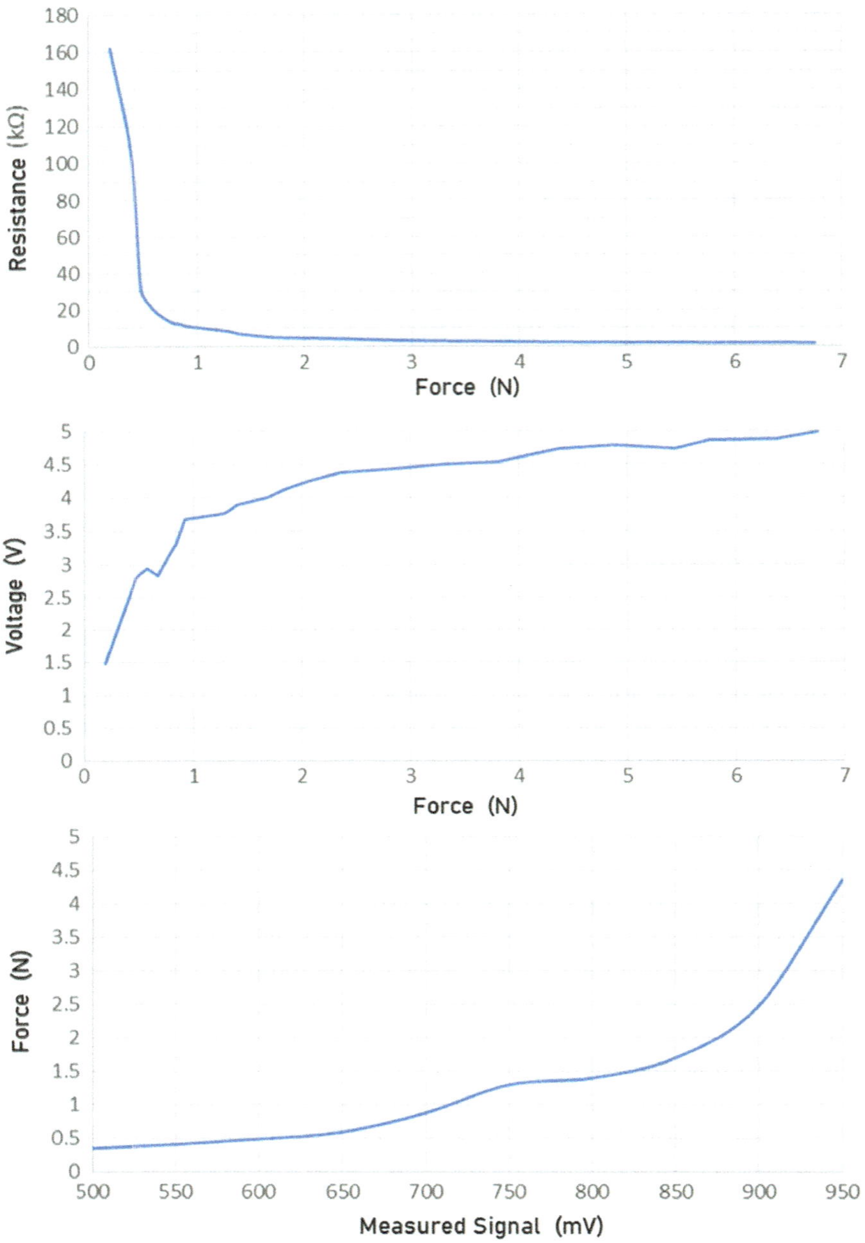

Fig. 12 General characteristics of the FSR sensor

Current time (millisecond) = Number of milliseconds since the program started.
Start Time = Current Time
Period = 60000 ms (1 min)/number of required cycles.
If (Current Time − Start Time ge period) Then move Actuator to the right position

6 Embedded System Design

A microcontroller type ATmega2560 has been used as the main controller in the device. It has 54 digital input/output pins 15 of them can be used as PWM outputs, 16 analog inputs, three timers, 4 UARTs with a 16 MHz crystal oscillator. It is used as the embedded microcontroller that scans, measures and controls all variables in the system. A low-cost Wi-Fi module (type ESP8266) is directly connected to the main microcontroller. This module is provided with a full TCP/IP stack and microcontroller capability for wireless communication with the health center or the specialist to follow the patient's condition and provide medical advice for cardiac resuscitation.

The proposed portable device can be used in ambulances or sub-health centers as well as in hard-to-reach places, so that the battery must be rechargeable and work for an appropriate period of time.

6.1 Ventilator Interfacing

The general layout of the hardware design of the main control unit of the mechanical ventilator is illustrated in Fig. 13. An embedded microcontroller (type ATmega2560) is used for sensing, measurement and control. Five analog sensors are connected directly to the microcontroller, these sensors are; current sensor, pressure sensor,

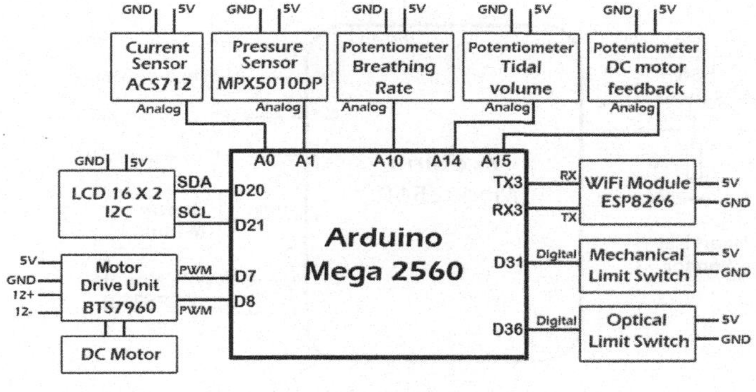

Fig. 13 Hardware design of the main control unit

arm positioning senor, tidal volume sensor, and breathing rate sensor. Four output lines from the microcontroller are connected to the motor drive unit to control the motor speed and direction.

6.2 CPR Interfacing

The hardware design of the CPR unit is given in Fig. 14. The embedded microcontroller has been used to scan FSR and heart rate sensors to calculate some parameters used for CPR monitoring. A buzzer module and three LEDs are used to indicate the state of pressure and force applied on the patient's chest when providing CPR. The microcontroller is directly connected to the ESP8266 WiFi unit for wireless communication with the health center or the specialist to follow the patient's condition and provide medical advice for cardiac resuscitation. This module is provided with a full TCP/IP stack and microcontroller capability for wireless communication with the health center or the specialist to follow the patient's condition and provide medical advice for cardiac resuscitation.

6.3 Operator Interface

The operator can select the required values suitable for the patient by adjusting the breathing rate and tidal volume for each breath directly from the device. The device can be accessed by the specialist or a medical center to monitor the case of the patient and also to modify the required parameters. The set-points include tidal volume and

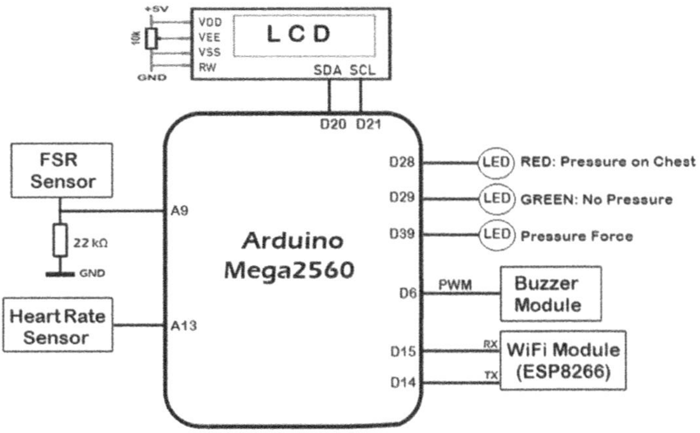

Fig. 14 Hardware design of the CPR unit

breathing rate. The ventilator LCD monitors breathing rate, tidal volume and minute ventilation. The CPR LCD monitors the heart rate, number of comprises on chest (0–30), number of cycles (1–5), CPR condition (good or not), total time for each cycle, and the rate of beat per minute. For each 30 comprises, the cycle will be increased by 1, during which the microcontroller checks the rate of comprises per minute (CPM). It is "Good" if it is between 100 and 120. Otherwise, the paramedic will adjust the CPM rate. After completing one full cycle, the LCD screen will display the total time and rate of beats per minute. Real-time monitoring helps to begin assisting the patient to recover (without using traditional manual methods) specifically in places that are difficult for ambulances to reach.

6.4 *Wireless Communication*

This module, as shown in Fig. 15, will connect to the Wi-Fi and start sending the data to the web hosting FTP using Linux server (WHM). All data and patient monitoring readings will be stored in the database (MySQL). The specialist will interact with the data through a graphical user interface. The ESP8266 acts as a client that makes an HTTP POST request to a PHP script to insert data (sensor readings, user input and device status) into a MySQL database. The goal of this part is to visualize the readings from anywhere in the world by accessing our own server domain in

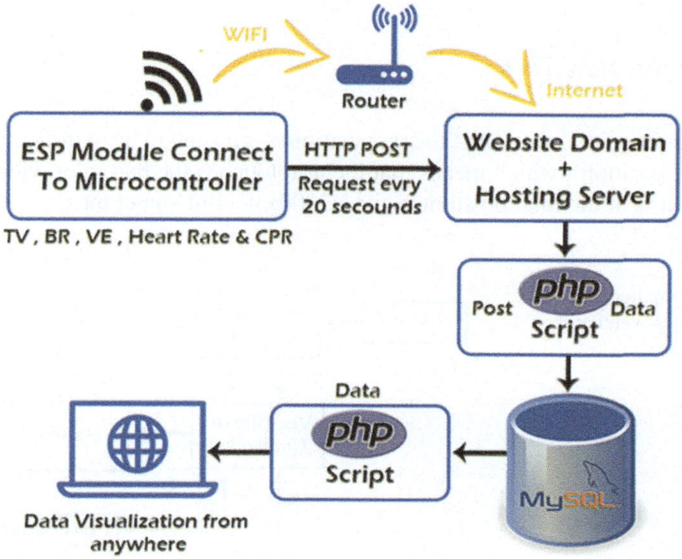

Fig. 15 Wireless communication layout

the internet (www.philadelphia.edu.jo). The specialist will directly update the set points of the ventilator to improve the patient's breathing situation (Al-Mutairi and Al-Aubidy 2020).

7 Real-Time Computer Control

Real-time control of the mechanical ventilator is based on two parameters; the tidal volume (TV) and the breathing rate (BR), as illustrated in Fig. 16. These two parameters are basically covering physiological parameters of human body related to gas exchange in mechanical ventilation. The controller will adjust the position of the DC motor based on the operator set points and real-time measurement of TV and BR. The proposed system allows the specialist from a health center to update the values of the ventilator set points if the critical case of the patient requires such adjustment.

Upon receiving an emergency signal from the device, the specialist will check the patient's condition by monitoring the received signals including; the heartbeat, tidal volume and breathing rate. Therefore, the specialist will directly update the set points of the ventilator to improve the patient's breathing situation.

The embedded microcontroller calculates minute ventilation (VE) and required arm position. The minute ventilation is the total volume of air entering the lungs in a minute, and calculated directly by the microcontroller when BR and TV are measured.

7.1 *Calibration Task*

At the beginning of operation, the arm position is adjusted to be at the starting point (the initial position), which means that the arm touches the bag to provide 100 mL, and when it is at the final position it gives 2000 mL. But sometimes, to ensure high

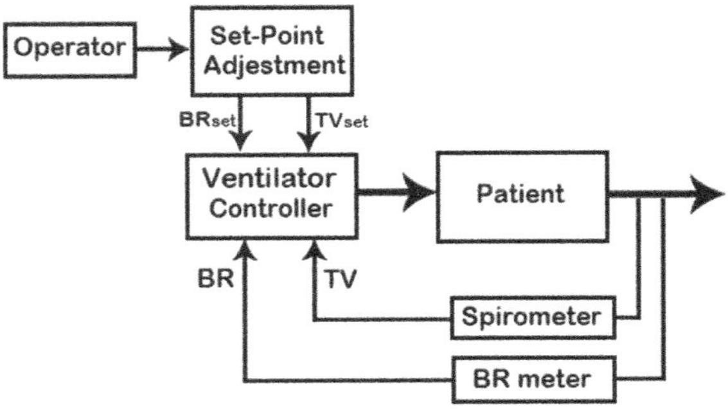

Fig. 16 Computer control of the mechanical ventilator

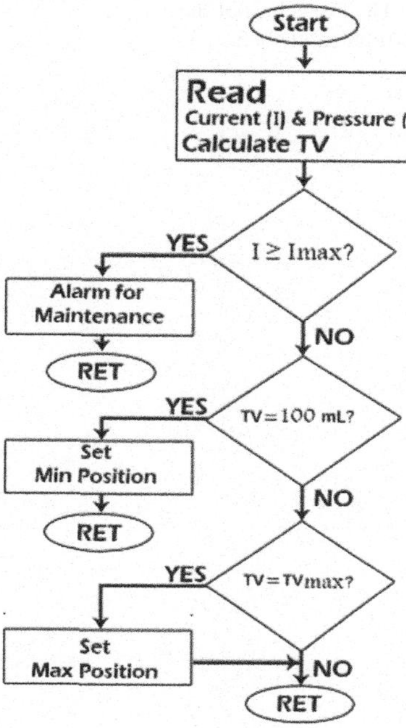

Fig. 17 Flowchart of the real-time calibration task

accuracy, it is advisable to adjust these positions to facilitate the treatment process by the calibration task, as shown in Fig. 17.

The calibration process is based on two feedback signals generated from current sensor and pressure sensor. The rated value of the motor current is about 7 amps for different speeds. When the motor arm exceeds the bag valve mask, the current value becomes 9 amps, therefore the device needs to be calibrated or serviced. The pressure sensor is used to calculate the minimum and maximum positions of the arm. The lowest position is determined by the pressure sensor to generate a volume of 100 mL, while the maximum position is indicated when generating the largest possible volume (2000 mL).

7.2 *Control Task*

Figure 18 shows the flowchart of the controller algorithm and can be summarized in the following steps:

- Calculate the required TV value based on the initial and final values of the actuator positions (calibration stage). It is expected that at this actuating position, the value will be closed to that required by the specialist.

Fig. 18 Flowchart of the real-time control task

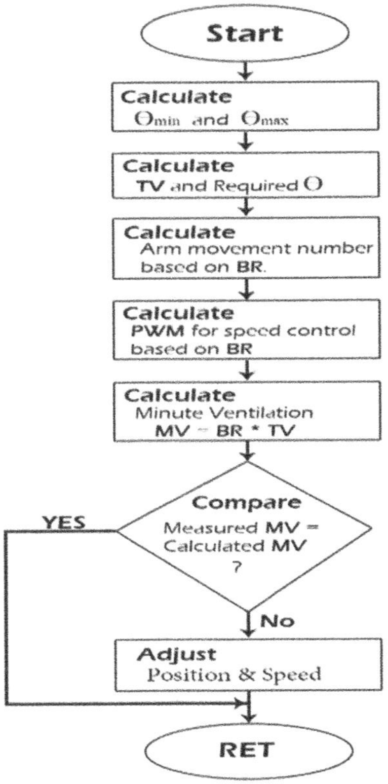

- Calculate the number of times: the TV output is executed each minute based on the BR value.
- Calculate the required actuating signal (PWM signal) to control the motor speed to ensure smooth breathing.
- Calculate minute ventilation (VE) by multiplying actual measured values of BR and TV.
- Compare the input setting with actual measured values from the system, and accordingly adjusts the actuating position and motor speed.
- The controller checks the current sensor, as well as the limit switches for safety operation. In case the current is high or one of the keys is set, it means that the device requires tuning and maintenance.

A general proportional integrated derivative controller (PID) has been used to adjust the position of the motor by providing the motor drive unit with the required PWM signal. The controller gain parameters are obtained by off-line simulation so that it gives the best time response with minimum overshoot, rise time, and almost zero steady-state error:

$$u(t) = K_P e(t) + K_i \int_0^t e(\tau)d\tau + K_d \frac{de(t)}{dt} \tag{3}$$

Any mechanical failure, that can be detected through the limit switches or sensors (current and pressure), will cause an interrupt, and the microcontroller will execute an interrupt service routine.

8 Testing and Validation

The implemented system is based on web application (PHP, CSS and HTML), with computer or smart phone browsing version. Three pages are used to manage the operation of the implemented system. The first page "Dashboard", shown in Fig. 19, is used for general information and to monitor any new high-risk cases with alert.

The "My Patients" page, see Fig. 20, is used to add patients (by adding their ID) or delete them and update their medical history. It updates automatically if there is any abnormal condition with alarm. It also allows adding descriptions and display of all patients in the database.

The "Control devices" page is used to control any ventilator by specifying the patient's name, as illustrated in Fig. 21.

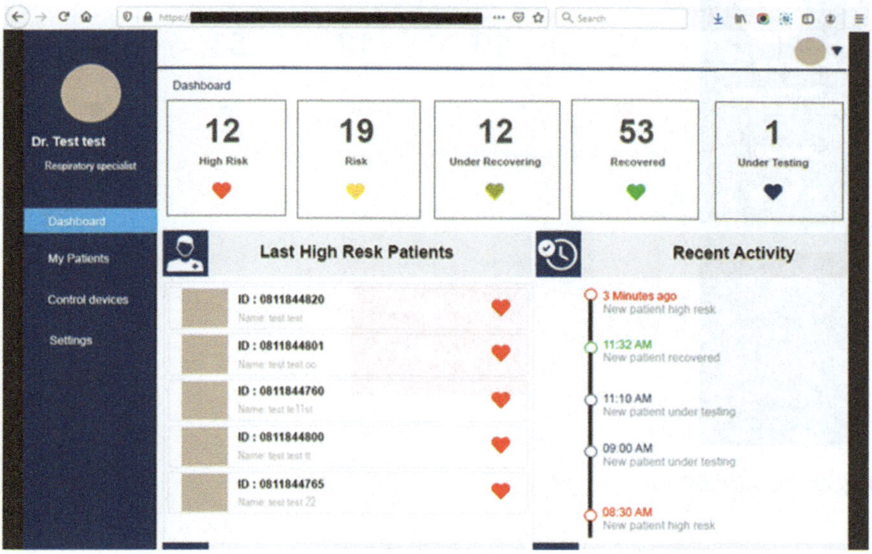

Fig. 19 "Dashboard" page

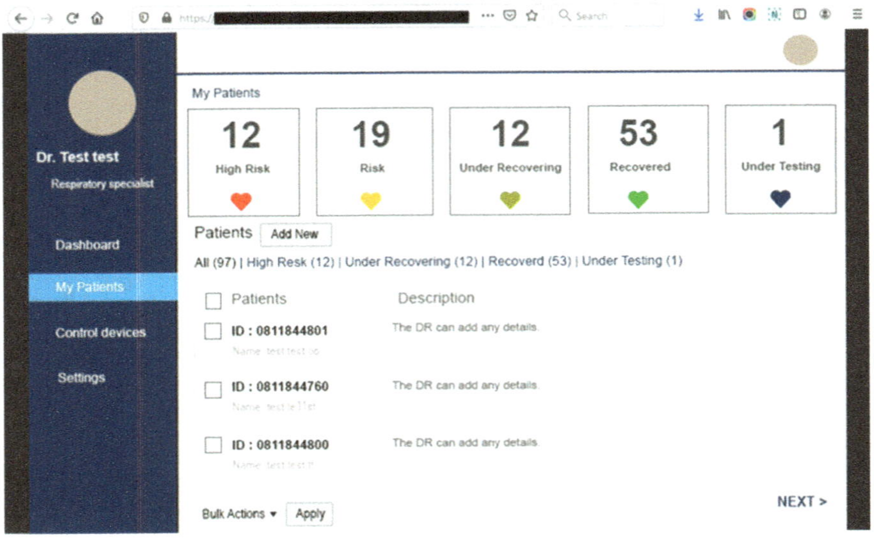

Fig. 20 "My Patients" page

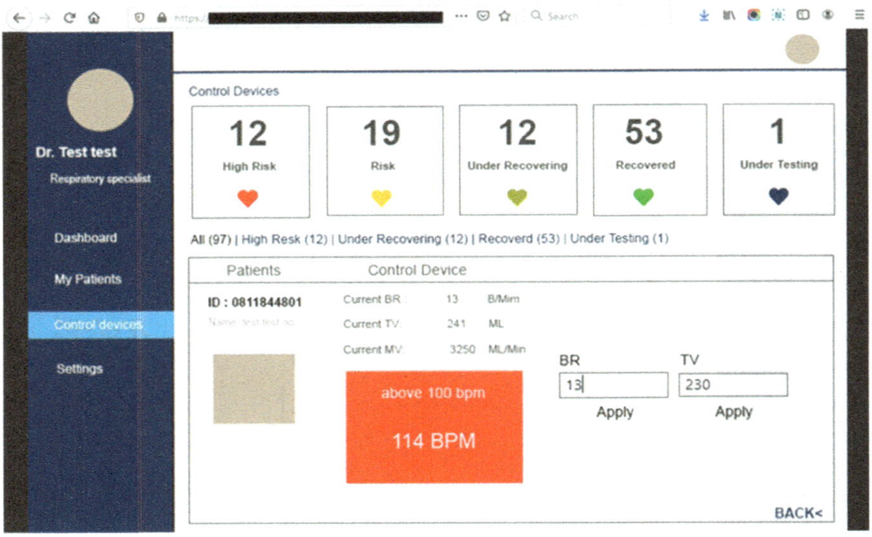

Fig. 21 "Dashboard" page

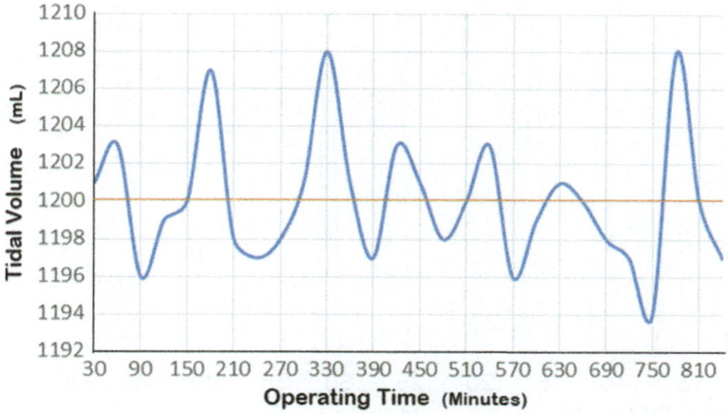

Fig. 22 Tidal volume test

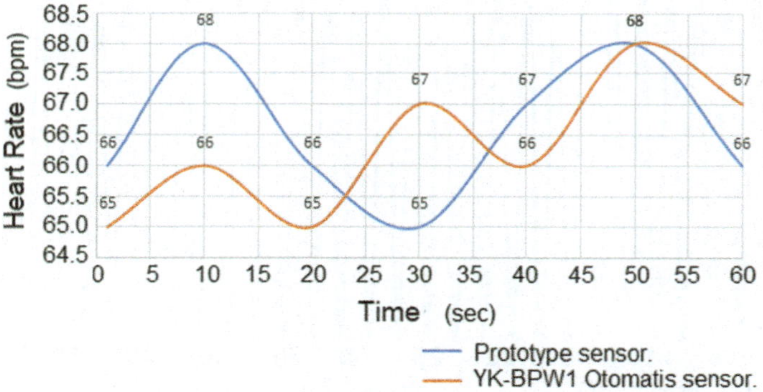

Fig. 23 Heart rate sensor test

The device was tested by running it continuously for 24 h and intermittently for more than 300 h. All data from the sensors and control data generated from the microcontroller were recorded in Excel files. Most of the results obtained were predictable in various possible patient scenarios. Some errors appeared in the initial stages when testing the device, and most of them did not affect the main work of the device, but affected its accuracy. Figure 22 shows the performance of the ventilator when the tidal volume set-point is 1200 mL. The reason for fluctuations in tidal volume (less than 1% which is acceptable) is due to the specifications of the low-cost pressure sensor used in the prototype.

The operation and performance of the heart rate sensor was also examined by comparing it to a cheap commercial device (type YK-BPW1), and the results were expected as shown in Fig. 23. The breathing actions generated from the embedded microcontroller for the implemented ventilator when the breathing rate is 12 bpm and 30 bpm are illustrated in Fig. 24.

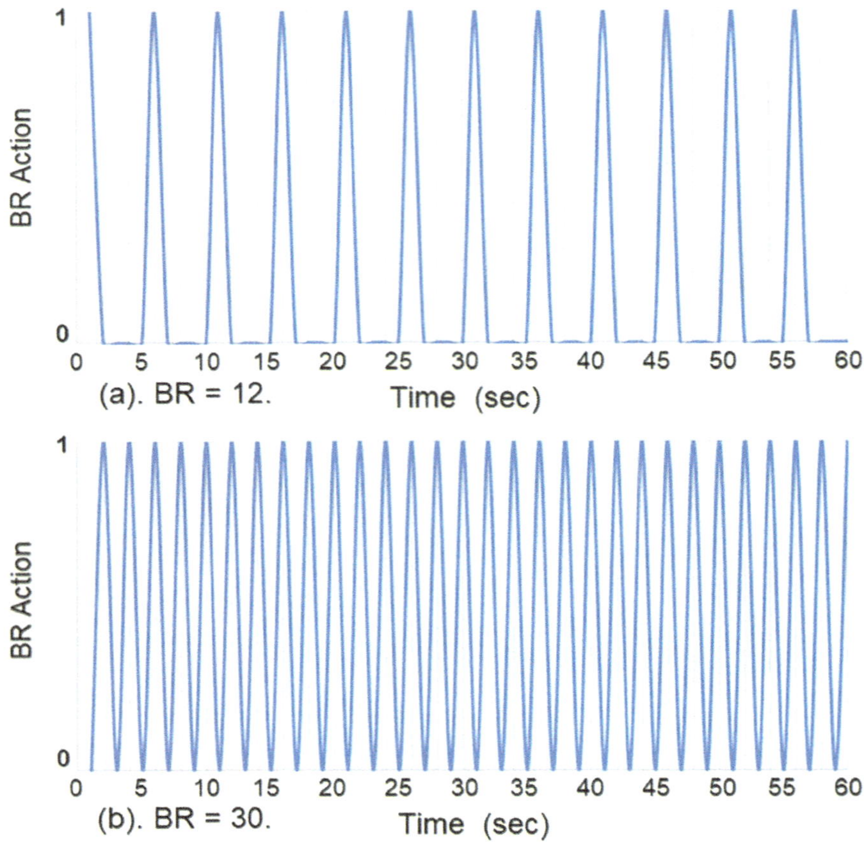

Fig. 24 Breathing rate test

The operation of the main driver (DC motor) is dependent on the respiratory rate set-point, as shown in Fig. 25. In this case, the duty cycle of the PWM signal is used to control the speed of the DC motor to obtain smooth breathing. The time response to tidal volume when the respiratory rate is 30 bpm is shown in Fig. 26. It is clear that the volume of the tides changes to follow the required set point (500 mL). When the tidal volume inlet changes to 1000 mL, the rise time is roughly the same. By using the a real-time control algorithm, the motor arm can return faster to the initial position after achieving the desired output of the tidal volume. This achieves the desired goal, because when a patient needs more tidal volume, this means a lower respiratory rate. Any change in the volume of the tide during the movement of the motor will not be reflected in the operation of the system, since the embedded microcontroller applies the change after the motor has returned to the initial position.

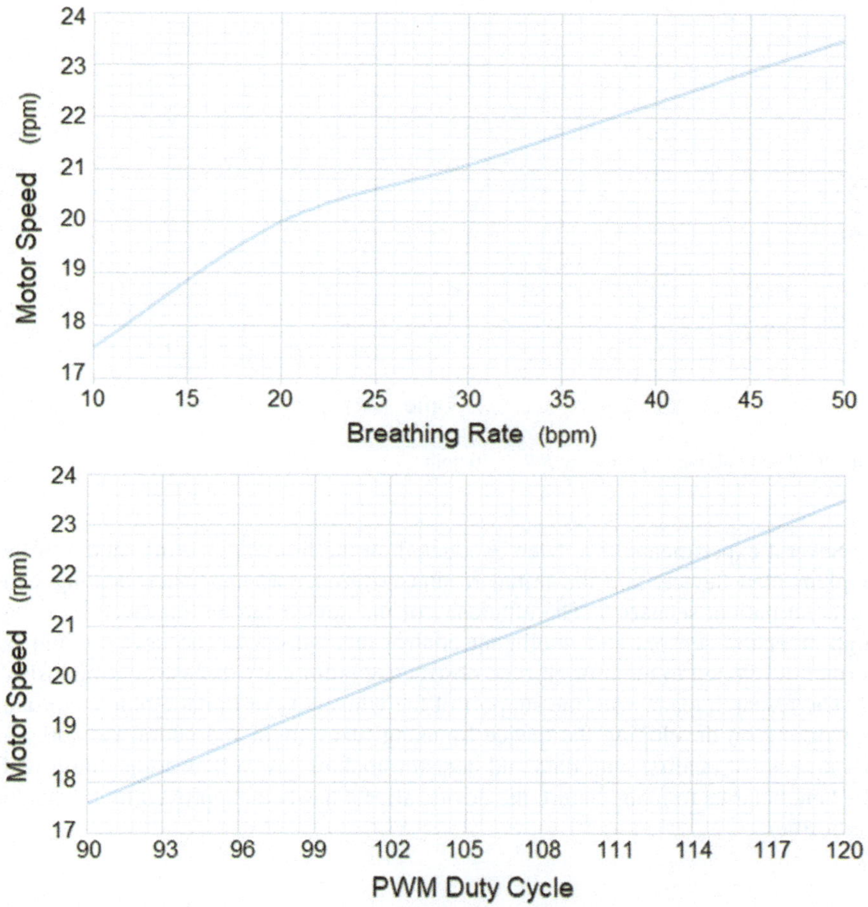

Fig. 25 Motor speed control based on BR

9 Conclusion

This chapter covers the application of reverse engineering concepts in the design and construction of a ventilator with CPR supervisory control and real-time monitoring of a patient's heart rate, as well as enabling a specialist or health center to access the device wirelessly. The proposed device offers the following features:

- Portable and can be used easily at home, ambulances or hard-to-reach areas.
- It does not have major risks to the patient's life.
- The cost of the first prototype is less than 800 USD.
- It can be easily accessed wirelessly by a specialist or medical center to update its parameters and provide medical advice.

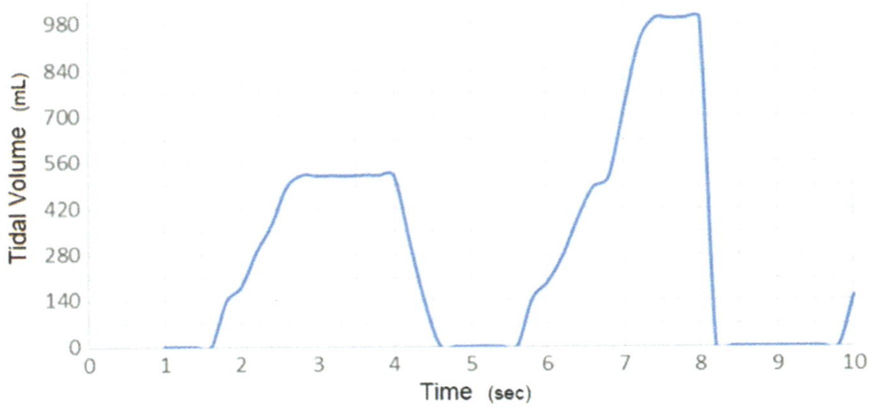

Fig. 26 Tidal volume response for $BR = 30$ bpm

Several experiments have been conducted on the first engineering model of the implemented device, and the initial results were encouraging when dealing with different patient scenarios. The efficiency of the device can be improved by using high-precision sensors and intelligent control methodology. The device could be further developed to monitor chronic obstructive pulmonary disease continuously.

The research team, with the support of the University of Philadelphia in Jordan, is working on developing the device by using fuzzy logic and neural networks in the process of dealing with data and decision-making, as well as cooperating with the beneficiaries to manufacture the device after it has been approved by the health authorities.

References

Al Husseini, A. M., Lee, H. J., Negrete, J., Powelson, S., Servi, A., Slocum, A., et al. (2010). Design and prototyping of a low-cost portable mechanical ventilator. In *Design of Medical Devices Conference (DMD2010)*, April, USA (pp. 1–9).

Alian, A. A., & Shelley, K. H. (2014). Photoplethysmography. *Best Practice and Research Clinical Anaesthesiology*, *28*(4), 395–406.

Al-Mutairi, A. W., & Al-Aubidy, K. M. (2020). Design and construction of a low cost portable cardiopulmonary resuscitation and ventilator device. In *17th IEEE International Multi-Conference on Systems, Signals, Devices (SSD'2020)*, Tunisia, 20–23 July.

Atkins, D.L., Berger, S., Duff, J.P., Gonzales, J.C., Hunt, E.A., Joyner, et al. (2015). Part 11: Pediatric basic life support and cardiopulmonary resuscitation quality. 2015 American heart association guidelines update for cardiopulmonary resuscitation and emergency cardiovascular care. *Circulation*, *132*(18 Suppl 2), 519–525.

Barrett, K. E., Barman, S. M., Boitano, S., & Brooks, H. (2012). Ganong's review of medical physiology. In *LANGE Basic Science* (24th ed.). New York: McGraw Hill.

Beardsell, I. (2009). *Get through MCEM part A: MCQs*. London: Royal Society of Medicine Press.

Chatburn, R.L. (2004). Computer control of mechanical ventilation. *Respiratory Care, 49*(5), 507–517.

Fludger, S., & Klein, A. (2008). Portable ventilators. *Continuing Education in Anaesthesia Critical Care & Pain, 8*(6), 199–203.

Ghafoor, M. J., Naseem, M., Ilyas, F., Sarfaraz, M. S., Ali, M. I., & Ejaz, A. (2017). Prototyping of a cost effective and portable ventilator. In *2017 International Conference on Innovations in Electrical Engineering and Computational Technologies (ICIEECT)*, April, Pakistan.

Guler, H., & Ata, F. (2014). Design of a fuzzy-LABVIEW-based mechanical ventilator. *Computer Systems Science and Engineering, 29*(3), 1–20.

Jackson, C.D., & Muthiah, M.P. (2020). What is positive-pressure ventilation and how did its use develop for mechanical ventilation? Medscape. Accessed 10 Jul 2020.

Kaynar, A. M., & Sharma, S. (2020). What are the differences between positive-pressure and negative pressure ventilators for the treatment of respiratory failure. Medscape. Accessed 7 Apr 2020.

Mandal, A. (2019). What is a positive pressure ventilator? News Medical Life Sciences. Accessed 28 Jun 2019, https://www.news-medical.net/health/What-is-a-Positive-Pressure-Ventilator.aspx

Schaublin, J., Derighetti, M., Feigenwinter, P., Petersen-Felix, S., & Zbinden, A. M. (1996). Fuzzy logic control of mechanical ventilation during anaesthesia. *British Journal of Anaesthesia, 77*(5), 636–641.

Speedy Publishing, L. L. C. (2015). *CPR and first aid care*. Speedy study guides. MDK Publisher.

Sun, Y., Kohane, I., & Stark, A. R. (1994). Fuzzy logic assisted control of inspired oxygen in ventilated newborn infants. In *Proceedings of the Annual Symposium on Computer Application in Medical Care*, USA (pp. 757–761).

Thayer, K. (2018). How does reverse engineering work? IEEE Global Spec. Retrieved February 26, Available online (20 May 2020) at: https://insights.globalspec.com/article/7367/how-does-reverse-engineering-work

Villaverde, A.F.,& Banga. J. R. (2014). Reverse engineering and identification in systems biology: Strategies, perspectives and challenges. *Journal of the Royal Society Interface, 11*(91), 20130505.

Wang, C., Zhang, G., & Wu, T. (2016). A model-based decision support system for mechanical ventilation using fuzzy logic. *International Journal of Simulation: Systems, Science & Technology, 17*(36), 27.1–27.7.

Woollam, C. H. (1976). The development of apparatus for intermittent negative pressure respiration. *Anaesthsia, 31*(5), 537–547.

Virtual Reality and Augmented Reality Technologies for Smart Physical Rehabilitation

Octavian Postolache, João Monge, Ricardo Alexandre, Oana Geman, Yu Jin, and Gabriela Postolache

Abstract Virtual Reality and Augmented Reality technologies provides new experiences to the users during physical rehabilitation training sessions increasing the engagement for improved physical outcomes. These relatively new technologies are the starting point on the development of new services that can be defined as "Physical Therapy at Home" or "Remote Physical Therapy". The VR and AR serious game can be considered complementary of the regular training in clinics, where mechanical equipment is commonly used. By comparison with classical technology for training the new VR and AR based solutions that are part of IoT Physical Rehabilitation ecosystem assure data storage for individual exercises performed by the users. Data analysis is commonly implemented in this kind of architecture, the obtained results can be used by physiotherapist to perform and objective evaluation of patient physical outcome and to tailor the training plans. The user interaction with virtual scenarios or virtual objects in real scenario imply the usage of human computer interfaces characterized by different degree of unobtrusively. Thus, wearable, environment embedded or remote sensing solutions together different type of IoT

O. Postolache (✉) · J. Monge · R. Alexandre
Department of Information Science and Technologies, ISCTE-Instituto Universitario de Lisboa, Lisbon, Portugal
e-mail: opostolache@lx.it.pt

J. Monge
e-mail: joao_monge@iscte-iul.pt

R. Alexandre
e-mail: rjfae20@gmail.com

O. Postolache · J. Monge · G. Postolache
Instituto de Telecomunicacoes, Lisbon, Portugal
e-mail: pgpostolache@gmail.com

O. Geman
Health and Human Development Department, Stefan cel Mare University (USV), Suceava, Romania
e-mail: oana.geman@usm.ro

Y. Jin
Shanghai Maritime University, Shanghai, China
e-mail: nora.jin0908@gmail.com

O. Kanoun and N. Derbel (eds.), *Advanced Systems for Biomedical Applications*,
Smart Sensors, Measurement and Instrumentation 39,
https://doi.org/10.1007/978-3-030-71221-1_8

physical rehabilitation architectures are considered in this chapter. Different AR and VR scenarios were implemented using Unity game development software for different computation platform. Different types of interfaces based on new technologies that assure the user interaction with AR or VR rehabilitation environment are considered including wearable smart sensors, environment embedded sensors, mobile device and big displays.

Keywords Smart sensors · Virtual reality · Augmented reality · Physical rehabilitation · Smart interfaces

1 Introduction

The ageing phenomena it is a reality already alerted by the united nation through the latest statistics. Thus, the global population aged 60 years or over numbered 962 million in 2017, more than twice as large as in 1980 when there were 382 million older persons worldwide. The number of older persons is expected to double again by 2050, when it is projected to reach nearly 2.1 billion (United Nations 2020). Ageing phenomena is also related to the increase costs for healthcare services including physical rehabilitation services which contribute to extension of quality of life for the elderly, taking into account the direct relation between motor capabilities and health status. Historically, rehabilitation has described a range of responses to disability, from interventions to improve body function to more comprehensive measures designed to promote inclusion that is what all of us consider including the elderly population.

Physical therapy methods are applied on people affected by motor impairments, with the motor capabilities recovering objective. The elderly are the most affected population group by motor disabilities. They can be affected by different types of motor impairments that means difficulties to perform simple daily tasks such as picking up an object eating alone or even dressing. Such consequences may restrict personal activities and avoid the full participation of the elderly in the community that generally affect the well-being of this group of population.

Worldwide, cerebrovascular accidents (stroke) are the second leading cause of death and the third leading cause of disability. According with Internet Stroke Center (Internet Stroke Center 2020) highlights in USA about three-quarters of all strokes occur in people over the age of 65 while the risk of having a stroke more than doubles each decade after the age of 55. Important number of the people that suffered a stroke event requires intensive intervention aimed mainly at maintaining and improving of motor skills. Several studies (Béjot et al. 2016; Timsit 2016) suggest stroke as the most significant cause of disability in adults. Medical reports showed that in the first six months after stroke, 49% of patients needed help with bathing, 31% of patients needed help with dressing, and 33% of patients needed help with feeding. Integrated medical care and rehabilitation program for stroke patients aimed at the recovery of functional independence and social participation in daily life. The new technologies were involved in the last years on physical rehabilitation revolution. Thus, Virtual Reality (VR) and AR (Augmented Reality) technologies and therapeutic serious games have been plenary applied as new intervention tools of smart rehabilitation for post-stroke physical therapy (Prashun et al. 2018; Reis et al. 2019).

The health community has shown an increasing interest in therapeutic approaches based on VR and AR therapeutic serious games. These games are designed and implemented to preserve the fun or entertainment for the gamer, but also to assure the improvement of physical and/or motor skills for people during the training session (Susi et al. 2007). Examples of already existing solution on the market can be mentioned. Thus, can be Physio are virtual reality rehabilitation platform for physical therapy that aims to provide a captivating experience for patients of any age (Serious Game Market 2020).

The possibility of programming the serious game scenario according to the type and the difficulty of the rehabilitation tasks associated to the therapeutic plans is one of the strength of AR or VR based rehabilitation. These tasks can be reliably imposed by the physiotherapists in complementary mode with classical physical rehabilitation methods over a period of several weeks or months, and the outcome of the participant can be evaluated taking into account the fully digitalization of the VR or AR Rehab services (Mealy 2020).

The interactions with VR and AR scenarios associated with the serious games involve different devices that characterized by different sensing technologies such as optical, MEMS, piezoelectric, piezo resistive, capacitive. These sensing devices are integrated on wearable modules that can assures the user—serious game interaction. The wearables are able to revolutionize the healthcare services including physical rehabilitation. The individuals can be the subjects of remote monitoring based on wearable, such as smart gloves (Alexandre et al. 2019) but also using remote sensing expressed by Kinect, Leap Motion Controller or Structure Occipital sensors (Ferreira et al. 2017; Lourenço et al. 2018; Yu et al. 2019) as parts of personalized health systems. Providing information about patient rehabilitation status these technologies provide better communication between physiotherapist and patient and also between physiotherapists for a better rehabilitation plan based on examples and data based experience for similar rehabilitation needs. A good therapeutic communication confers a higher possibility to obtain valid informed consent, positive clinical outcomes, higher levels of patient satisfaction, and higher levels of patients' compliance with rehabilitation programs and decreases the levels of patient frustration.

This chapter is organized such as: Sect. 2 presents the reported architectures in the field of serious game for physical rehabilitation including wearable and also remote sensing, Sect. 3 and alternative and a new Theragames approach based on augmented reality. The last section of the chapter includes the conclusions a set of ideas regarding the future of physical rehabilitation in a digital society.

2 Virtual Reality Serious Game

The interest and evolution of the virtual reality serious game rehabilitation approach has been emerging in the last years. Thus, patients and general public are nowadays motivated to interact with VR scenarios as part of physical rehabilitation training (Cary et al. 2014). One of the solution for the interaction between the user and VR scenario is based on wearable sensors mainly expressed by IMU (Oarde et al. 2014) or remote sensing One of the natural user interface that was reported as appropriate for VR serious game for physical rehabilitation was Kinect sensor. The sensor provides information about 3D human body joints positions and can be used to calculate trajectories of different part of body including lower limb or upper limb during the training session. Additional information can be extracted using wearable sensors such as EMG that are commonly used to extract information about muscle activation. The interaction with VR scenario elements can be carried out using 3D-IMU that provides values of angles between different body parts information that can be transmitted by different objects of VR including the avatar. In the following parts different types of interfaces associated with VR serious games are considered.

2.1 Wearable Interface for VR Serious Game

The wearable interface is materialized by low-cost hardware and software components that assures highly interactive therapeutic games in VR scenarios. The implemented natural interface is expressed by a set of smart gloves that assures physical rehabilitation monitoring for of hands and hands' fingers. Additionally, a headband device that includes a 3D-IMU is used to measuring the rotation angles and linear accelerations of the head with the aim of integrating the patient into First-person Controller in VR scenarios. The IMU information can be also used to evaluate the posture of the user during the therapeutic game sessions. This system provides evaluation metrics of the patient's rehabilitation that can be used to extract the physical rehabilitation outcomes. The general architecture of the system that includes smart gloves and headband for user—VR interaction is presented in Fig. 1.

The above presented framework is characterized by a set of four blocks that include hardware and software components. The framework assure the patient access to a set of therapeutic serious games characterized by natural interaction using the developed wearable devices and VR scenarios that were created using Unity3D with scripts developed in C #. The developed software is associated with computation platform that is also the client of the implemented system. This block, "1" receive through Bluetooth the data from the wearable smart sensors that were developed using Arduino nano platform, IMUs, force and flexion sensors. The data from the client (block 2) is transmitted to the server which stores the settings and the results of the serious therapeutic games.

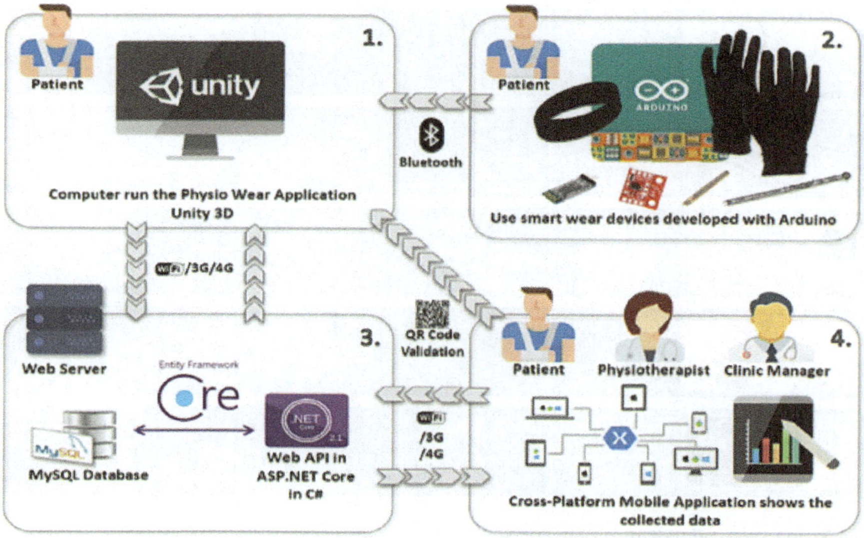

Fig. 1 VR serious game framework for physical rehabilitation

The software technologies used on the server side are MySQL database and the ASP.NET Core in Web API developed in C# that allows communication between applications. A mobile application was developed that assure data management of patient data and health professionals' data as so as the visualization of the data associated with VR serious games. The architecture of the system can be also represented by different layers that materialize and IoT system (Fig. 2).

The IoT architecture includes a Sensor Layer that is materialized by a set of IMUs (MPU-9250), 3D IMUs distributed on the level of the smart gloves and a set of analog sensors (FlexSensors 2.2" and FlexiForce A201) and conditioned circuit that assure the measurement of applied force and flexion for each finger. Figure 3 shows the developed smart gloves prototype.

The Edge Layer consists of Arduino Nano computation platform that receive the data from IMU through I^2C (Inter-Integrated Circuit). At the same time the ADC and Arduino Nano analog multiplexer are used to acquire the signals from the analog sensors for force and flexion. Using a Bluetooth transceiver associated to the edge layer the data is transmitted to Gaming Platform Layer.

The Gaming Platform Layer inspects the data generated by the VR serious game itself, and include: game' scores, angles of hand movement, fingers' feedback force. These data associated with the VR game are processed on the client side level and stored in a remote server that materialize the Cloud Database Layer. The Internet connectivity is based on Ethernet or Wi-Fi connection.

The VR serious game was developed on the Unity platform (Borromeo 2018). Two VR games for upper limb rehabilitation were consider: Cans Down challenge and the Coffee Pong challenge. Smart gloves and headband are assure the user monitoring

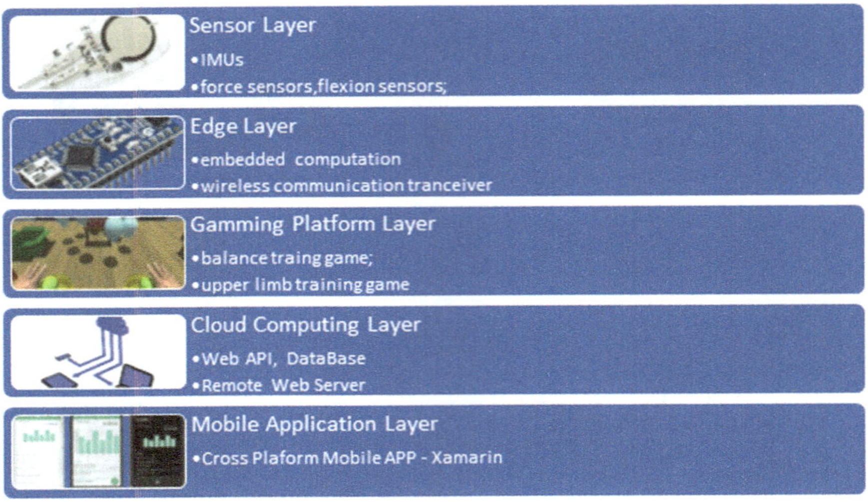

Fig. 2 IoT physical therapy architecture

Fig. 3 Smart gloves and headband prototype as VR serious game user interface

during rehabilitation exercise. Thus, the upper limbs and finger motion are recorded during interaction with virtual objects as part of game action (Fig. 4).

The Cans Down game goal is to knock down the stack of cans by throwing the tennis balls that appear in the golden goblets. The game requires exigent upper limb movement. Thus, the user need to grasp a ball, when two fingers of one hand should touch it simultaneously. To throw the ball, the hand must be open at the end of the movement. A new ball is generated automatically whenever another ball touches the ground. For each can that falls on the ground is gained one point. During the game the kinematic and dynamic parameters associated with hand motion are stored in a database to be used for evaluation of upper limb motion capabilities and finger forces and finger flexion.

Fig. 4 Therapeutic serious games, cans down challenge and coffee pong challenge, respectively

The Coffee Pong also aims to create a rehabilitation exercise of the upper limbs including the fingers. Thus, finger movements are acquired, during the grab action of virtual objects and shoot them in a certain direction. This game allows the patient to take the available ball on the ping-pong table with one hand and take the ping pong racket with the other. The goal is to get the balls to knock down the coffee cups on the other side of the table, producing joint movements. For each ball played, a new ball is placed on the table. If the patient drops the racket on the floor, a new racket will be placed in the initial position. For each cup of coffee tipped the patient gains one point.

The Cloud Database Layer will store all data, from the user profile accounts created by the Admin' Clinics or by the medical staff in the Mobile Application, as well as all workout monitoring data from the Gaming Platform. The Mobile App Layer performs profile validation on the Gaming Platform, through a unique QR Code generated, for each patient. This QR Code will contain all the information necessary to configure the Game and the parameters to be evaluated by the physiotherapist in each exercise session. This layer is also responsible for the data processing, data analysis and data visualization. The data communication between the mobile device and cloud is performed using Wi-Fi or 4G connectivity.

2.2 NUI Kinect—Virtual Serious Game

The advances in the field of virtual serious games are also highlighted on the level of the user interfaces that are used, providing natural interaction with VR scenarios. Thus, can be underlined the importance of Nintendo Wii^{TM} (Nintendo, Kyoto, Japan) system and later of the Microsoft Kinect^{TM} system with their body controlled type of interaction, allowed using them for studies in physical rehabilitation for different purposes such as: balance control (Borghese et al. 2013), fall risk reduction (Baranyi et al. 2013), gait (Muñoz et al. 2018; Tannous et al. 2016), muscle strength (Jani et al. 2017) and also upper limbs rehabilitation (Chen et al. 2016; Geman et al. 2019; Postolache et al. 2019; Wang et al. 2019).

The Microsoft Kinect system, as natural user interface (NUI), provides real-time detection of body joints through Microsoft Software Development Kit (SDK) that permit to access the features (e.g. x, y, z position programmatically). This data can be used for body motion analysis assessing anatomical landmark position and angular displacement data during commonly performed clinical tests of postural control. Based on this sensor the interaction with Virtual Reality Serious Game (VR) is carried out and adding Internet connectivity, remote physical rehabilitation services can be developed including also remote exercise session planning and session analysis by the physical therapist.

The Kinect commercial games provide high level of interaction however data that is provided is mainly expressed by the user scores. Thus, important information related to user joints coordinates associated with user motion during the gaming periods are neglected. Using this information several games were reported in the literature such as Theragames or serious games for physical rehabilitation (Postolache et al. 2016). Additionally, through tailored game scenarios the motivation of the users can be increased that may lead to sustainable levels of exercise adherence.

Several implementation for upper limb and lower limb rehabilitation are reported in the literature (Cary et al. 2014; Postolache et al. 2015; Wang et al. 2019). The general architecture of the VR serious game system is based on client server architecture with a remote server for the database storage. Additionally a mobile APP or a web APP are developed for the game's parameters setup and presentation of results of the exercise sessions to the physical therapists. The remote server can be used for the physical therapy serious game and by mobile application. The Postolache et al. (2019) reported the development of Application Programming Interface (API) running on an Apache Server (v2.4.7) with PHP v5.5.9 that assure extended functionalities regarding the data management.

A general representation of users' interaction flow associated with Physical Rehabilitation Framework is presented in Fig. 5.

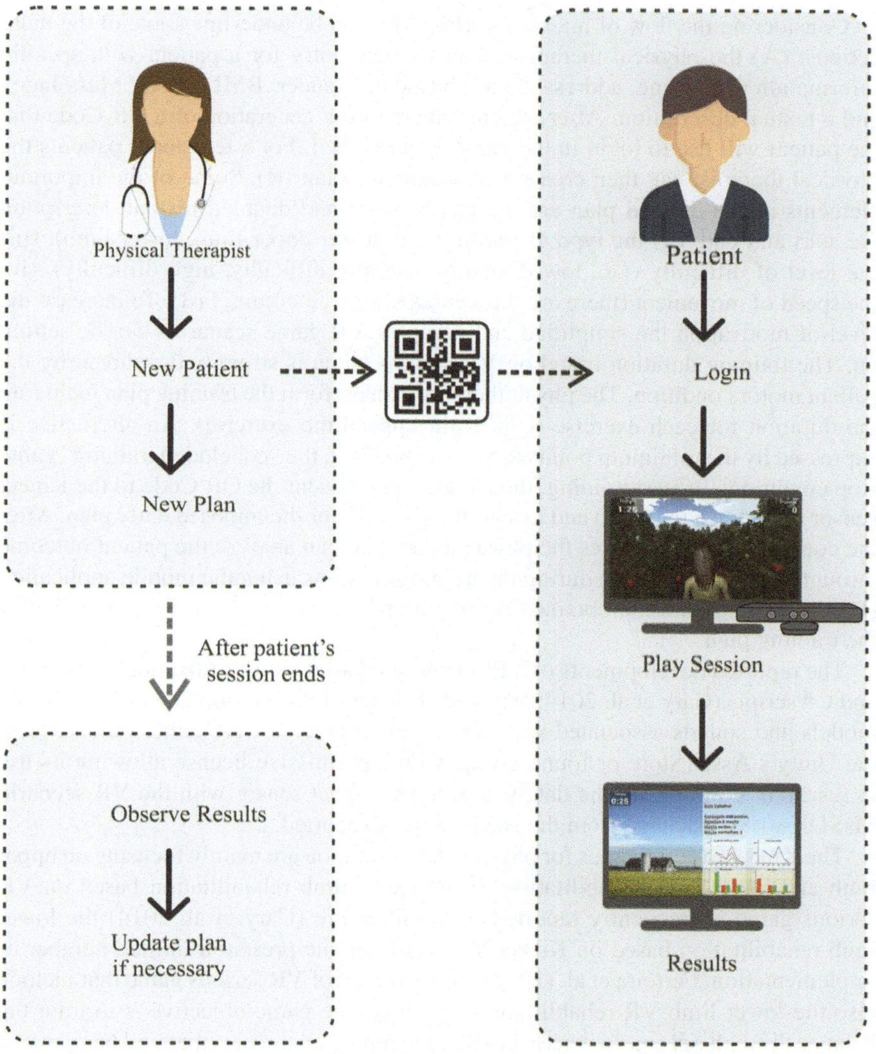

Fig. 5 Flow of interaction with the mobile application (physical therapist) and the kinect serious game (patient)

Considering the flow of interaction (Fig. 5) it can be underline some of the main actions: (A) the physical therapist creates a new entry for a patient with specific information (e.g. name, address, email, birthdate, gender, BMI—Body Mass Index and a textual description. After patient data record a generation of a QR Code that the patient will use to login in the game is performed. For a registered patients the physical therapist can then create a new exercise plan (B). Some of the important elements of the defined plan are: (i) plan's name and detailed textual description the start and end, (ii) the type of training (i.e. lower upper limb, lower limb), (iii) the level of difficulty (i.e., low difficulty, medium difficulty, high difficulty), (iv) the speed of movement (there are three modes: slow, medium, fast). To increase the level of motivation the simplified and complex VR game scenarios can be setting up. The training duration based on VR serious game is strongly dependent by the patient motor condition. The physiotherapist can perform the training plan including the duration for each exercise (e.g. 2 min upper limb exercise). An alternative is expressed by the minimum points that corresponds to the "concluded training" game stop condition. To start training, the patient may present the QR Code to the Kinect sensor to perform login (C) and to start the exercise for the imposed daily plan. After the concluding the exercises the physical therapist can analyze the patient outcome through calculated metrics during the training sessions using the mobile application (D). According with the obtained results the physiotherapist can decide to change the training plan.

The reported developments of VR serious are based on the utilization of Unity3D and C# scripts (Cary et al. 2014; Wang et al. 2019). In some implementation the 3D models and sounds associated with virtual environments were either purchased in the Unity's Asset Store or found online with a permissive license allowing its use in research. To integrate the data provided by Kinect sensor with the VR scenario MsSDK v1.8.1 package from the Asset Store is reported.

The Kinect serious games for physical rehabilitation are mainly focusing on upper limb and lower limb rehabilitation. If the upper limb rehabilitation based on VR serious game is frequently reported in the literature (Cary et al. 2014) the lower limb rehabilitation based on Kinect VR serious game present a limited number of implementation. Ferreira et al. (2017) reported a set of VR serious game that include also the lower limb VR rehabilitation scenario. The game objective is to train the balance through a scenario where the user is stepping on the tiles that will be spawned on the game. The user can perform the lateral movements Fig. 6a or front and back steps as shown below in Fig. 6b. The game is configured using a menu, where game duration, tiles spawn range (extended range will conduct to bigger angles between both legs), tiles speed, and even the score objective allowing the physiotherapist to impose the training plan. According with the patients' needs it is also possible to choose which leg motion will be trained. The game mechanic consists in stepping the tiles as they appear, reaching the objective score within the time limit imposed. By stepping on the tiles, the patients will train their lower limb mobility. The game may also be used for cognitive rehabilitation, thus the user might select only the imposed tiles (e.g. green tales) stepping the different color tales (e.g. red tales) that lead to losing points.

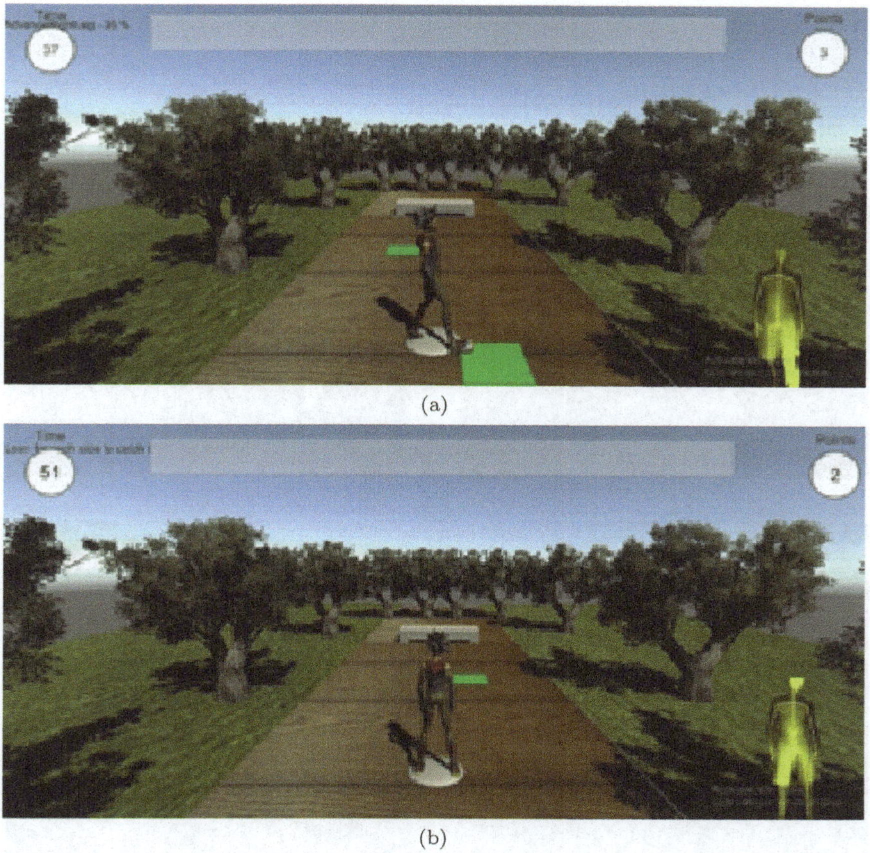

Fig. 6 Lower limb rehabilitation serious game-step on tile

During the training based serious game several metrics are on-line calculated and stored in the database. Some of the metrics are:

- Time between tale-on tale-off (times to step on the tile and return to initial position);
- Average velocity of each movement;
- Angle between legs when the step is performed;
- Distance between feet when tile is stepped;
- Number of steps for whole training period.

Advanced processing of the database stored data can be carried out on the server side level.

Regarding the upper limb rehabilitation after presentation of the QRcode to the Kinect sensor camera the patient record may be are presented on the screen followed by Objectives screen. In this screen, the objective of the particular game session is

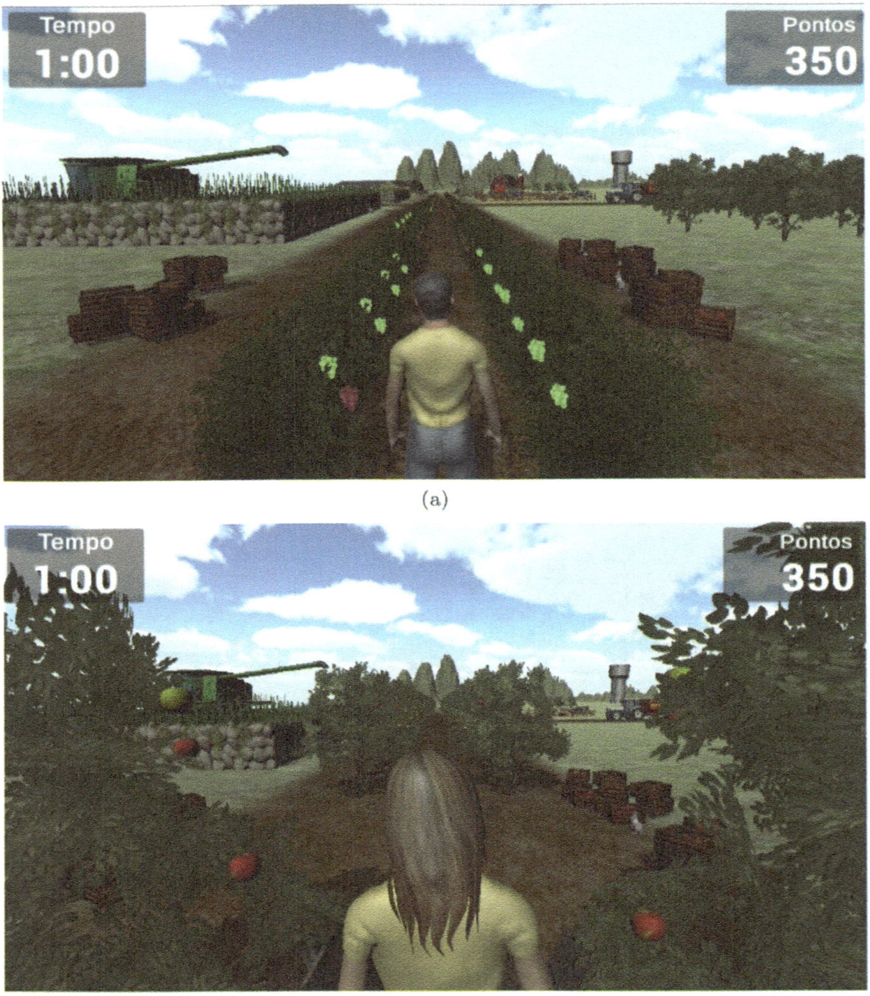

Fig. 7 Game screen for: **a** lower angles level; **b** higher angles level

presented in accordance to the configurations made by the physical therapist made by Physio APP.

The serious game for upper limb rehabilitation presents as objective to catch fruits associated with angles ("Low angles" or "High angles") that might be reached by left hand, right hand or both hands during the game based training session. In case "Low angles" setting imposed by physiotherapist through personalized training plan are marked by raspberries while "High angles" motion are associated with apples. The game screen is presented in Fig. 7.

The serious game screens presented above underline the avatar personalization (gender based) of the game according user's registration. On the top right corner, the current score ("pontos") is presented and is updated automatically when the patient picks a fruit. When the countdown ends, the avatar will start to move in a straight line, according to the movement speed defined. The patients' hand motion activity for fruit harvester is detected by the Kinect and reproduced by avatar. The developed game associated with each fruits a collision box that surrounds them in shape and the avatar's hands have collision boxes, too. When the avatar hand enters in contact with a fruit, that fruit is "picked" and increase the score (50 points for green fruits and 100 points for red fruits). Since the Kinect also detects user body positional changes (moving left or right for example) on the movement, a zone where the patient can move and pick fruits was defined. To increase the user motivation the VR scenario is characterized by the environment includes more or less visual elements (e.g., birds flying when the participant approaches the tree/bush they are on; bunnies, cats, dogs running, horses, cows; goats and pigs). All the visual elements were accompanied by corresponding auditory elements to increase the immersion level of the patient to the defined context.

When the participant reaches the required points (e.g. 1000 points) or the countdown timer reaches zero in the case of training by time a result screen with graphical information regarding the score achieved, the number of fruits (red and green) picked, per hand are presented. The results screen also presents the corresponding angles made by arms during the fruits harvesting session. Additionally, the progress the hand motion capability during several training sessions (e.g. 5 sessions). Thus, several game sessions of 1 min were performed and the values of the angles associated with body motion during the training was stored in the database for off-line data processing and data analysis.

In Tables 1 and 2 are presented the values of the measured posture angles and the statistics for two female and two male volunteers (Wang et al. 2019). Values of minimum, maximum, average and standard deviation are calculated for whole the training session, where the physical training objective focused on upper limb rehabilitation that is translated on fruit pickup on the player side.

As it can be observed in the above tables, during the training session the users are performed the left arm and right arm motion activity that will affect the posture angles. Based on the measured body angle metrics it is possible to evaluate the posture and the stability of the volunteers as a training outcome. Using a set of thresholds for low level of stability the accidents during the training sessions can be avoided based on previous alarm generation.

Correlations between the body posture angles and the values of angles associated with the position of the fruits in the trees can be extracted.

Table 1 Body posture angles metrics (met) during the fruits harvesting game performed with the left hand by two female volunteers F1 and F2; met-metrics. le = left elbow, ls = left shoulder angle, re = right elbow, rs = right shoulder; n = neck angle; s = spine angle

V	met	le	ls	re	rs	n	s
F1	min	151	104	146	30	96	114
	max	179	155	179	147	141	139
	av	168	120.5	171	109.1	119	126.9
	std	5.4	9.37	7.58	29.58	9.28	5.31
F2	min	119	73	120	43	108	115
	max	155	125	170	140	132	139
	av	138	97.95	143.5	107.5	122.2	122.6
	std	10.3	17.67	14.39	21.12	5.9	5.87

Table 2 Body posture angles metrics (met) during the fruits harvesting game performed with the left hand by two male volunteers M1 and M2, met-metrics)

V	met	le	ls	re	rs	n	s
M1	min	121	56	121	33	111	123
	max	177	122	179	122	140	131
	av	165.3	106.7	164.5	83.77	23	125.2
	std	11.12	11.58	14.04	20	6.29	1.91
M2	min	113	57	115	22	122	123
	max	175	126	159	87	138	30
	av	154.3	90.64	143	44.96	131.5	125.6
	std	17.63	18.93	12.61	21.98	4.14	1.47

2.3 NUI Leap Motion Controller—VR Serious Game

The VR serious game that focuses on hand and finger motion rehabilitation is based on a natural user interface expressed by a motion and depth controller Leap Motion Controller (LMC), (Kavian and Ghomsheh 2020; Ultraleap 2018), that materialize the natural user interface with the VR scenarios. The use of serious games in conjunction with the LMC and VR allows the patient with physical and/or motor physical rehabilitation needs, to perform a dynamic interaction increasing the motivation allowing different activities to be performed repeatedly over a pre-set period.

The LMC can detect the movement of the hands when they are on the device's field of vision, the movements being recognized by the device and translated into actions associated with the implemented VR scenarios. The LMC streams data at a variable acquisition rate of up 120 Hz. The device is highly sensitive to hand small amplitude movements, due to its ability to map the entire zone over it, being able to detect small movements of the whole hand. According to company reports, the controller uses two monochrome cameras and three IR (InfraRed) LEDs (wavelengths = 850 nm) to rebuild the 3D scene and track the positions of the hands and fingers (Fig. 8).

The device software and VR is based on device SDK (Software Development Kit) that provides data access through direct hand and finger mapping detection.

The main advantages of this sensor compared to other motion sensors, also applied to rehabilitation are: provides natural user interfacing and through remote sensing (no need for the device to be attached to person's body, which can cause discomfort on certain occasions); the provided SDK for the development of applications.

The implemented VR serious game aims is to pick up the cubes and to introduce according with the color in the corresponding box to earn points. During the game time cubes appear on the scene, for each correct cube the patient earns a point and for each failure loses a point. Figure 9 illustrates the CollectCube game interface.

Three levels of difficulty were considered: easy, medium and hard. The EASY level is characterized by single box (green) and over time appear green cubes; the MEDIUM scenario included two boxes (green and red) and over time randomly appear cubes with green or red color and HARD level is characterized by a scenario that includes three boxes (green, red and blue) and over time randomly appear cubes with green, red or blue color. When the physiotherapist creates a training plan for the patient, the avatar hands are considered according with the patient gender or patient age. The physiotherapist can chose which hand to be used for a specific training (Fig. 10).

A mobile device is used together a mobile application to assure the interaction between the physiotherapists and the system. Thus the mobile application is used by physiotherapist to perform queries regarding patient data and to create training plans after the registration of the patients. The system also provides to the physiotherapists the possibility to visualize the training results in order to estimate the patient outcome in objective manner. The developed mobile application can generate PDF reports that include information of the exercises that were already done. Referring the training plans the physiotherapists can check whether a training is active or expired, change existing settings, view results, and sort the existing trainings, for example by time order or view only the active trainings.

The LeaPhysio system for physical rehabilitation of the upper limbs was evaluated using 8 healthy volunteers (4 female occupational therapists (age range 29–31 years) and 4 healthy volunteers (1 female and 3 male, with the age range 25–48 years). All participants were played the CollectCube APP above presented and the selected training was for both hands the training session being selected of 3 min. At the end of the training session, hand and finger data were collected by the application for off-line analysis. Thus, in Fig. 11 a set of charts associated with both hands training for two participants are presented. The charts are clearly underline the evolution in time of Z coordinate of the volunteers' palm. The Z position refers to the depth axis, for example, during the game the participants need to move the hand in depth (Z axis) to place the cubes in the respective boxes (Postolache et al. 2017).

The presented values for palm Z position are expressed in cm. A value equal to 0 means that the participant has a hand over the leap motion controller, values greater than 0 means that the hand is moved to the front (towards the boxes) and negative

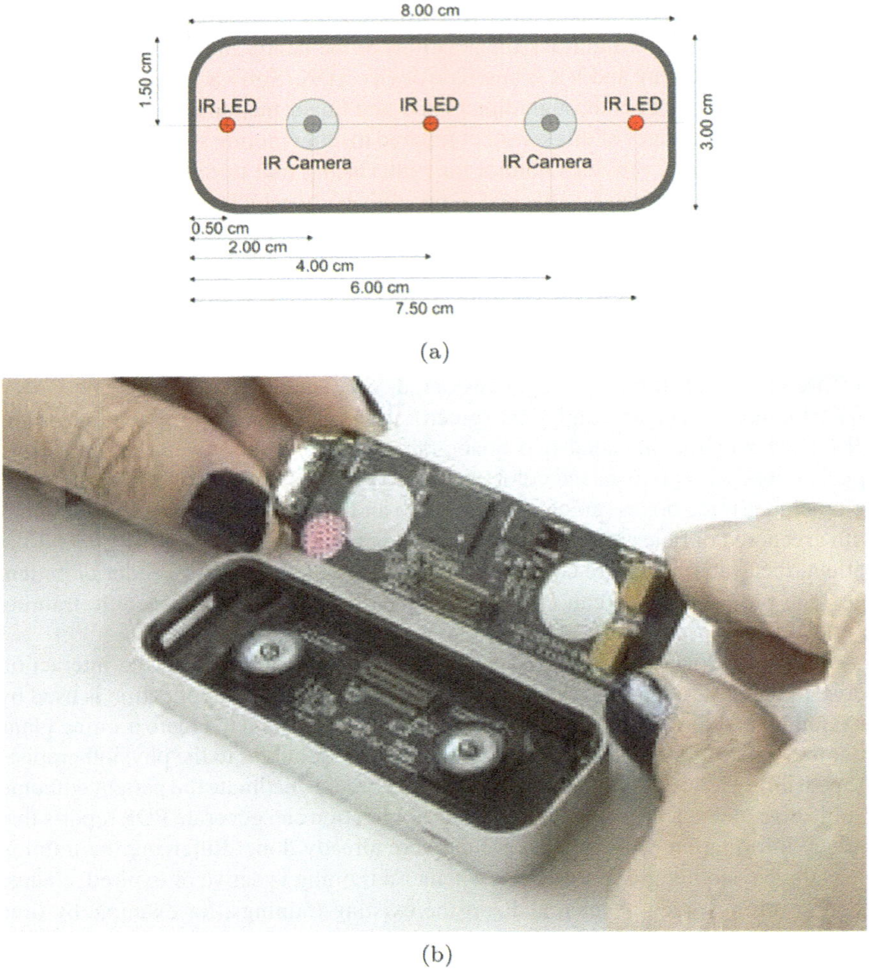

Fig. 8 Leap Motion Controller: **a** main block diagram; **b** LMC implementation

values means that the hand is moved to the back part. The participant2 obtained a better score for values greater than 0, compared with the participant1, meaning that the hands of participant3 was closer to the boxes. The participant1 used more the right hand, while the participant2 had a similar performance in both hands.

The VR serious game with interactions between the volunteer and the training scenario using Kinect of Leap Motion Controller are reliable and appropriate solutions to complement the daily used technology in physical therapy clinics. The visualization of the VR scenarios was always performed using high dimensions display that provide semi immersive interaction. The natural user interfaces and the semi-

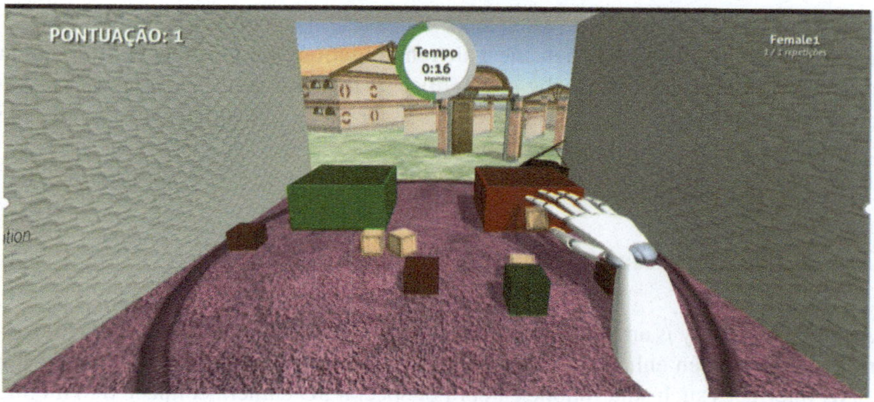

Fig. 9 CollectCube game interface

Fig. 10 Hand configuration for the patient to be used during a specific training

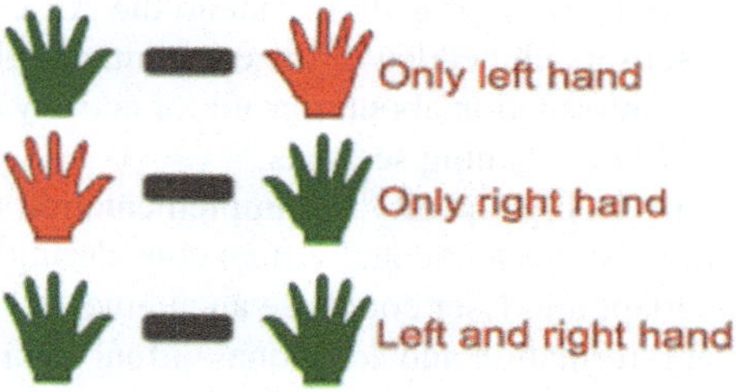

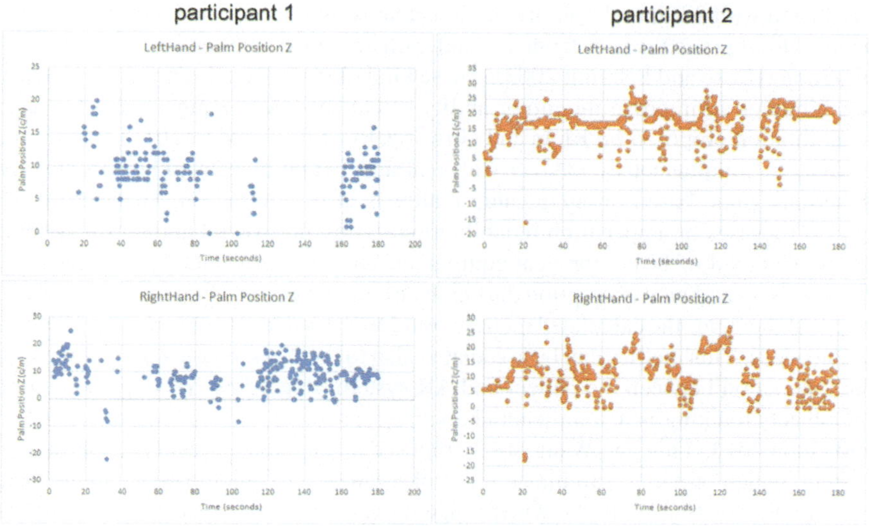

Fig. 11 Left and right hand motion evolution according with the Z coordinates of palm for two participants

immersive training based on VR mainly focusses on upper limb rehabilitation or upper limb rehabilitation when the gait rehabilitation can be practiced for much reduced distances. To extend the capabilities of this alternative systems for physical rehabilitation augmented reality devices associated with AR scenarios were also considered and presented in this chapter.

3 Augmented Reality Serious Game

Augmented reality is an interactive experience that presents you with a real environment that has been enhanced or altered by computer-generated information. Using a AR interfaces such smartphone camera or special 3D camera connected to mobile computation platform, it allows the user to experience the real world with additional data or images overlaid on top of it in the form of three-dimensional scenarios, interactive models, or text. To extend the AR experience of the users' wearable sensors or sensors embedded in the environment objects can be considered to provide specific information about user motor activity and health status of the user during daily activity or training sessions.

In the high quality AR implementation the users will not have a feeling of separation between real and virtual considering the real time processing of virtual to real overlapping. User could see an augmented scene that is characterized by real-world the information and additional virtual information at the same time.

As part of smart rehabilitation technologies AR serious games can provide data associated with AR based training sessions that is used to calculate metrics that can be used to improve the user feedback and motivation for an effective physical rehabilitation (Hesse and Schmidt 2003). These kind of games are designed for physical rehabilitation to improve the motor ability of patients. A training system for hand rehabilitation proposed on Klein and Gilda (2013), Liu and Mei (2017) enables the patients to simultaneously interact with real and virtual objects and environments. A serious game based on augmented reality is also reported on Alexandre et al. (2019) focusses on upper-limb rehabilitation. Using the AR rehabilitation systems the user got rid of heavy medical equipment, got more motivation and reduce the mental stress and pain perception during the training process. Due to the diversity of physical therapies the calculated metrics could be extended according to the rehabilitation needs. This metrics can be associated also to the AR scenarios optimizing also the interaction between the patient and AR system. Regarding AR implementation the objects tracking is a significant challenge associated with AR scenarios. Klein and Gilda 2013; Ying and Wang 2017 proposed an augmented reality rehabilitation system, which provides arm image for stroke patients, to guide patients to do some rehabilitation actions with the wider range movement. In this research the patient's arm tracking is presented as the result of recognizing some specific areas, such as the bottom of the face, shoulder area and intersection between shirt and neck. Parts of body tracking and used object tracking can be performed by marking objects. Thus upper limbs training system for stroke patients is reported in Microsoft (2020). A virtual cup on the table is considered for upper limb exercises. Thus the patients are motivated to perform a set of imposed actions, based on a set of marked pictures on the table. Additionally a camera was utilized to capture markers of the picture and to represent AR objects in front of the patients by marker detection and pose estimation. The reported work is focusing on static therapy training as part of the patient's body motion recovery, however can be mentioned limitation on data recordings of patients' motion parameters and physical characteristics as so as related AR visual effects.

3.1 AR Sensing

To mixt real and virtual worlds the AR needs the help of a few components expressed by camera, sensors, computer vision, and a graphical interface (e.g. smart phone). Basic AR implementations reports the usage of regular camera embedded on the mobile devices (e.g. smart phone) however the usage of depth camera (3D scanner) improve the AR experience through the accurate localization of objects placed on real environment. Referring to the AR interfaces for Theragames it can be mentioned Microsoft HoloLens or Oculus Quest or 3D Occipital Structure sensor.

Microsoft HoloLens 2 offers the most comfortable and immersive mixed reality experience available, with industry-leading solutions that deliver value in minutes-all enhanced by the reliability, security, and scalability of cloud and AI services from Microsoft (Androidcentral 2020). Move freely, with no wires or external packs to get in your way. The HoloLens 2 headset is a self-contained computer with Wi-Fi connectivity assuring higher mobility and avoid the necessity of external computation platform.

The Oculus Quest interface is characterized by a tracking system used in the Oculus Rift S, named Oculus Insight. On the Quest, the system relies on four wide angle cameras located on each corner of the headset to spatially track the headset. The main specifications are display panel: OLED with 1440×1600 per eye resolution, 72 Hz refresh rate, embedded computing based on Qualcomm Snapdragon 835 processor and 4 GB RAM; 6 degrees of freedom head and hand tracking. The hand motion associated with AR game scenarios particularly Theragames is followed by two touch controllers (Khoshelham and Elberink 2012).

The fluidity of interactions and the reduced latency during the AR serious game is mainly dependent by the embedded computation platforms associated with the systems such as HoloLens and Oculus Quest. The embedded computation platform is a big strength when it is considered the software compatibility but it cannot updated during the time as it can be considered for the AR systems where the computation platform is external such for the 3D Occipital Structure sensor.

Regarding the Structure depth sensor, it is based on a laser-emitting diode infrared (IR) radiation range projector and an IR camera. The infrared projector emits a single beam which is split into multiple beams by a diffraction grating to create a constant pattern of speckles projected onto the real world. The lighting points projected on

Table 3 3D structure occipital characteristics

Characteristics	Values
Min distance	0.40 m
Max distance	3.5 m
Accuracy	Up to 0.5 mm
Precision	1% *d
Field of view	H: 58°; V: 45°
Illumination	IR structured light projector

the surface of object are acquired by the infrared camera and stored together the distance between system and the object to Structure system memory. To extract the depth information a comparison with a reference distance is used. Thus when the measured distance is smaller or larger than the reference distance, data point on the infrared image will changed according to the difference obtained above (Burke and Morrow 2008). Additionally the used smart phone iPhone's RGB camera send raw data (raw RGB) to a Structure system on chip for processing (Monge and Postolache 2018). The processed data point on the surface of object shows different colors on the depth image depending on the distance between IR camera and real object. Several characteristics of the Structure sensors are presented in the Table 3.

Important advantage of the Structure sensors is the fact that can be mounted on the headset assuring high portability by comparison with Kinect sensor which is a fixed sensor characterized also by high power consumption by comparison with Structure sensor. For the structure sensor the battery permits a 4 h active sensing autonomy.

3.2 AR Serious Games for Physical Rehabilitation

A category of AR serious game for physical rehabilitation are using mobile device camera and specific software, developed in ARKit or Unity to add virtual objects in a real scenario. The team was working to develop interactive physical therapy session where the patient receives the information from a virtual physiotherapist or personal motor rehabilitation trainer (Fig. 12). The visual interaction with virtual trainer is performed through smart phone camera associated with a headset (Yu et al. 2019). The system didn't provide the tracking capabilities, however it can present the results associated with performed training exercises in interactive mode for higher user engagement. To increase the information content that can plenty characterize the performed training the system include wearable EMG that deliver information about muscle activity. The EMG measurement WBAN are distributed on the upper and lower limbs according to the muscle activity monitoring needs.

Considering the physical rehabilitation the tracking represents an important characteristic associated to a system for AR physical training. In this context a 3D Occip-

Fig. 12 Left and right hand motion evolution according with the Z coordinates of palm for two participants

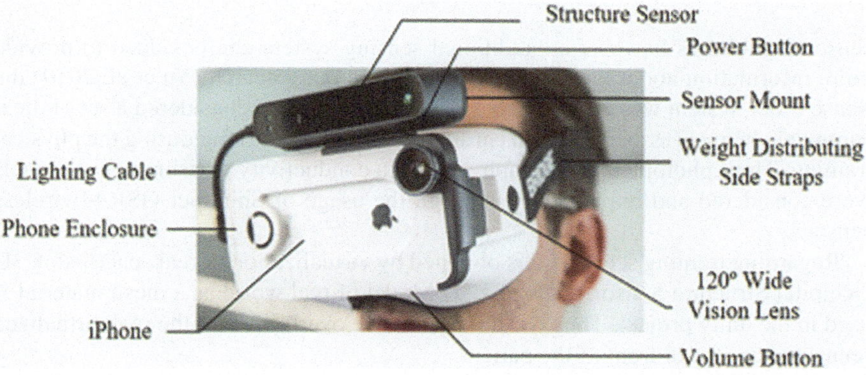

Fig. 13 AR system based 3D occipital structure sensor and iPhone 8—bridge headset

ital Structure sensor that is depth camera is reported by Yu Jin et al. (2019). The Structure sensor is connected to iPhone 8 that is a part of headset that support also the sensor (Fig. 13). Structure sensor gives to the mobile device the ability to capture real world in three dimensions that is denominated virtualization of real environment, virtualization.

Using the AR system based on 3D Occipital Structure Sensor and iPhone 8 an iOS serious game "PhysioMaze" was developed. The used 3D gaming technology for game implementation was Unity 3D. Additionally structure bridge engine SDK was considered. The iOS App associated with the serious game provides a rehabilitation training plan for stroke patients. This serious game focusses on patient balance and requires patients to walk in a maze according to a specified route. Some constraints are imposed to the patient under the training plan. The body balance assessment is included, thus the patients can't touch wall, preserving their balance. The player position in the AR scene is mainly obtained using the Occipital Structure

Fig. 14 Interface of "Maze" AR serious game for lower limb physical rehabilitation scenario

sensor capabilities however an additional sensing system can be added to provide more information about the gait rehabilitation. In our research (Yu et al. 2019) the Sense Floor system was considered. The authors were also considered a set of measurement channels associated with cardiac and stress monitoring during the physical training. Thus, photoplethysmography and skin conductivity measurement channels were considered and implemented through the usage of Shimmer GSR+ wireless sensor.

Regarding training scenario it is obtained by virtualization of real space using 3D Occipital Structure Sensor. Thus, the 3D Model of real world as a mesh material is used in the unity project. Thus, virtual objects are overlapped on the real virtualized scene and appear in front of the patient.

In Fig. 14 the implemented AR physical training scenario is presented. In this scenario the "Maze" serious game is associated with 2 m by 4 m area according with size of sense floor.

The patient under training wearing the Bridge Headset and Shimmer sensor will walk into this virtual maze. According to the training plan that fit to the patient rehabilitation needs a specific maze pathway is considered including signs on the ground and the patient deviations from the right path are calculated (Fig. 15). The maximum errors considering the training session data for two patients (p_1, p_2) were 0.588 m for p_1 and 0.476 m for p_2.

To improve the user balance, they are not allowed to touch the walls during training according to the serious game rules. The number of walls collision will affect the general score. The current score is recorded and displayed on the upper right corner of the screen. Additional message are displayed such "Go into the maze and start training" or "Go outside the maze from the exit".

The general score is mainly obtained through objects pickup actions. Thus, gold rings, diamonds, coins, stars are collected to increase the session score. Countdown starts when training started, warning sounds play when user was fail the objective to finish the maze in time according to the training plan (e.g. recorded time > 180 s).

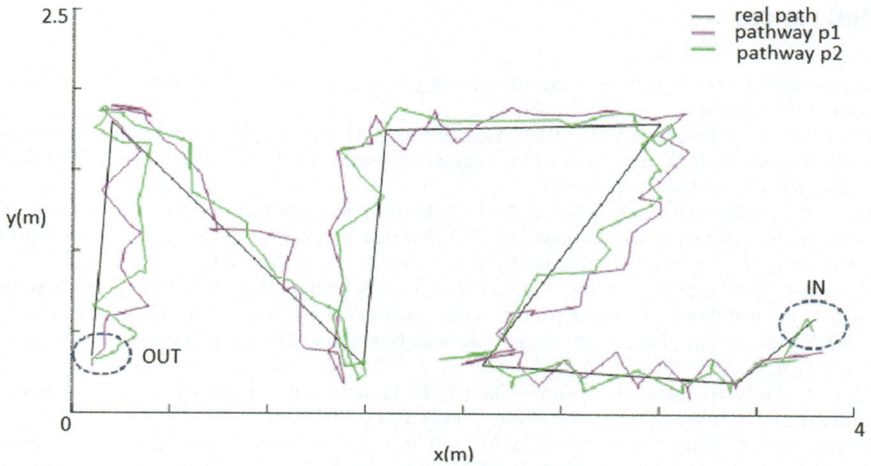

Fig. 15 "Maze" game. Training paths: imposed by training plan and performed pathways by two patient (p_1, p_2)

The implemented AR serious game highlights the possibility to increase the balance and gait capability in the patients increasing, in the same time, the engagement of the patients.

4 Conclusion

The main purpose of this chapter is to describe therapeutic gaming systems based on sensors, in the context of AR or VR scenarios. The used software technologies for the AR or VR components is expressed by Unity real time development platform. IoT architectures for rehabilitation systems are considered, underlining the remote sensing solutions materialized by Microsoft Kinect, Leap Motion Controller or 3D Occipital Structure Sensors. Additionally wearable sensing solutions expressed by IMU sensors, force and flexion sensors as part of smart gloves interface are presented. Based on described smart rehabilitation systems, the physiotherapist can follow the patient during the prescribed training plan for days, weeks or months. Taking into account the data storage based on implemented IoT, the serious game data for the active patients is processed to extract the rehabilitation model and to predict the rehabilitation outcomes. The developed system may be the starting point for new era of smart healthcare services allowing the users to perform in-home training session according to the imposed physical rehabilitation plan. The quality of service will assure of high level of patients motivation and reduced rehabilitation period.

References

Androidcentral. (1111). Oculus quest: Everything you need to know. https://www.androidcentral.com/oculus-quest.

Alexandre, R., & Postolache, O. (2016). Wearable and IoT technologies application for physical rehabilitation. In *IEEE International Symposium on Sensing and Instrumentation in IoT Era ISSI*, Shanghai, China (Vol. 1, pp. 1–6).

Alexandre, R., Postolache, O., & Girão, P. M. (2019). Physical rehabilitation based on smart wearable and virtual reality serious game. In *IEEE International Instrumentation and Measurement Technology Conference I2MTC*, Auckland, New Zealand, May (Vol. 1, pp. 1–6).

Baranyi, R., Willinger, R., Lederer, N., Grechenig, T., & Schramm, W. (2013). Chances for serious games in rehabilitation of stroke patients on the example of utilizing the Wii Fit Balance Board. In *IEEE 2nd International Conference on Serious Games and Applications for Health (SeGAH)* (pp. 1–6).

Béjot, Y., Bailly, H., Durier, J., & Giroud, M. (2016). Epidemiology of stroke in Europe and trends for the 21st century. *La Presse Médicale*, *45*(12), Part 2, 391–398.

Borghese, N. A., Mainetti, R., Pirovano, M., & Lanzi, P. L. (2013). An intelligent game engine for the at-home rehabilitation of stroke 704 patients. In *IEEE 2nd International Conference on Serious Games and Applications for Health (SeGAH)*, May (Vol. 705, pp. 1–8).

Borromeo, N. A. (2020). Hands-on unity 2020 game development ed. Packt.

Burke, J. W., & Morrow, P. J. (2008). Vision based games for upper-limb stroke rehabilitation. In *International Machine Vision and Image Processing Conference (IMVIP)* (pp. 159–164).

Cary, F. C., Postolache, O., & Girão, P. M. (2014). Kinect based system and serious game motivating approach for physiotherapy assessment and remote session monitoring. In *International Conference on Sensing Technology - ICST*, Liverpool, United Kingdom, September (Vol. 1, pp. 1–5).

Chen, P., Du, Y., Shih, C., Yang, L., Lin, H. T., & Fan, S. C. (2016). Development of an upper limb rehabilitation system using inertial movement units and kinect device. In *International Conference on Advanced Materials for Science and Engineering (ICAMSE)* (pp. 1–6).

Ferreira, D. F., Oliveira, R. O., & Postolache, O. (2017). Physical rehabilitation based on kinect serious games. In *International Conference on Sensing Technology - ICST*, Sydney, Australia, December (Vol. 1, pp. 1–6).

Geman, O., Postolache, O., & Chiuchisan, I. (2019). Mathematical models used in intelligent assistive technologies: Response surface methodology in software tools optimization for medical rehabilitation (pp. 83–110). Berlin: Springer Book.

Hesse, S., & Schmidt, H. (2003). Upper and lower extremity robotic devices for rehabilitation and for studying motor control. *Current Opinion in Neurology, 16*, 705–710.

Internet Stroke Center. (2020). Stroke statistics (2020). http://www.strokecenter.org/patients/about-stroke/stroke-statistics/. Accessed Aug 2020.

Jani, A. B., Bagree, R., & Roy, A. K. (2017). Design of a low-power, low-cost 751 ECG & EMG sensor for wearable biometric and medical application. In *IEEE Sensors*, Glasgow, UK, November (pp. 1–3).

Kavian, M., & Ghomsheh, A. (2020). Monitoring wrist and fingers range of motion using leap motion camera for physical rehabilitation. In *International Conference on Machine Vision and Image Processing (MVIP)* (pp. 1–6).

Klein, A., & Gilda, A. (2013). A markeless augmented reality tracking for enhancing the user interaction during virtual rehabilitation. In *XV Symposium on Virtual and Augmented Reality* (pp. 117–124).

Khoshelham, K., & Elberink, S. O. (2012). Accuracy and resolution of Kinect depth data for indoor mapping applications. *Sensors, 12*(2), 1437–1454.

Liu, J., & Mei, J. (2017). Augmented reality-based training system for hand rehabilitation. *Multimedia Tools and Applications*, 14847–14867.

Lourenço, F., Postolache, O., & Postolache, G. (2018). Tailored virtual reality and mobile application for motor rehabilitation. In *IEEE International Instrumentation and Measurement Technology Conference I2MTC*, Houston, United States, May (Vol. 1, pp. 1–6).

Mealy, P. (2020). Virtual reality vs. augmented reality: Strengths and weaknesses. Dummies a Willey band. https://www.dummies.com/software/virtual-reality-vs-augmented-reality-strengths-and-weaknesses/. Accessed Aug 2020.

Microsoft. (2020). Hololense 2. https://www.microsoft.com/en-us/hololens/hardware. Accessed Sept 2020.

Monge, J., & Postolache, O. (2018). Augmented reality and smart sensors for physical rehabilitation. In *International Conference and Exposition on Electrical and Power Engineering - EPE*, Iasi, Romania, October (Vol. 1, pp. 1–6)

Muñoz, J. E., Villada, J. F., Casanova, S., Montoya, M. F., & Henao, O. A. (2018). Dynamic systems theory in human movement exploring coordination patterns by angle-angle diagrams using kinect. In *10th International Conference on Virtual Worlds and Games for Serious Applications (VS-Games)*.

Oarde, E. D., Libatique, N. C., Tangonan, G. L., Sotto, D. M., & Pacaldo, A. T. (2014). Digital motion analysis system for rehabilitation using wearable sensors. In *International Conference on Humanoid, Nanotechnology, Information Technology, Communication and Control, Environment and Management (HNICEM)* (pp. 1–6).

Postolache, O., Cary, F. C., Duarte, N. D., & Girão, P. M. (2015). Physiotherapy assessment based on kinect and mobile APPs. In *International Conference on Information, Intelligence Systems and Applications - IISA*, Corfu, Greece, July (Vol. 1, pp. 1–6).

Postolache, O., Girão, P. S., López, A., Ferrero, F. J., Dias Pereira, J. M., & Postolache, G. (2016). Postural balance analysis using force platform for K-theragame users. In *IEEE International Symposium on Medical Measurements and Applications (MeMeA)* (pp. 1–6).

Postolache, O., Lourenço, F. L., Dias Pereira, J. M., & Girão, P.M. (2017). Serious game for physical rehabilitation: Measuring the effectiveness of virtual and real training environments. In *IEEE International Instrumentation and Measurement Technology Conference I2MTC*, Torino, Italy, May (Vol. 1, pp. 1–6).

Postolache, O., Teixeira, L., Cordeiro, J., Rodrigues, M., Lima, L. L., Arriaga, P. A., Girão, P. M., Postolache, G. (2019). Tailored virtual reality for smart physiotherapy. In *IEEE Advanced Topics in Electrical Engineering - ATEE*, Bucharest, Romania, March (Vol. 1, pp. 1–6).

Prashun, P., Hadley, G., Gatzidis, C., & Swain, I. (2010). Investigating the trend of virtual reality-based stroke rehabilitation systems. In *14th International Conference on Information Visualisation* (pp. 1–6).

Reis, E., Postolache, G., Teixeira, L., Arriaga, P., Lima, M. L., & Postolache, O. (2019). Exergames for motor rehabilitation in older adults: An umbrella review. *Physical Therapy Reviews, 24*(3–4), 84–99.

Susi, T., Johannesson, M., & Backlund, P. (2007). Serious games-An overview. School of Humanities and Informatics, University of Skovde, Skovde, Sweden.

Serious Game Market. (2020). VR serious games redefining physical therapy. https://www.seriousgamemarket.com/2017/07/vr-serious-games-redefining-physical.html. Accessed Aug 2020.

Tannous, H., Istrate, D., Ho Ba Tho, M. C., & Dao, T. T. (2016). Feasibility study of a serious game based on kinect system for functional rehabilitation of the lower limbs. *European Research in Telemedicine*, *5*(3), 97–104.

Timsit, S. (2016). Stroke at the beginning of the XXIst century. *La Presse Médicale*, *45*(12), Part, 2, 389–390.

Ultraleap. (2020). Leap motion controller. https://www.ultraleap.com/product/leap-motion-controller/. Accessed Aug 2020.

United Nations. (2017). World population ageing. United Nation highlights. https://www.un.org/en/development/desa/population/publications/pdf/ageing/WPA2017_Highlights.pdf. Accessed Aug 2020.

Wang, L., Liu, J., & Lan, J. (2019). Feature evaluation of upper limb exercise rehabilitation interactive system based on kinect. IEEE Access, *7*.

Ying, W., & Wang, A. (2017). Augmented reality based upper limb rehabilitation system. In *13th IEEE International Conference on Electronic Measurement & Instruments (ICEMI)* (pp. 426–430).

Yu, J., Monge, J., Postolache, O., & Niu, W. (2019). Augmented reality with application in physical rehabilitation. In *International Conference on Sensing and Instrumentation in IoT Era (ISSI)* (pp. 1–6).

Control of Lower Limb Exoskeletons for Gait Rehabilitation Purposes

Boutheina Maalej, Rim Jallouli–Khlif, and Nabil Derbel

Abstract Mobility deficiency is one of the main impairments for children with Cerebral Palsy (CP). Various approaches and techniques are used for rehabilitation purposes in order to improve the life quality and the autonomy of these young patients. Nowadays, beyond conventional techniques, such as manual stretching and massage, robotic-based gait training therapy is considered as one of the most effective key tools to compensate and rehabilitate functional skills of people with cerebral palsy. In the literature, several works are dealing with different aspects regarding these robotic rehabilitation devices, such as design, optimization, actuation and control. In this chapter, two control schemes based on sliding mode control with integral action are developed. The proposed control solutions have been numerically implemented and tested in different simulation scenarios to show their effectiveness and robustness.

Keywords Cerebral palsy · Kids rehabilitation · Adaptive sliding mode with integral action · Lower limb exoskeleton

B. Maalej (✉)
ENIS, Control and Energy Management Laboratory, Digital Research Center of Sfax, Clinical Investigation Center, University of Sfax, Sfax, Tunisia
e-mail: maalej.boutheina@gmail.com

University of Gabes, Gabes, Tunisia

R. Jallouli–Khlif
Control and Energy Management Laboratory, Higher Institute of Computer Science and Multimedia of Sfax, Sfax, Tunisia
e-mail: rim.jalloulikhlif@isims.usf.tn

N. Derbel
ENIS, Control and Energy Management Laboratory, Digital Research Center of Sfax, Sfax, Tunisia
e-mail: nabil.derbel@enis.tn

O. Kanoun and N. Derbel (eds.), *Advanced Systems for Biomedical Applications*, Smart Sensors, Measurement and Instrumentation 39,
https://doi.org/10.1007/978-3-030-71221-1_9

1 Introduction

Advances in Robotic technologies for rehabilitation have been significantly developed in the last six decades. In fact, exoskeletons have progressed from the stuff of science fiction to the commercialization (Dollar and Herr 2008). The basic intension is to develop assisting technologies for disabled persons. In fact, studies carried on animals and humans affirm that the brain reorganization is tightly related to the intensity and the magnitude of the sensory motor activities (Bayón et al. 2016a). Thus, since the early age, physical activities are prior to avoid muscles spasticity (Sumathy and Renjith 2014). The cerebral palsy, as defined by an international panel in the mid-2000s, is described by some permanent disorders inducing activity limitations due to non-progressive disturbances during the development of fetal or infant brain (Graham et al. 2016). The cerebral palsy is considered as the first neurological disease affecting both balance and motor disability (Bax et al. 2005). Multiple factors are considered as direct causes for cerebral palsy (Nelson and Ellenberg 1986) like:

- congenital deformity,
- fetal growth limitations (Blair and Stanley 1993),
- inflammation during the fetal and neonatal period (Grether and Nelson 1997),
- untreated maternal hypothyroidism,
- perinatal stroke,
- and premature birth, which is considered as the most important risk factor: up to 15% of more risk for kids born before 28 weeks of gestation (O'Shea et al. 2009).

Frequently, motor disabilities caused by cerebral palsy (muscle weaknesses, contracture in joint levels and spasticity) are joined by disturbances of sensation, cognition, communication, perception and behavior disorders (Balci 2016). Therefore, therapy approaches should follow the child's necessities. Notably, the rehabilitation should focus on improving the locomotor function by training (Barbeau and Rossignol 1994) to provide an appropriate spinal cord neurons restructuring. The principal objective is to improve the patient's autonomy in daily life activities.

Medical costs are consequent factors to take into account. In fact, about 500.000 children in US are affected by cerebral palsy, and the number is higher in Europe.

Mainly two groups of rehabilitation approaches are used:

- standard therapy based on physical and occupational therapy as walking, standing, stretching activities and flexibility,
- robot-assisted therapy that aims to compensate or rehabilitate the motor skills of cerebral palsy patients in a controlled and safe way (Bayón et al. 2016b).

Robot-assisted therapy is considered to be an alternative solution to complete the treatment (Meyer-Heim and Hedel 2013). Several studies and companies are now interested in disabled persons, especially for children (e.g. Lokomat, Andago, Armeo, etc.) because fundamental abilities and skills should be developed during the first age, which increases the rate of the rehabilitation success. For lower limbs rehabilitation, some systems are already used by people having reduced mobility, such as walkers

(NF-Walker, Innowalk, Innowalk-Pro) (Bayón et al. 2016b) and exoskeleton robots (Lokomat, GT-1 RehaStim) (Maalej et al. 2018). In a study including 33 CP patients all around 7 years old, dealing with 40 sessions of 20 min, 24% succeeded to walk without assistance and 15.3% of the patients rest had a gait pattern improvement (Smania et al. 2011).

Taking into consideration multitudes of variable parameters around the posture variations and the physical deformities of kids suffering from cerebral palsy, the control approaches of such systems need to get a great importance in order to follow the assistance as needed, the user's intention detection, the modularity, the safety, the stability and the robustness against parametric variations (Maalej et al. 2020; Patané et al. 2017).

This chapter presents two control laws for a gait rehabilitation robot device dedicated to children. The first approach is a sliding mode control with an integral action, implemented in the nominal case and in the presence of parametric variations and external perturbations. Then, considering the case where the system parameters are ill-known, unknown or they vary in time, we propose to implement the adaptive sliding mode control with an integral action.

This chapter is organized as follows. Section 2 introduces both methods of rehabilitation: the standard therapy and the robot assisted one. Section 3 deals with the proposed control approaches for lower limb exoskeletons dedicated to young patients. Section 4 presents the simulation results which prove the validity of these approaches. Finally, Sect. 5 presents the conclusion and future works.

2 Rehabilitation Challenge

Cerebral Palsy takes part of the most frequently disabilities at early age and induces heavy charges not only on families and children, but also on health, educational and social institutions. It commonly consists of a disorder of posture and movement associated to an alteration in the immature brain (Bayón et al. 2016b). Treatments for CP patients are related to their specific pathology and differ from physical rehabilitation to medication and even surgery. Several techniques and approaches are used for physical rehabilitation of children with CP. Currently, they are basically classified as standard approaches without any equipment, and approaches using new technologies through specified robots.

2.1 Standard Rehabilitation

Standard Rehabilitation therapy is very important for persons suffering from different kinds of disorders, as delayed motor development, namely starting from the early age. Even if recovery from these disorders is complex and it could be a long process (Munera et al. 2017), children can improve and maintain their quality of life, promote their social reintegration and achieve the maximum independence by learning the appropriate activities and exercises. Physicians cooperate with the child, family, and members of a multidisciplinary team, to arrange a composite care system offering the maximal profit to each child. It consists of physical exercises where physiotherapists focus on the rehabilitation of the functional capacities of children. In fact, since the child capacities are relatively low, therapists try to correct deformities and improve the patient's motor skills, as young children have greater brain plasticity than adults. For this reason, an intervention at an early age is highly recommended to change the motor patterns.

During last decades, there have been different therapeutic practices used for children suffering from cerebral palsy (Cuccurullo 2004):

- *Proprioceptive Neuromuscular Facilitation (PNF)*: is based on the theory of spiral and diagonal movement, through stimulation of nerve, muscle and sensory receptors to facilitate movement patterns that will have better functional pertinence.
- *Bobath approach*: is probably the most used Neuro Developmental Technique (NDT) to normalize tone, restrict primitive reflex patterns and improve automatic reactions.
- *Sensory motor approach/Rood approach*: focuses on cutaneous sensorimotor to activate the muscle tone and promote voluntary motor activity.
- *Brunnstrom approach/Movement therapy*: is opposite to Bobath approach, it encourages the use of abnormal movements, and consider them a basic process of recovery till normal ones are reached.
- *Motor relearning program/Carr and Shepherd Approach*: is influenced by Bobath approach, and based on learning general strategies to surround motor difficulties, in order to find the problem solving at each task.

Standard physical therapy, even it helps patients to enhance their motor skills, it offers moderate results that are not sufficient in some cases. For this reason, researchers have been studied other tools for more efficient gait rehabilitation therapy, notably the robot assisted therapy.

2.2 Robotic Devices in Rehabilitation

The use of robotic trainers for patients with mobility limitations has been developed in the last decades since new generations of robotic devices offer to patients an active participation in task specific activities. This is thanks to the development of new therapeutical control strategies and to the modularity of new exoskeletons, achieving the treatment adaptation to the patient's needs (Bayón et al. 2018). Among the history, the earliest like exoskeleton devices are dated on 1890 (Ali 2014). At the end of 1960, the Mihailo Pupin Institute exoskeleton was a pioneer in gait assistance (Ali 2014). Due to technological constraints, it wasn't really a practical solution. In 2001, Lokomat device has been used in several hospitals and rehabilitation centers.

Nowadays, technology advances afford a wide expansion in the field of exoskeletons for rehabilitation purposes. Different devices are already commercialized, such as ReWalk, eLEGS, MindWalker, Hal, REX, Vanderbilt exoskeleton (Ali 2014). Each one has a specific mechanism regarding its control. This evolution in gait training robotic devices is thanks to their benefit outcomes on patients life, while they allow the patient to achieve a set of movements in a limited time (Balci 2016). In general, they can improve endurance, balance and coordination, strength, motor planning, and confidence (Maalej et al. 2019). A new study has been undertaken on 28 children (6–16 years) with different levels of CP according to the gross motor function classification system "GMFCS" (2–4 years), in order to investigate the effects of robotic rehabilitation, notably on spasticity and motor functions (Filiz and Arslan 2018).

Researchers and engineers in mechatronic and robotic fields are working on the implementation of joints' control laws that support children morphology variations, as limb lengths and masses, ensuring a perfect gait tracking. New robotic trainers, like CPWalker platform (Bayón et al. 2018), can be easily controlled through an interface which controls and monitors the exercises online. The device cost is also a pioneer constraint that should be considered in order to make a better quality of life affordable for CP children and their families.

In addition, robot guided therapy permits the option of home treatment (Balci 2016), which may offers comfortable conditions especially for children, by avoiding repetitive difficult displacements to therapeutic centers.

Achieving functional walking ability is an important goal for children. Robotic systems (Rifai and Amirat 2016) can be motivating and challenging. In fact, lower-limb exoskeletons are highly efficient in reducing the hard work of the physiotherapists. Moreover, they can guarantee a high number of repetitions (3300 steps per session vs 50–800 steps using standard therapy) as well as longer therapeutic sessions.

In this chapter, we are interested in the rehabilitation of kids aged from two to thirteen years old. The difference of kids age leads to a huge difference of morphology (limb lengths, masses, etc.) which proves the necessity of good control aspects of the exoskeletons' joints.

3 Proposed Control Approaches for Exoskeleton

In this section, we propose to control the lower limb exoskeleton using sliding mode controllers. First, in order to cancel the static error in the case of external disturbances and in the case of ill-known system parameters, the sliding mode control with integral action is developed. Then, in case where system parameters (vector θ) are ill known, or unknown or they vary in time, an adaptive approach increases the robustness and improves the dynamical behaviour of the controlled system.

The dynamic model of the lower limb exoskeleton is established using Lagrange formulation. The proposed model which includes the exoskeleton and the kid legs is as follows:

$$M(q)\ddot{q} + C(q,\dot{q})\dot{q} + G(q) = \tau \tag{1}$$

where:

- $M(q) \in \mathbb{R}^{n\times n}$ is the inertia matrix
- n is the number of degrees of freedom
- $C(q,\dot{q}) \in \mathbb{R}^n$ is the vector of the Coriolis, centrifugal forces
- $G(q) \in \mathbb{R}^n$ represents the gravitational forces
- $\tau \in \mathbb{R}^n$ is the vector of torques generated by actuators
- $q = [q_1\ q_2\ \cdots\ q_n]^T \in \mathbb{R}^n$ is the position vector
- $\dot{q} = [\dot{q}_1\ \dot{q}_2\ \cdots\ \dot{q}_n]^T \in \mathbb{R}^n$ is the velocity vector
- $\ddot{q} = [\ddot{q}_1\ \ddot{q}_2\ \cdots\ \ddot{q}_n]^T \in \mathbb{R}^n$ is the acceleration vector

The control of such complex systems has been extensively studied in the literature. Researchers have proposed to use variable structure systems (Maalej et al. 2019; Slotine and Li 1987). Others have proposed to combine sliding mode control with intelligent approaches (Alimi and Derbel 1995; Frikha et al. 2010). Moreover, the theory of fractionary systems has been implemented in multiple applications thanks to its robustness (Yousfi et al. 2013, 2014).

Variable structure systems have known a significant enhancement since its appearance. They are based on sliding mode controllers applied to complex nonlinear systems even in case of parametric variations or external perturbations. In fact, the dynamical behavior of the system becomes insensitive to all kinds of perturbations. The sliding control approach consists of driving the system to the sliding surface and then remaining on it in order to converge to the desired trajectory. Consequently, the control contains two terms. The first one is responsible to drive the system to the sliding surface and the second one is dedicated to remain the system on it. However, it results the chattering phenomenon on the control which can damage the actuators. For that, there are several modifications to cancel this drawback.

In the sequel, we will implement the proposed control approaches on a 2-degrees of freedom exoskeleton ($n = 2$).

3.1 Sliding Mode Control with an Integral Action

Let us define the sliding function as follows:

$$s = \dot{e} + 2\lambda e + \lambda^2 \int e\,dt = \left(\frac{d}{dt} + \lambda\right)^2 \left[\int e\,dt\right] \tag{2}$$

where: $e = q - q_d$ is the tracking error and q_d is the desired trajectory of normal kids. Hence, in the case where the system remains on the sliding surface where

$$s = 0 \tag{3}$$

we obtain:

$$\dot{s} = 0 \tag{4}$$

then we obtain successively:

$$\ddot{q} - \ddot{q}_d + 2\lambda(\dot{q} - \dot{q}_d) + \lambda^2(q - q_d) = 0 \tag{5}$$

$$M^{-1}(\tau - C\dot{q} - G) - \ddot{q}_d + 2\lambda(\dot{q} - \dot{q}_d) + \lambda^2 e = 0 \tag{6}$$

The expression of the equivalent control is as follows:

$$\tau = \tau_{eq} = C\dot{q} + G + M\ddot{q}_d - 2\lambda M(\dot{q} - \dot{q}_d) - \lambda^2 Me \tag{7}$$

If $s \neq 0, \dot{s} \neq 0$, the high frequency discontinuous term, $\Delta\tau$ is added to the equivalent control:

$$\tau = \tau_{eq} + \Delta\tau \tag{8}$$

with:

$$\Delta\tau = -MK \text{ sign } (s) \tag{9}$$

with K a diagonal positive, definite matrix. Then:

$$\dot{s} = M^{-1}\Delta\tau \tag{10}$$

In order to ensure the stability of the closed-loop system, let us consider the following Lyapunov function:

$$V = \frac{1}{2}s^T s \tag{11}$$

This choice guarantees the minimization of s and the convergence to zero of the error e (remaining on the sliding surface). Then, the differential of V with respect to time:

$$\dot{V} = s^T\dot{s} = s^T M^{-1}\Delta\tau = -s^T K \text{ sign } (s) \leq 0 \tag{12}$$

3.2 Adaptive Sliding Mode Control with an Integral Action

In reality, parameters are not well-known. An adaptive approach becomes required. Hence, let us consider the vector θ of unknown parameters. Then, the control law of the sliding mode controller becomes:

$$\tau = \tau_{eq}(\widehat{\theta}) + \Delta\tau(\widehat{\theta})$$

$\widehat{\theta}$ is an estimation of θ.

The derivative of the sliding surface $\dot{s}$ is expressed as:

$$\dot{s} = \ddot{e} + 2\lambda\dot{e} + \lambda^2 e \;=\; M^{-1}(\tau - C\dot{q} - G) - \ddot{q}_d + 2\lambda(\dot{q} - \dot{q}_d) + \lambda^2 e$$

In the sequel, we can write:

$$\dot{s} = F(\theta) + G(\theta)\tau(\theta)$$

Functions F and G depend also on the state vector of the system.

The equivalent control and the increment control with well known parameters are:

$$\tau_{eq}(\theta) = -[G(\theta)]^{-1}F(\theta) \tag{13}$$

$$\Delta\tau(\theta) = -[G(\theta)]^{-1}K \text{ sign } s = [G(\theta)]^{-1}H$$

H is independent from θ.

However, with the ill-known vector θ, they become:

$$\tau_{eq}(\widehat{\theta}) = -[G(\widehat{\theta})]^{-1}F(\widehat{\theta}) \tag{14}$$

$$\Delta\tau(\widehat{\theta}) = [G(\widehat{\theta})]^{-1}H$$

Then:

$$\dot{s} = F(\theta) + G(\theta)[G(\widehat{\theta})]^{-1}\left[H - F(\widehat{\theta})\right]$$

We can write:

$$\dot{s}(\theta) = \dot{s}(\widehat{\theta}) + \sum_i \frac{\partial \dot{s}(\theta)}{\partial \theta_i}\widetilde{\theta}_i + O(\widetilde{\theta})^2$$

where: $\widetilde{\theta} = \theta - \widehat{\theta}$. Thus:

$$\dot{s} = H + \sum_i L_i\widetilde{\theta}_i + O(\widetilde{\theta})^2$$

where:

$$L_i = \frac{\partial F(\theta)}{\partial \theta_i} + \frac{\partial G(\theta)}{\partial \theta_i}[G(\widehat{\theta})]^{-1}\Big[H - F(\widehat{\theta})\Big]$$

Consider the Lyapunov function:

$$V = \frac{1}{2}s^T s + \frac{1}{2}\sum_i \frac{1}{\rho_i}\widetilde{\theta}_i^2$$

Its differential with respect to time is:

$$\begin{aligned}
\dot{V} &= s^T\dot{s} + \sum_i \frac{1}{\rho_i}\widetilde{\theta}_i\dot{\widetilde{\theta}}_i \\
&= s^T\Big[-K\ \text{sign}\ (s) + \sum_i L_i\widetilde{\theta}_i\Big] + \sum_i \frac{1}{\rho_i}\widetilde{\theta}_i\dot{\widetilde{\theta}}_i + O(\widetilde{\theta})^2 \\
&= -s^T K\ \text{sign}\ (s) + \sum_i \underbrace{\Big[s^T L_i + \frac{1}{\rho_i}\dot{\widetilde{\theta}}_i\Big]}_{=0}\widetilde{\theta}_i + O(\widetilde{\theta})^2 \\
&= -s^T K\ \text{sign}\ (s) + O(\widetilde{\theta})^2 \leq 0
\end{aligned}$$

with the adaptation law:

$$\dot{\widetilde{\theta}}_i = -\rho_i s^T L_i = -\rho_i L_i^T s \tag{15}$$

This confirms the stability of the overall system.

Assuming that θ is constant or it presents slow variations: $\dot{\theta} \simeq 0$, then:

$$\dot{\widehat{\theta}}_i = \rho_i s^T L_i = \rho_i L_i^T s$$

4 Simulation Results

In this section, simulation results of sliding mode and adaptive sliding mode controllers with integral actions are presented. In the sequel, we have proposed to use the hyperbolic tangent function instead of the sign function, in order to eliminate the chattering phenomenon:

$$\tanh(x) = \frac{e^x - e^{-x}}{e^x + e^{-x}} = \frac{2}{1 + e^{-2x}} - 1 \tag{16}$$

Figures 1 and 2 represent the evolution of positions, speeds, applied torques and the evolution of the position errors of hip and knee joints, respectively, using sliding mode controller with integral action (while using hyperbolic tangent function).

Figure 3 shows the sliding functions. It is clear that the system is on the sliding surface.

In order to test the robustness of the proposed sliding mode control, two scenarios have been established. First, in case of parametric variations (addition of 50% of the masses and lengths). Figures 4 and 5 show that the system follows the desired trajectory for the hip and the knee joints. Figure 6 presents the evolution of the sliding functions. It is obvious that the system remains to the sliding surface.

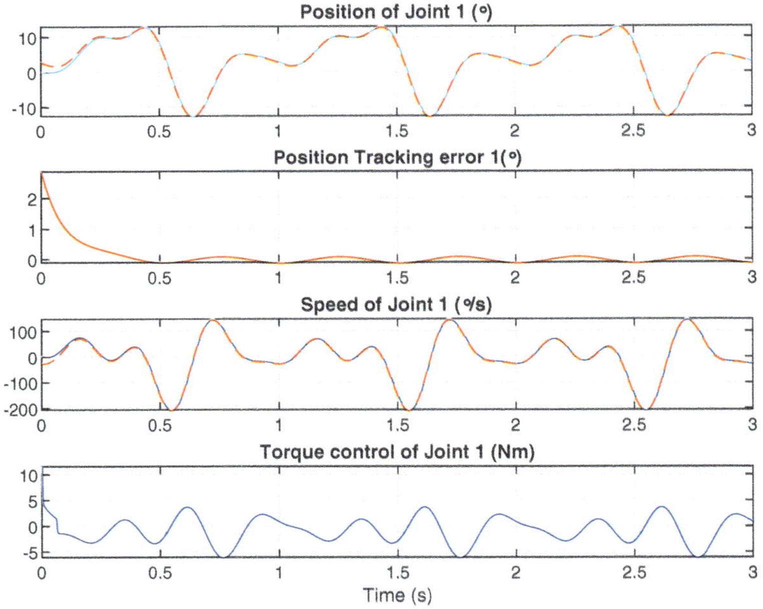

Fig. 1 Sliding mode with integral action of the hip joint (joint 1) with hyperbolic tangent function

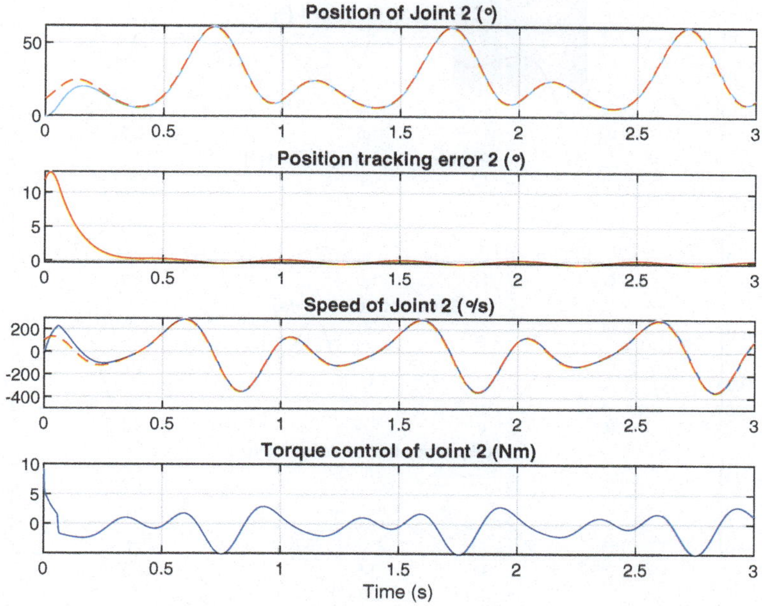

Fig. 2 Sliding mode with integral action of the knee joint (joint 2) with hyperbolic tangent function

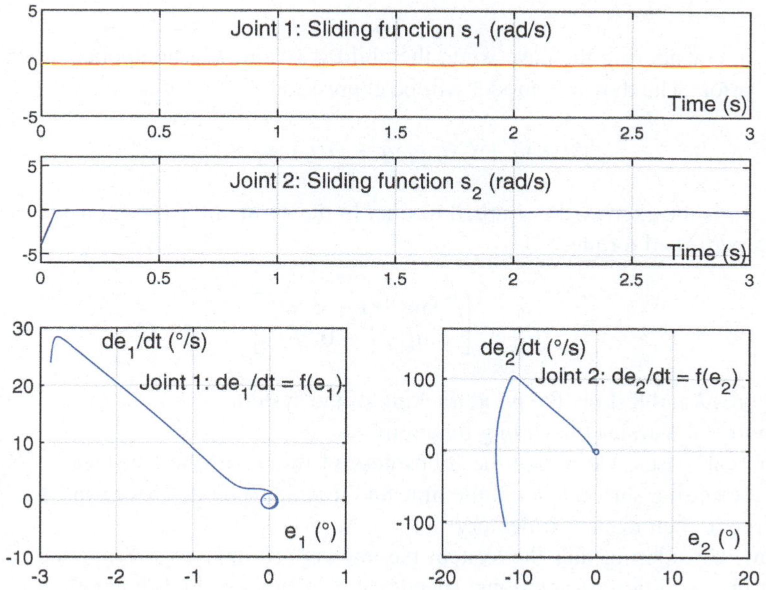

Fig. 3 Sliding functions with hyperbolic tangent function

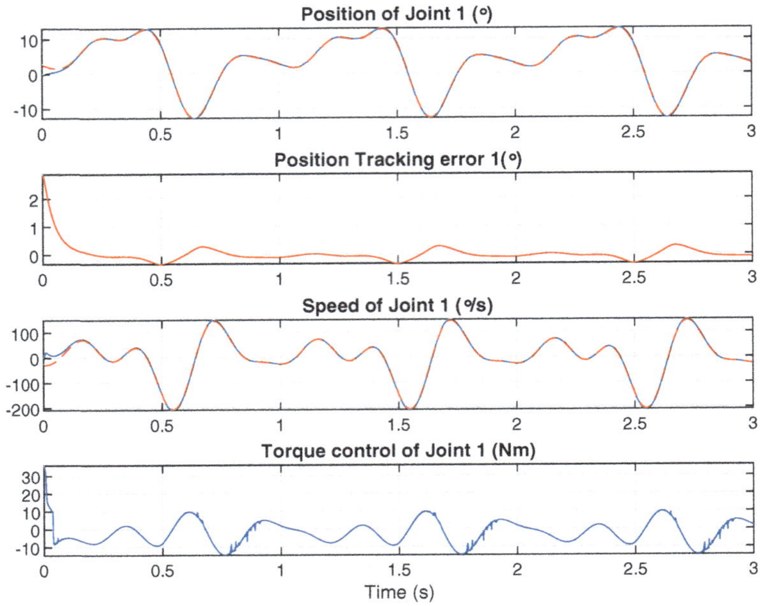

Fig. 4 Influence of 50% of masses and length uncertainties on the hip joint (joint 1) tracking with sliding mode control with integral action

Then, considering the case when the child is scared and he applies an external perturbation. The dynamic model will be expressed by:

$$M(q)\ddot{q} + C(q, \dot{q})\dot{q} + G(q) = \tau + \tau_p \tag{17}$$

where τ_p is the exercised perturbed torque. In the simulations, we consider the following perturbed torque:

$$\tau_p = \begin{bmatrix} 2\sin(8\pi t + 0.85) \\ \sin(8\pi t - 0.35) \end{bmatrix}$$

Figures 7 and 8 show the good tracking of the system.

Figure 9 illustrates the sliding functions.

For both cases, we notice the robustness of this controller. In fact, the system reach the sliding surfaces in a finite time and it remains on it. This results that joints converge to their desired trajectory.

Now, considering that the system parameters are unknown. Figures 10 and 11 depict the evolution of positions, speeds, applied torques and the evolution of the position errors of hip and knee joints, respectively, using the adaptive sliding mode controller with integral action, using the hyperbolic tangent function in the case

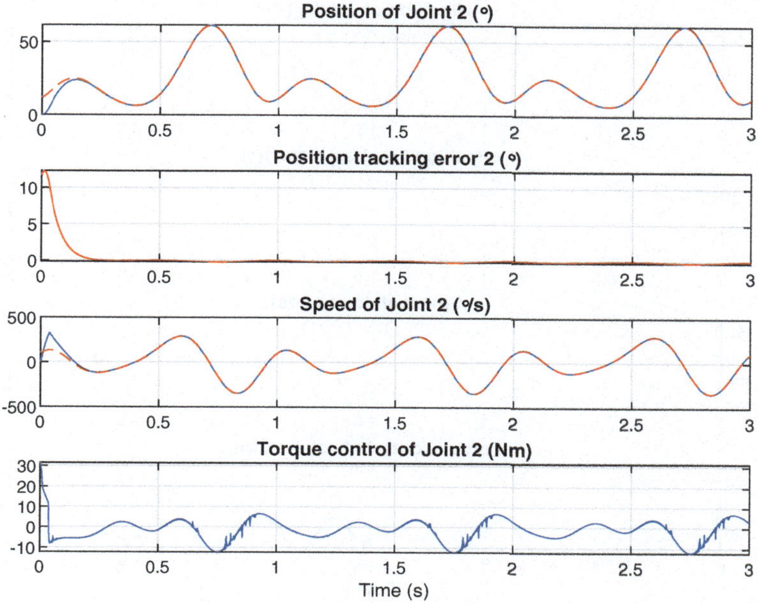

Fig. 5 Influence of 50% of masses and lengths uncertainties on the knee joint (joint 2) tracking with sliding mode control with integral action

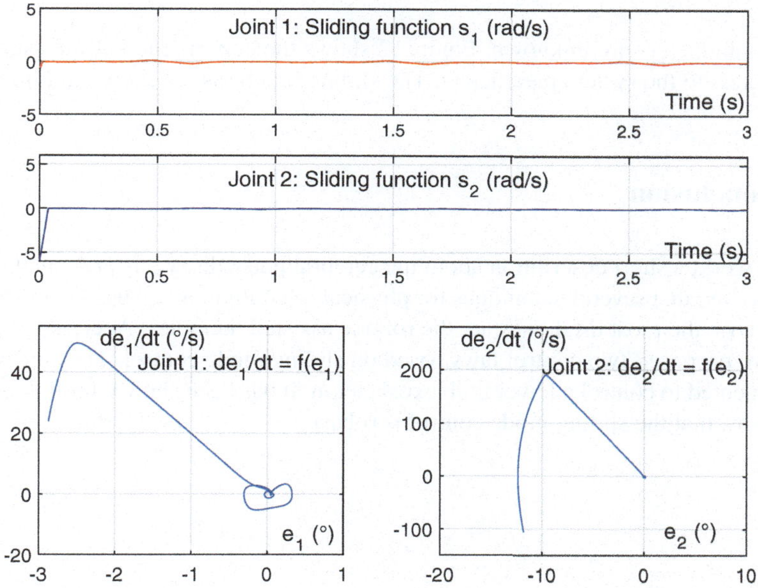

Fig. 6 Sliding functions in case of 50% of parametric variations

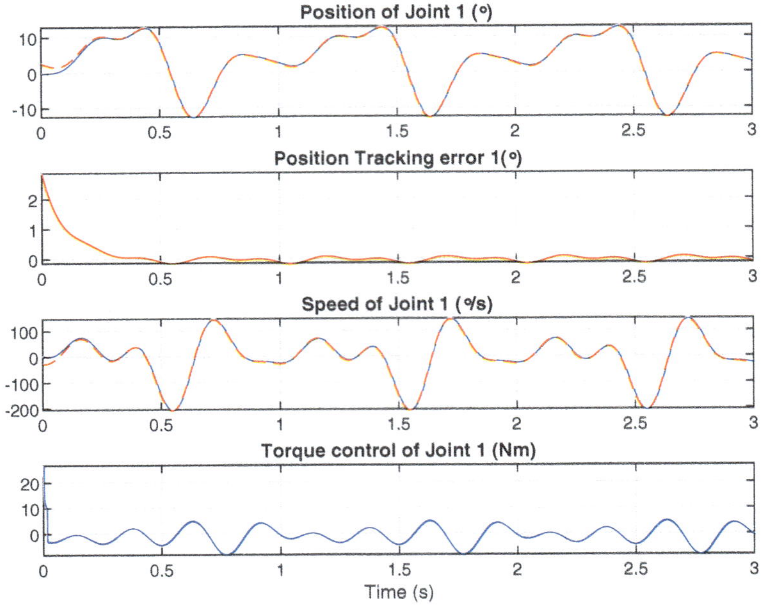

Fig. 7 Influence of an external perturbation on the hip joint (joint 1) with sliding mode control with integral action

where parameters are unknown. Figure 12 shows the convergence of the estimated parameters to the system parameters. The sliding functions are shown in Fig. 13.

5 Conclusion

In this work, a short description about the cerebral palsy disease is presented. It has been proved that several techniques for physical rehabilitation are used in the world. Nowadays, the most used ones are the robotic assisted therapies. According to this, we have proposed two control laws based on sliding mode control. They have been implemented to control a lower limb exoskeleton. It has been shown through several scenarios, that the sliding mode control is robust.

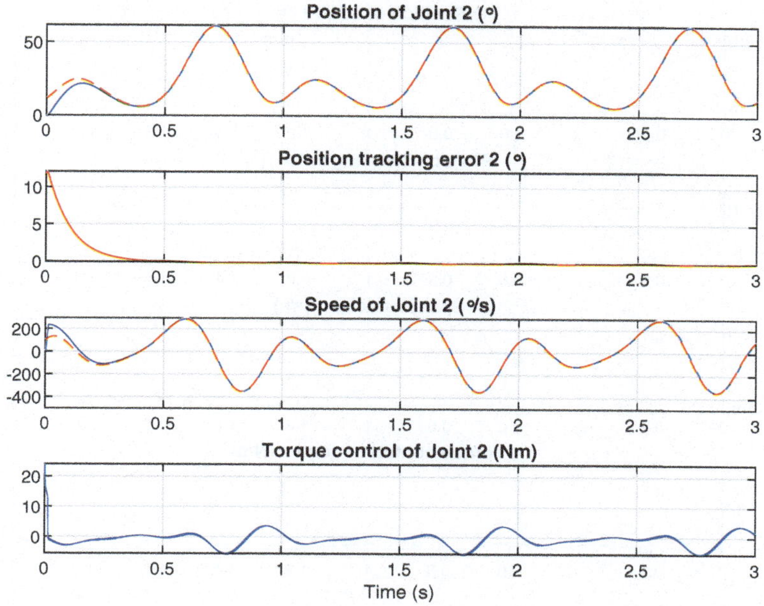

Fig. 8 Influence of an external perturbation on the knee joint (joint 2) with sliding mode control with integral action

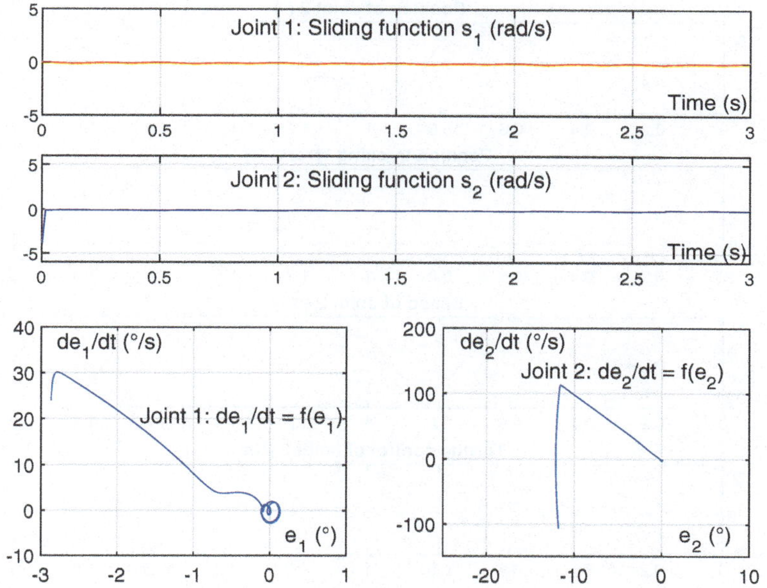

Fig. 9 Sliding functions in case of external perturbations

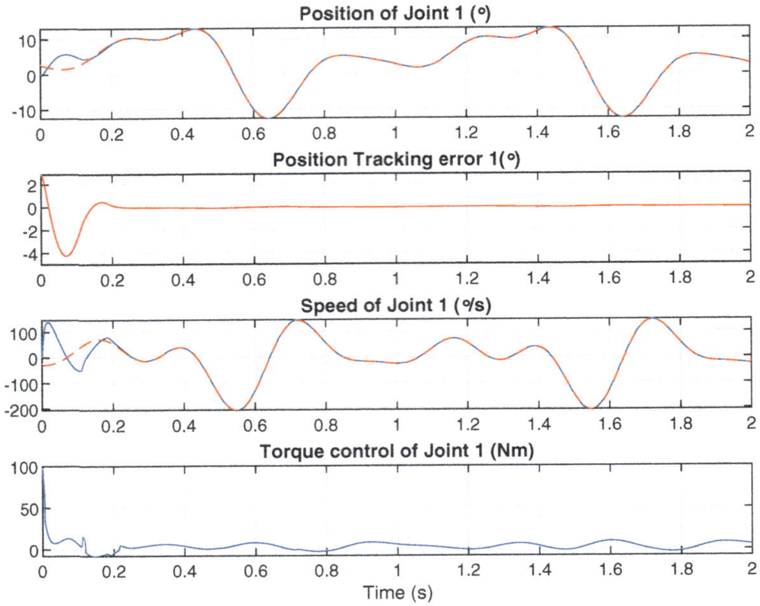

Fig. 10 Adaptive sliding mode with integral action of the hip joint (joint 1) with hyperbolic tangent function

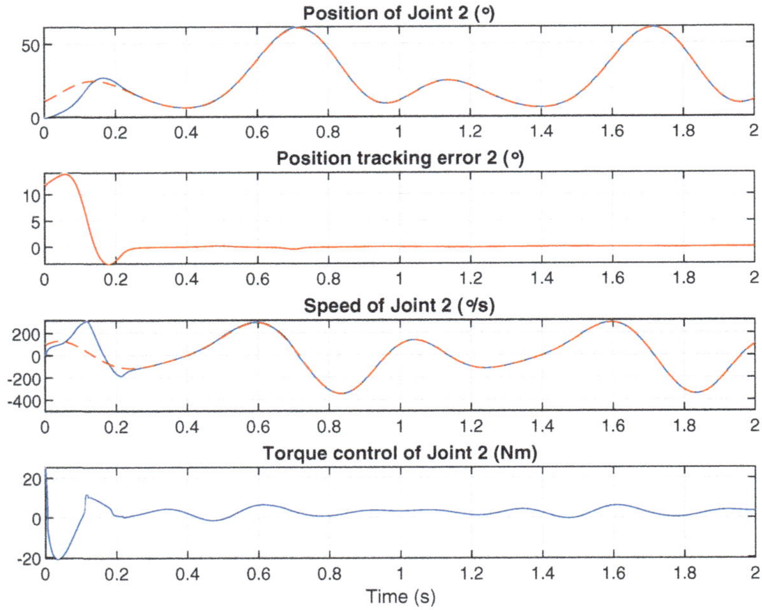

Fig. 11 Adaptive sliding mode with integral action of the knee joint (joint 2) with hyperbolic tangent function

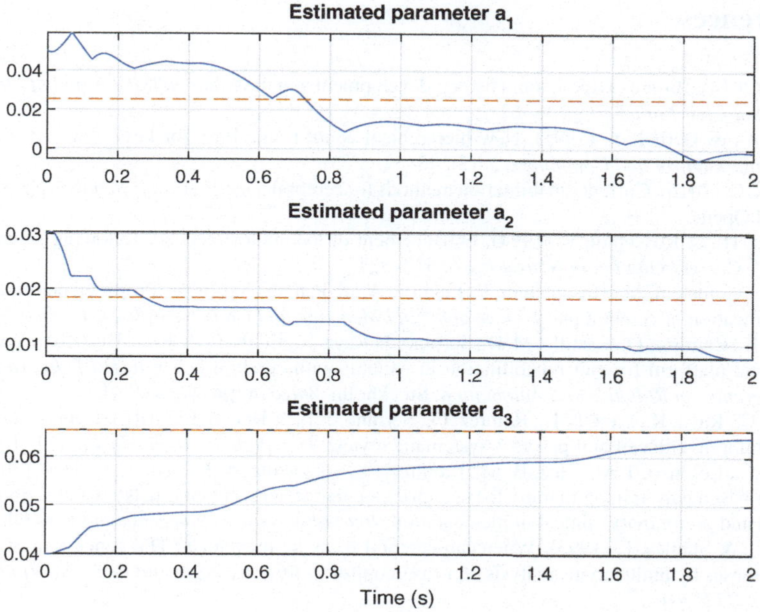

Fig. 12 Evolution of the estimated Parameters for the adaptive approach of sliding mode with integral action

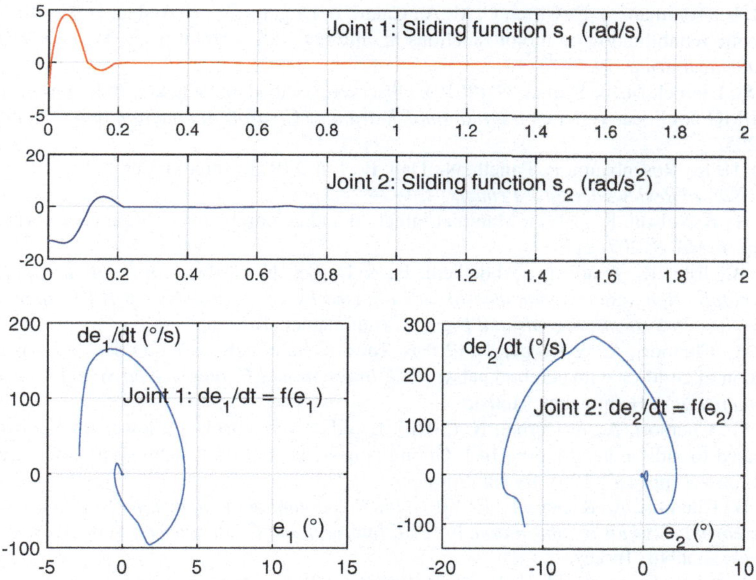

Fig. 13 Sliding functions: Adaptive sliding mode with hyperbolic tangent function and unknown parameters

References

Ali, H. (2014). Bionic exoskeleton: History, development and the future. *IOSR Journal of Mechanical and Civil Engineering*, 58–62.

Alimi, A., & Derbel, N. (1995). New hierarchical control algorithm for large scale time delay-systems. *Control and Computers*, *23*(2), 48–52.

Balci, N. C. (2016). Current rehabilitation methods for cerebral palsy. In *Cerebral palsy* (pp. 51–78). IntechOpen.

Barbeau, H., & Rossignol, S. (1994). Enhancement of locomotor recovery following spinal cord injury. *Current Opinion in Neurology, 6,* 517–524.

Bax, M., Goldstein, M., Rosenbaum, P., Leviton, A., & Paneth, N. (2005). Proposed definition and classification of cerebral palsy. *Developmental Medicine and Child Neurology, 47,* 571–576.

Bayón, C., Ramirez, O., Castillo, M. D., Serrano, J., Raya, R., Belda-Lois, J., et al. (2016). Cpwalker: Robotic platform for gait rehabilitation in patients with cerebral palsy. In *IEEE International Conference on Robotics and Automation*, Stockholm, Sweden (pp. 3736–3741).

Bayón, C., Raya, R., Lara, S. L., Ramrez, O., Serrano, J. I., & Rocon, E. (2016). Robotic therapies for children with cerebral palsy: A systematic review. *Translational Biomedicine, 7,* 1–10.

Bayón, C., Lorenzo, T. M., Saiz, B. M., Ramirez, O., Somarriba, A. P., Lara, S. L., et al. (2018). A robot-based gait training therapy for pediatric population with cerebral palsy: Goal setting, proposal and preliminary clinical implementation. *Journal of Neuroengineering and Rehabilitation.*

Blair, E., & Stanley, F. (1993). When can cerebral palsy be prevented? The generation of causal hypotheses by multivariate analysis of a case-control study. *Paediatric and Perinatal Epidemiology, 3,* 272–301.

Cuccurullo, S. (2004). *Physical medicine and rehabilitation board review.* New York: Demos Medical.

Dollar, A. M., & Herr, H. (2008). Lower extremity exoskeletons and active orthoses: Challenges and state-of-the-art. *IEEE Transactions on Robotics, 24,* 144–158.

Filiz, M. B., Toraman, N. F., Çiftçi, C. M. A., Çakır, T., Doğan, ŞK., & Arslan, H. (2018). Effects of robotic rehabilitation on motor functions in children with cerebral palsy. *Meandros Medical and Dental Journal, 19,* 211–218.

Frikha, S., Djemel, M., & Derbel, N. (2010). Observer based adaptive neuro-sliding mode control for MIMO nonlinear systems. *International Journal of Control, Automation and Systems, 8*(2), 257–265.

Graham, H. K., Rosenbaum, P., Paneth, N., Dan, B., Lin, J. P., Damino, L., et al. (2016). Cerebral palsy. *Nature Reviews-Disease Primers, 2,* 1–24.

Grether, J., & Nelson, K. (1997). Maternal infection and cerebral palsy in infants of normal birth weight. *JAMA, 3,* 207–211.

Maalej, B., Jribi, R., Ayadi, N., Abdelhedi, F., & Derbel, N. (2018). *On a robotic application for rehabilitation systems dedicated to kids affected by cerebral palsy*. In *IEEE International Conference on Systems, Signals and Devices*, Hammamet, Tunisia.

Maalej, B., Chemori, A., & Derbel, N. (2019). Towards an effective robotic device for gait rehabilitation of children with cerebral palsy. *IEEE International Conference on Signal, Control and Communication*, Hammamet, Tunisia.

Maalej, B., Chemori, A., & Derbel, N. (2020). L1 adaptive control of a lower limb exoskeleton dedicated to kids' rehabilitation. In J. Ghommam, N. Derbel, & Q. Zhu (Eds.), *New trends in robot control* (pp. 107–129). Berlin: Springer.

Maalej, B., Chemori, A., & Derbel, N. (2019). *Intelligent tuning of augmented L1 adaptive control for cerebral palsy kids rehabilitation.* In *IEEE International Conference on Systems, Signals and Devices*, Istanbul, Turkey.

Meyer-Heim, A., & van Hedel, H. J. (2013). Robot-assisted and computer-enhanced therapies for children with cerebral palsy: Current state and clinical implementation. *Seminars in Pediatric Neurology, 20,* 139–145.

Munera, M., Marroquin, A., Jimenez, L., Lara, J. S., Gomez, C., Rodriguez, S., et al. (2017). Lokomat therapy in Colombia: Current state and cognitive aspects. In *Proceedings of International Conference on Rehabilitation Robotics*.

Nelson, K., & Ellenberg, J. (1986). Antecedents of cerebral palsy. Multivariate analysis of risk. *New England Journal of Medicine, 2,* 81–86.

O'Shea, T. M., Allred, E. N., Dammann, O., Hirtz, D., Kuban, K. C. K., Paneth, N., et al. (2009). The ELGAN study of the brain and related disorders in extremely low gestational age newborns. *Early Human Development, 11*, 719–725.

Patané, F., Rossi, S., Del Sette, F. F., Taborri, J., & Cappa, P. (2017). Wake-up exoskeleton to assist children with cerebral palsy: Design and preliminary evaluation in level walking. *IEEE Transactions on Neural Systems and Rehabilitation Engineering, 25,* 906–916.

Rifai, H., Ben Abdessalem, M. S., Chemori, A., Mohammed, S., & Amirat, Y. (2016). Augmented L1 adaptive control of an actuated knee joint exoskeleton: From design to real-time experiments. In *IEEE International Conference on Robotics and Automation* (pp. 5708–5714).

Slotine, J.-J. E., & Li, W. (1987). On the adaptive control of robot manipulators. *International Journal of Robotics Research*, 49–59.

Smania, N., Bonetti, P., Gandolfi, M., Cosentino, A., Waldner, A., Hesse, S., et al. (2011). Improved gait after repetitive locomotor training in children with cerebral palsy. *American Journal of Physical Medicine & Rehabilitation, 90,* 137–149.

Stavsky, M., Mor, O., Mastrolia, S. A., Greenbaum, S., Than, N. G., & Erez, O. (2017). Cerebral palsy trends in epidemiology and recent development in prenatal mechanisms of disease, treatment, and prevention. *Frontiers in Pediatrics, 21,* 1–10.

Sumathy, G., & Renjith, A. (2014). A survey of vision and speech stimulation for cerebral palsy rehabilitation. In *International Conference on Control, Instrumentation, Communication and Computational Technologies* (pp. 1315–1319).

Yousfi, N., Melchior, P., Rekik, C., Derbel, N., & Oustaloup, A. (2013). Path tracking design based on Davidson-Cole prefilter using a centralized CRONE controller applied to multivariable systems. *Nonlinear Dynamics, 71*(4), 701–712.

Yousfi, N., Melchior, P., Lanusse, P., Derbel, N., & Oustaloup, A. (2014). Decentralizd CRONE control of nonsquare multivariable systems in path-tracking design. *Nonlinear Dynamics, 76*(1), 447–457.

Indoor Scene Simplification for Safe Navigation Using Saliency Map for the Benefit of Visually Impaired People

Marwa Chakroun, Sonda Ammar Bouhamed, Imene Khanfir Kallel, Basel Solaiman, and Houda Derbel

Abstract In computer vision and artificial intelligence fields, the construction of a performant system for scene interpretation is very important, especially for autonomous navigation applications. Currently used systems bring a lot of information that is sometimes not necessary to ensure safe navigation. In this chapter, we propose an approach that simplifies the observed scene as much as possible while ensuring autonomous navigation. Two main steps are involved: (i) highlighting salient regions in order to suppress background objects and (ii) employing the morphology operations to simplify scene interpretation. This approach is proposed for the benefit of visually impaired people (VIP) in order to navigate safely in their indoor environment only by having the necessary knowledge of the scene.

Keywords Saliency map · Scene simplification · Scene interpretation · Indoor environment · Vision system · Visually impaired people

M. Chakroun (✉) · S. A. Bouhamed · I. K. Kallel · H. Derbel
Control and Energy Managment Laboratory (CEM Lab), National School of Engineers of Sfax, University of Sfax, Sfax, Tunisia
e-mail: chakroun.marwa37@gmail.com

S. A. Bouhamed
e-mail: sonda.ammar@gmail.com

I. K. Kallel
e-mail: imen.khanfir.kallel@gmail.com

H. Derbel
e-mail: hd.derbel@gmail.com

M. Chakroun · S. A. Bouhamed · I. K. Kallel · B. Solaiman
Image and Information Processing Department (iTi), IMT Atlantique/Institut TELECOM, Technopôle Brest Iroise, Plouzané, France
e-mail: basel.solaiman@imt-atlantique.fr

O. Kanoun and N. Derbel (eds.), *Advanced Systems for Biomedical Applications*, Smart Sensors, Measurement and Instrumentation 39,
https://doi.org/10.1007/978-3-030-71221-1_10

1 Introduction

Artificial vision is a technological field for the application of computer vision to industrial needs, including process control, automatic assessment, and robot guidance (Rosati et al. 2011). Due to the increasing progress of robotics and industrial automation, machine vision research continues to develop. According to Marr's paradigm (1982), a vision system is a succession of procedures that transform information from one level to a higher level of abstraction. As illustrated in Fig. 1, Marr has organized this succession into three levels of representation: (i) The low-level aim is to detect the significant intensity changes in an image and to extract basic characteristics (regions, edges, etc.). (ii) The intermediate level describes the links between the two spaces 2D (image) and 3D (real-world). (iii) The high-level deals with the complete description of a scene. In this paradigm, Marr does not take into account the role of prior knowledge or expert knowledge for cooperation between these three levels. Thus, the extraction of the semantic content of the scene becomes very difficult.

Artificial vision is a direct result of the evolution of image processing which in turn depends on the selected technology of the vision sensor. We present below the most used state-of-the-art vision sensors to acquire indoor scenes.

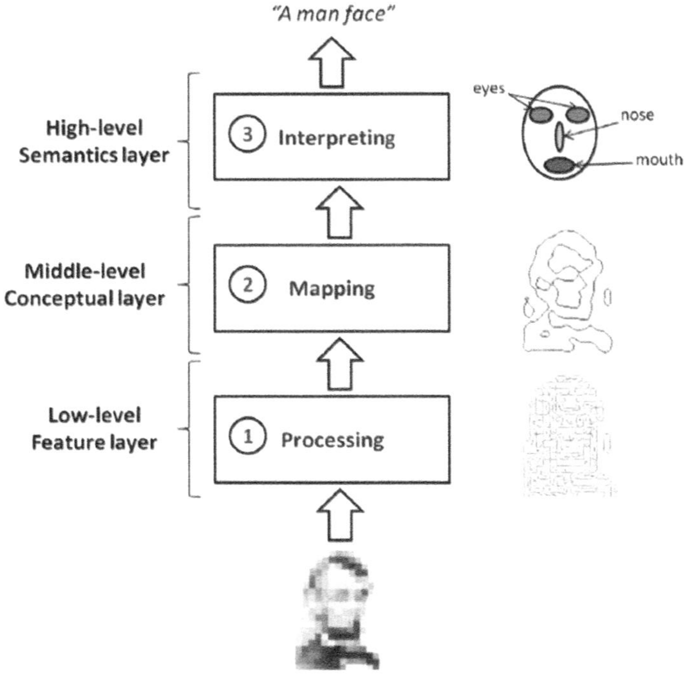

Fig. 1 Illustration of Marr's paradigm for a vision system (Marr 1982)

- **Stereovision systems:** They allow to calculate by triangulation, the depth of objects in 3D space, via their projections in plans of the two cameras. The positions' difference in the two images is called the disparity. The common problem of stereovision systems is the image processing time. Several solutions have therefore been proposed to reduce the computation time of the disparity image, in particular by employing the re-configurable systems mainly FPGAs (Lemonde 2005). Another problem that can be mentioned is self-calibration. The need to calibrate a stereo bench periodically after mounting it on a robot or a vehicle is considered one of the main problems related to stereovision. Indeed, the performance of a stereo bench depends entirely on the calibration. This is a difficult problem, especially when it must be solved in real-time. Accordingly, SolA (2007) proposed a limited calibration correction approach considering the known intrinsic parameters of the cameras.
- **Infrared Camera (IR):** Despite their low resolution and the poor contrast of provided images, IR cameras are used for vehicle navigation in difficult visibility conditions (Besbes et al. 2011; Matthies et al. 2007). In the literature, there is a lot of work based on IR cameras to detect objects in the night or smoky weather (Gond et al. 2008). IR cameras have different capacities and properties. For example, water absorbs more or less infrared waves depending on the frequency. In addition, the infrared images are not very textured, which prevents the use of traditional techniques for the detection of texture or visual cues. The cost of the infrared camera is still high which prevents their massive diffusion.
- **Monocular cameras:** They are characterized by their compactness, their low power consumption, ease of mounting which makes it easy to be replaced, and their light-weight. Since they are passive sensors, so they don't lead at any risk to the environment other than their involvement in e-waste as any other electronic component.

  The term scene refers to a subset of an environment in which multiple objects with a semantic meaning and spatial relationships connecting these objects are considered. Over the last three decades, several scene interpretation systems have been implemented in different fields of application (Hudelot 2005). These systems aim to identify and analyze this semantic content in order to extract new knowledge as humans do. The identification of the semantic content of a scene is achieved by the cooperation between the collected data and the different types of available knowledge. We propose to classify the scene interpretation systems according to the semantic levels on which the interpretation is performed. These levels are pixel-level, region-level, object-level, and scene-level. In this chapter, a particular interest is focused on both, pixel-level and region-level.

The scene interpretation process is strongly influenced by several factors:

- **The final objective:** The results of the interpretation depend strongly on the final objective aimed by the expert. The objective makes it possible to specify the semantic level of the interpretation and to adopt an appropriate strategy which focuses on the analysis of the semantic content corresponding to this level. Such a strategy makes it possible to exploit a large amount of data in a reasonable time.

- **Prior knowledge:** This knowledge, can be related to the expert, plays a key role in the interpretation process.

The remainder of this chapter is organized as follows. Section 2 summarizes some works of the state-of-the-art in the field of indoor modeling and interpretation. Sections 3 and 4 give overviews of saliency maps generation and morphological operations respectively. Section 5 presents the proposed scene interpretation process to provide VIP with safe navigation in their indoor environment. In Sect. 6, experimental results using real-indoor scenes are presented and discussed. Finally, Sect. 7 concludes the paper and proposes some perspectives.

2 Literature Review

Recent years have seen an increased request for indoor scenes modeling (Choi et al. 2015) thanks to its potential applications in planning, mapping, and navigation (Lehtola et al. 2017; Michailidis and Pajarola 2017; Pang et al. 2018; Wang et al. 2017). Lin et al. (2017) mentioned that there are actually five models for indoor modeling. The first is geometric modeling that applies exact coordinates to geometrically define the spatial structure and its elements. The second model is a grid model that presents the interior space using regular or irregular grid size. When the importance is more concerning interior path planning, the topological model comes in the third model. The semantic model is the fourth model which aims to present all different kinds of indoor objects. Due to some limitations of previous models, the hybrid model is presented to overcome these limitations. Generally, all the modeling methods mentioned above proceed one among these two approaches: Manhattan world assumption, or previous scanner knowledge. These approaches were performed to different datasets: Mobile laser scanner (MLS), Terrestrial laser scanner (TLS), RGB-Depth images, and RGB image point cloud.

Most of the available methods dealing with the point cloud can be categorized into three sets:

- **Linear primitive detection:** Current methods belonging to this set achieve good results in large-scale indoor modeling. Okorn et al. (2010) present a methodology for detecting lines using the Hough transform. However, this method is limited to Manhattan-World (MW) scenes and the output model is formed by a set of unconnected wall segments. A developed approach for modeling both MW and non-MW scenes is proposed by Oesau et al. (2014) by using cell decomposition after line fitting.
- **Planar primitive detection:** For this category, point clouds are classified into four classes: wall, floor, ceiling wall, and remaining points. Sanchez and Zakhor (2012) applies RANSAC to determine the first three classes. The result is a set of unconnected surfaces. Díaz Vilariño et al. (2016) define the first three classes by crossing planes according to their adjacency. Budroni and Boehm (2010) proposed

another planar primitive detection based on point density to identify walls in MW structures.

- **Volumetric primitive detection:** This category is generally limited to MW scenes applied in highly cluttered environments and implies less flexibility (Díaz Vilariño et al. 2014).
  Performance models based on point cloud can barely be compared since it typically relies on occlusions, the clutter-free environments or Manhattan-World structures (Becker et al. 2015; Khoshelham and Díaz Vilariño 2014) and on the topological, geometric, and semantic feature of the output model.

Simultaneous Localization and Mapping (SLAM) was frequently proposed to achieve indoor 3D modeling. Several sensors were integrated into SLAM robot to have efficient mapping topologies (Burgard and Hebert 2008), location estimation (Grisettiyz et al. 2005), feature extraction, and matching (Bay et al. 2006). Such sensors are well summarized in the following papers (Aulinas et al. 2008; Zaffar et al. 2018). Cadena et al. (2016) considered that the SLAM architecture system is formed by two contents which are the front-end and the-back end as shown in Fig. 2 of their paper.

To tackle the high processing time of SLAM algorithms, many works proposed to customize the architecture system. Bonato et al. (2007) and Nikolic et al. (2014) implement the most time-consuming part of the SLAM algorithm on FPGA (Field-Programmable Gate Array). Spampinato et al. (2011) used FPGA with 1800K system gates, 84 block RAM (18KB each) and 84 DSP 48A blocks to have a real-time SLAM algorithm. To reduce the SLAM robot cost, (Yap and Shelton 2009) used the low-cost ultrasonic sensor. Beevers and Huang (2007) used a microcontroller as the main processor and infrared rangefinders. Nevertheless, their SLAM is far from real-time since it requires a half a minute for every prediction step. As well, (Gifford et al. 2008) considered 6 IR sensors instead of using a costly rangefinder. The Complexity graph presents another problem for SLAM robots. Ismail (2019) showed an example of a complex graph in Fig. 3 of his paper. (Eade et al. 2010) developed a technique for reducing complexity graph by eliminating nodes and restraining connectivity between them. The author did not show if their proposed robot is a real-time platform. SLAM algorithms require a high processing power computer, i.e., desktop or laptop. Vincke et al. (2010) proposed an optimization architecture using several processors (DSP, RISC processor, GPU) and a co-processor for data preprocessing. Vincke et al. 2012 developed their SLAM algorithm in a multi-core embedded system exploiting SIMD architecture. Some works even use GPGPU to achieve low processing power systems.

Although all optimization methods, mentioned above, it is undesirable to develop a small robot using the SLAM algorithm without taking advantage of its high hardware/software computational power. Moreover, we can assure safe navigation to VIP by informing them only about the objects of interest presented in the scene. For that in this chapter, we opt for the generation of the saliency maps.

3 Saliency Map: Overview

The human visual system (HVS) is able to identify the most salient region in a fast and efficient manner, assuring us to easily focus our attention on these interesting regions (Cheng et al. 2015). Attention is the crucial link between perception and cognition. Hence, saliency detection has been widely used by researchers in cognitive neuroscience (Mannan et al. 2009), psychology (Wolfe and Horowitz 2004), and especially in computer vision (Cheng et al. 2013; Hou et al. 2012). Ma and Zhang (2003) mentioned that the attention's focus can be determined by two mechanisms either by stimulus (bottom-up approaches) or goal (top-down approaches). Bottom-up methods are closely linked to image characteristics such as color, edges, and gradient. These approaches are fast, unpredictable, feed-forward and involuntary in contrast to top-down methods that are slow, voluntary, task-driven and closed-loop. Salient object detection aims to imitate our HVS in finding the objects of interest in images. Hence, the extraction of salient objects can be a key to simplify the scene and help the VIP to navigate in their indoor environment by an easily understood scene interpretation.

3.1 Salient Detection Methods

Due to the lack of high-level knowledge, some works are based on what is called contrast prior to all their saliency methods (Liu et al. 2007; Yan et al. 2013). Other works (Jiang et al. 2013; Yang et al. 2013) use boundary prior. They consider that boundary regions are backgrounds to improve saliency computation. State-of-the-art results showed that such methods are more effective than contrast prior methods. However, these approaches may fail even when the object slightly touches the image border since they consider all image boundary as background. To this end, the robust background detection (RBD) proposed by Zhu et al. (2014) is used. This model is based on boundary connectivity measure that supposes that image patch as background only when the region it belongs to is deeply linked to the image boundary.

3.2 Salient Detection for the Benefit of VIP

Extracting key visual information has always been a very important issue for the VIP. Saliency detection is highly used to aware VIP of detected obstacles. Muthulakshmi and Ganesh (2012) proposed a method, based on an early model of eye fixation called Ullman and Koch's model (Koch and Ullman 1985) to track and recognize an object for VIP. Chen et al. (2011) propose an obstacle detection system from a saliency map based on visual characteristics such as color, direction, and intensity. The localization of the object on the road is realized after the thresholding of the saliency map. Wang et al. (2013) develop an application that detects from saliency maps doors and signage to support the VIP to navigate in unfamiliar indoor environments.

In our work, we will focus only on regions of interest instead of all objects presented in the scene to simplify as much as possible the scene.

4 Morphological Operations

Mathematical morphology is an extension of Minkowski's set theory (Serra 1983) and it is widely used as an image processing technique based on shape's analysis. Morphological transforms apply a structuring element to an image set, generating a new image at the same size.

4.1 Basis of Mathematical Morphology Transforms

The combination of morphological operations such as dilation, erosion, closing, and opening is useful to perform image analysis (Serra 1983; Terol-Villalobos 2001).

- **Erosion operation:** is generally applied to reduce objects in the image by enlarging the width of minimum regions and reducing peaks. Therefore, it may reduce positive noises but do just a little bit of negative ones. Eroding an image f by structuring element B is expressed by (1):

$$[\varepsilon_B(f)](x) = \bigwedge_{b \in B} f(x+b) \tag{1}$$

- **Dilation operation:** is used to stretch or shrink the input image by enlarging the width of maximum regions and increasing the valleys. Therefore, it may reduce negative impulsive noises but do just a little bit of positives ones. Dilating an image f by structuring element B is defined by (2):

$$[\delta_B(f)](x) = \bigvee_{b \in B} f(x+b) \tag{2}$$

- **Opening operation:** is simply performed by applying an erosion followed by dilation using the same structuring element for both operations. The idea behind the morphological opening is to eliminate objects that are smaller than the structuring element by applying erosion operation then recover the shape of remaining objects by dilation operation.

$$\gamma_B(f) = \delta_B[\varepsilon_B(f)] \tag{3}$$

- **Closing operation:** is an opening operation performed in reverse. In contrast to opening, it aims to recover the dilated image.

$$\phi_B(f) = \varepsilon_B[\delta_B(f)] \tag{4}$$

4.2 Opening and Closing by Reconstruction

Opening and closing by reconstruction are more useful than opening and closing operators. They permit the elimination of unwanted regions without significantly affecting the remaining objects of the image. Moreover, they help to recover structures that are partially damaged by dilation or erosion. The characteristic stems from the way the transformation is constructed through a geodetic dilation and erosion transform (Vincent and Dougherty 1994).

Geodesic dilation and erosion are based on two images: one image will be processed and defined as mark image and denoted by f and the second image will be considered as a mask image and denoted by g. Geodesic dilation and erosion of f with respect to g are expressed as follows respectively.

$$\delta_g{}^l(f) = \delta^l(f) \wedge g \tag{5}$$

$$\varepsilon_g{}^l(f) = \varepsilon^l(f) \vee g \tag{6}$$

where:
$\wedge$: means taking minimum value point by point
$\vee$: means taking maximum value point by point

5 Proposed Method

The proposed method automatically detects the object of interest and simplify indoor scene elements based on four main steps, as shown in Fig. 2.

- **Step 1:** Down-sampling and smoothing.
- **Step 2:** Saliency map generation based on RBD salient object detection model.
- **Step 3:** Saliency map simplification using morphological operations.
- **Step 4:** Region merging process in order to reduce over-segmentation problems.

In the following sub-sections, each step of the proposed method is detailed and discussed.

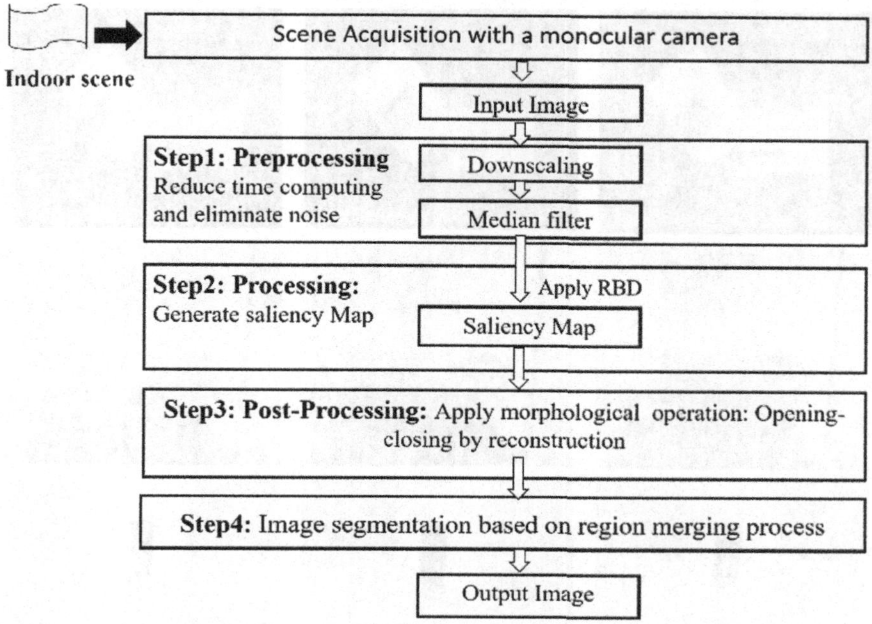

Fig. 2 Overall process of the proposed strategy

5.1 Step 1: Preprocessing: Downscaling and Noise Removing

Image pre-processing is crucial and common for any computer vision system. It aims is to improve image quality and enhance computational efficiency. In this paper, the pre-processing step is based on two sub-step namely downscaling and noise removing.

5.1.1 Downscaling

As shown in Fig. 3, the captured images are firstly downscaled to a quarter of their original size, in order to boost the computation speed of our proposed system. Therefore, the original image size that is 480×640 pixels is downscaled to 120×160 pixels. The size reduction does not affect system performances since the original size of the object of interest is widely bigger than 2×2 pixels.

5.1.2 Noise Removing

The downscaling averages a pixel group, which consequently involves a smoothing effect. Therefore, the downscaling helps to slightly reduce the image noise. In order

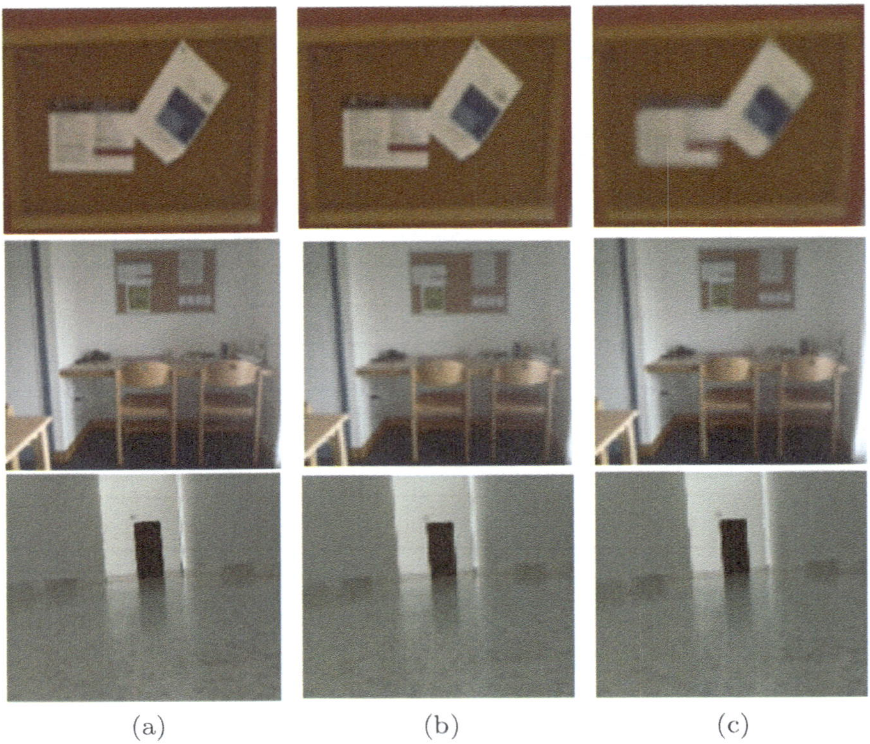

Fig. 3 Preprocessing steps **b** downscaling **c** median filtering

to increase the quality of the downscaled image and reduce the noise, we apply a median filter. This filter is chosen because it is fast and capable to produce an appropriate smoothing. Figure 3 shows the original images, the downscaling, and the median filtering results.

5.2 Step 2: Processing: Saliency Map Generation (1st Scene Simplification)

The saliency is defined as the most important region of the image. The image may have more than one salient area and, therefore, more prominent areas than others. Two kinds of models can be used to generate a saliency map namely fixation prediction model and object prediction model. Even though both saliency models types are supposed to be interchangeably applicable, their resulting saliency maps exhibit widely different characteristics because of the distinct objectives of salience detection. Figure 4 shows some examples of considered input images and their corresponding saliency map using the fixation prediction model (Fig. 4a) and the object prediction model (Fig. 4b). Obtained results, presented in Fig. 4, allow deducing that generated saliency map via fixation prediction models typically have prominent areas that look like illuminated spots while generated saliency maps via salient object detection models have smooth connected areas. In our study, we will take an interest in the use of an object prediction model since it affords more efficient saliency maps compared to those generated by a fixation prediction model. Using the object prediction model, as shown in Fig. 4 (scenes (1) and (2)), some hanging wall-objects presented in the input image are disappeared from the saliency maps. Figure 4 (scenes (3)) shows that some door details are disappeared as well. From the above-mentioned results, we can deduce that saliency map generation serves to a first simplification of the scene interpretation.

5.3 *Step 3: Post-processing: Opening-Closing by Reconstruction (2nd Scene Simplification)*

To achieve the desired result and the most accurate detection of objects of interest, the optimized saliency map has been followed with some morphological procedures. These operations are commonly a grouping of nonlinear procedures executed relatively in order to prepare pixels without altering their numeral values. As mentioned in Sect. 4, dilatation and erosion are the morphological key operators. Generally, dilation is used for rising the shapes and filling small holes and empty spaces while erosion is used for decreasing shapes and removing small protrusions. Combining dilation and erosion, two useful operators can be generated, namely opening and closing. The opening is the result of erosion followed by dilation which is generally used to remove slightly connections. The closing is the reverse process of opening and typically it is used to fill small holes.

Zhang and Sclaroff (2013) used the opening-closing by reconstruction only to enhance the contrast of the generated saliency maps. Bharath et al. (2013) used only opening by reconstruction without closing in order to find the regional maxima which will be then computed and used for segmentation.

In our work, we propose to use opening-closing by reconstruction to reduce the generated saliency map details in order to simplify the interpretation for the visually impaired people. Figure 5 shows that applying the two morphological operations namely opening-closing by reconstruction allows to: (i) smooth homogeneous region as shown in Fig. 5 (line (1) and (2)) and (ii) reduce the number of objects using the saliency map followed by opening-closing by reconstruction as shown in Fig. 5 (line (3) and (4)). From the obtained results, we can deduce that applying opening-closing by reconstruction to the saliency maps serves to a second scene simplification. On the first hand, non-fundamental information from the saliency map has vanished (e.g. the objects hanged on the table Fig. 5 (line (3), objects hanged on the wall as shown in Fig. 5 (line (4)) and on the other hand it reduces the over-segmentation problem

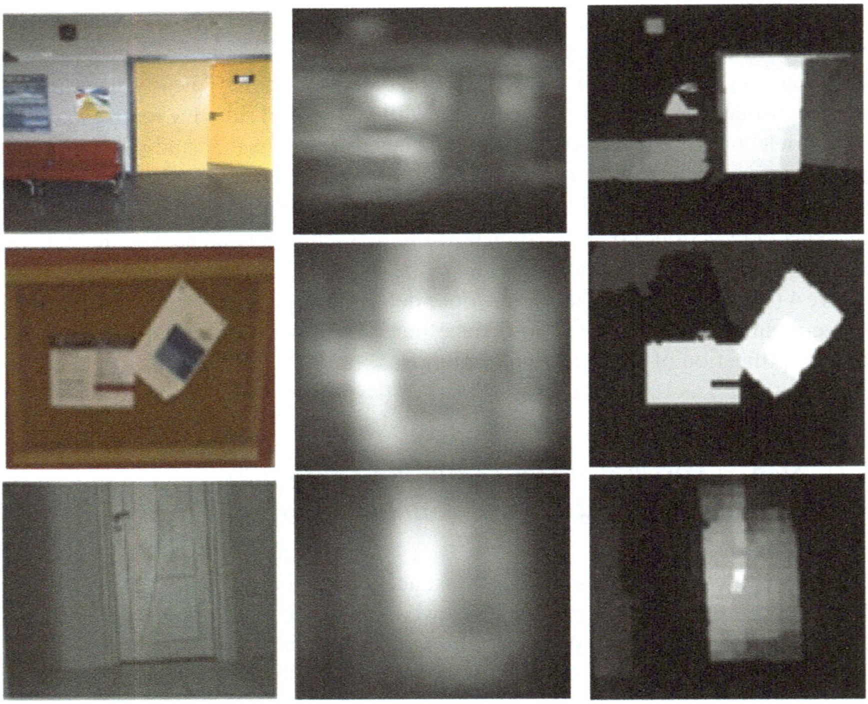

(a) (b) (c)

Fig. 4 Generation of saliency map of **a** input images by applying **b** a fixation prediction model and **c** an object prediction model

since it allows to have more homogeneous regions as shown in Fig. 5 (line (1) and (2)).

5.4 *Step 4: Image Segmentation Using Region Merging (3rd Simplification)*

The proposed strategy aims to simplify the observed scene as much as possible. Therefore, we opt to use a segmentation technique based on region merging process since it allows to yield segmentation results with reduced over-segmentation problems. Figure 6 shows some segmentation results applied to the post-processed saliency maps. From the obtained results presented in Fig. 6, we can observe how this segmentation method reduced the number of regions presented in the post-processed saliency maps by merging regions characterized by similar pixels' value. Therefore, this segmentation technique contributes as well to simplify the scene interpretation.

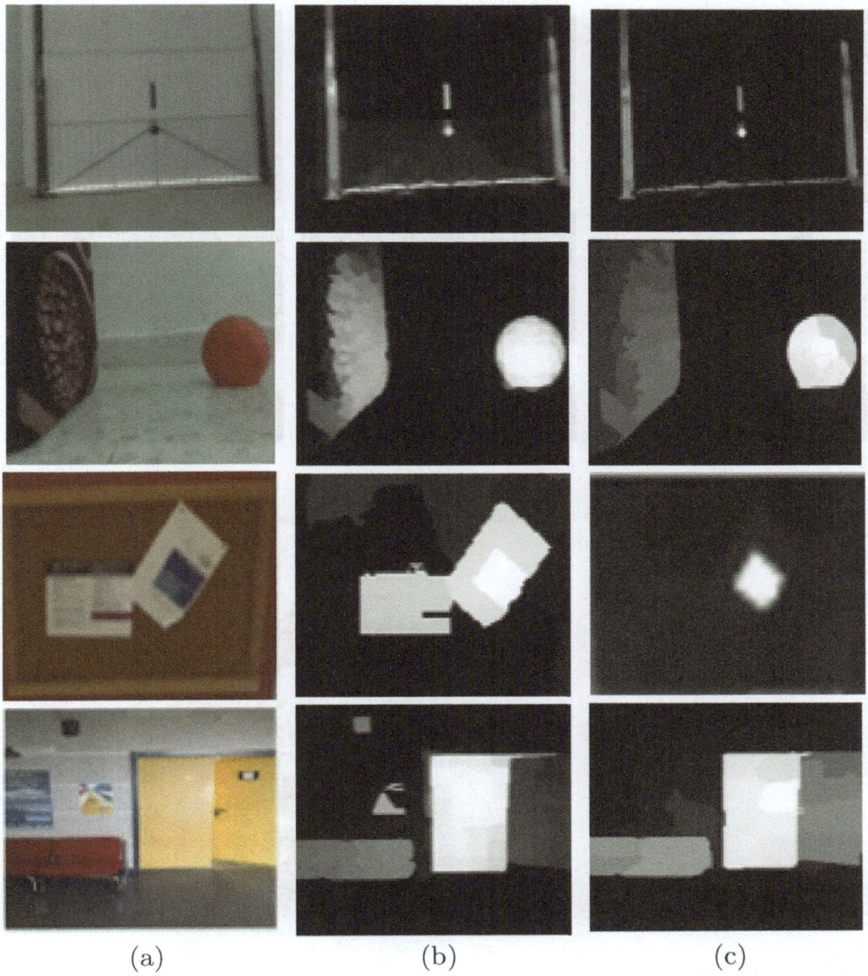

Fig. 5 Simplification details of **b** the saliency map using **c** the morphology operation opening-closing by reconstruction

6 Experiments

6.1 Dataset

To evaluate our proposed method, we built our own dataset using a monocular camera embedded in the NA-System device presented in Fig. 7, already designed in our laboratory. The constructed dataset includes 20 images corresponding to different scenes that have one or more objects located at different distances and/or orientations from

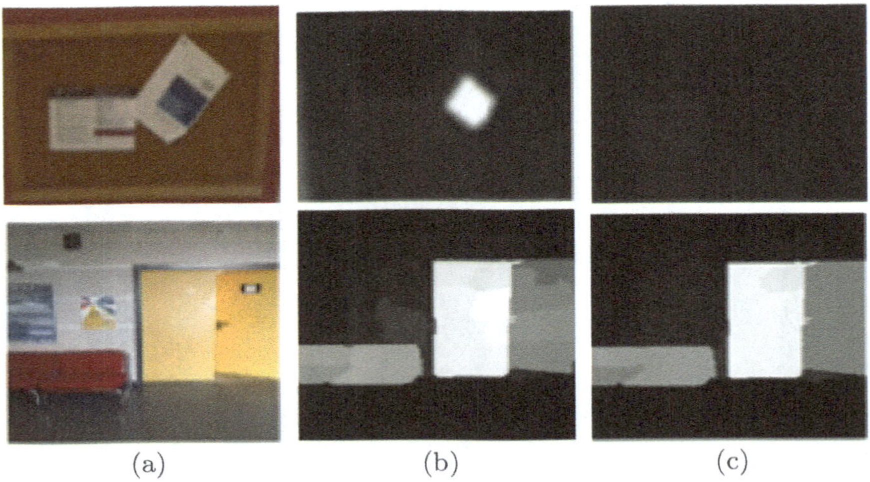

Fig. 6 Third simplification of **b** the saliency map followed by opening-closing by reconstruction by **c** merging homogeneous regions

Fig. 7 NA_System device

the device. In addition, the considered scenes involve various objects with different characteristics (e.g. size, shape, position, color, etc.). Likewise, different luminosity conditions have been taken into account. Since, our research work is related to indoor environment interpretation, our captures gather particular scene elements such as corridor (Fig. 8k), re-entrant corner (Fig. 8m), outgoing corner (Fig. 8i), wall (Fig. 8a, r), turn right (Fig. 8s), turn left (Fig. 8o), etc.

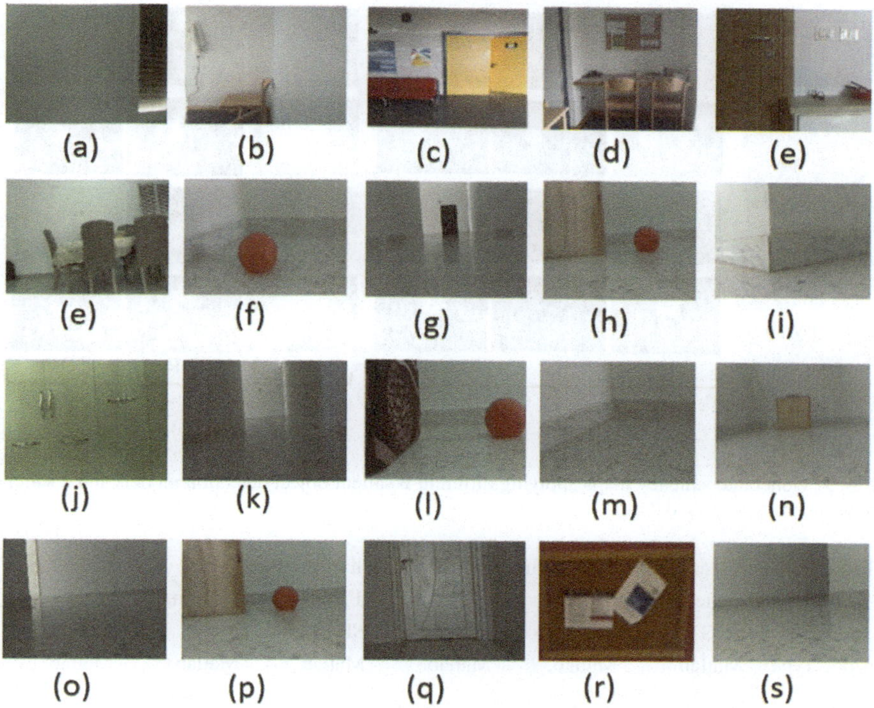

Fig. 8 Indoor scene examples

6.2 Results and Discussion

The proposed method aims to simplify as much as possible the indoor scene while ensuring autonomous navigation for visually impaired people. In our method, three scene simplifications are carried in: (i) processing step (ii) post-processing step and (iii) segmentation step.

Two kinds of saliency models are applied. For salient object detection model, we have tested SEG (Rahtu et al. 2010), AC (Achanta et al. 2009), RBD (Zhu et al. 2014), etc.). For salient fixation model, we have tested SIM (Murray et al. 2011), SR (Hou et al. 2012), COV (Erdem and Erdem 2013). The evaluation of saliency models is done according to the time processing and simplicity of their generated saliency maps.

In the following, the performances of the different models will be discussed in order to evaluate the ability of each model to present the considered scene within fewest objects of interest. Figure 9 shows the generated saliency maps applying different (a) salient object detection models and (c) salient fixation prediction models. The obtained results show that:

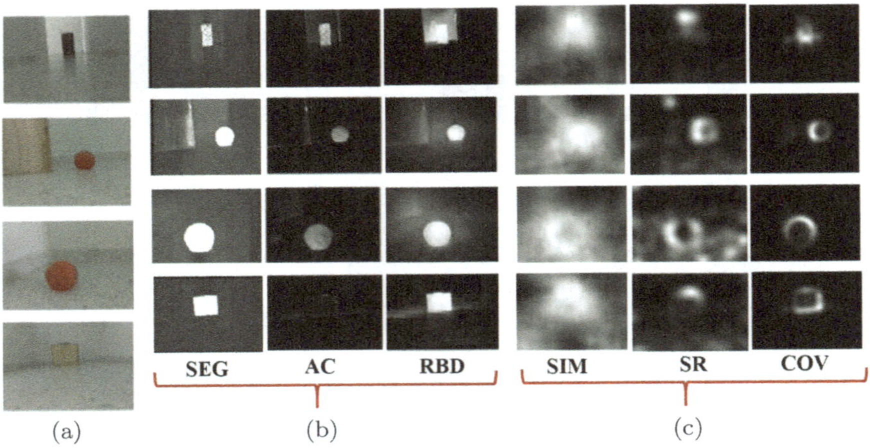

Fig. 9 Generated saliency maps applying different **b** salient object detection models and **c** salient fixation prediction models

Table 1 Comparison of processing time required for each method

	Salient object models			Fixation prediction models		
Model	SEG	AC	RBD	SIM	SR	COV
Code Type	Matlab $c++$	Matlab $c++$	Matlab	Matlab	Matlab	Matlab
Time (ms)	21.8	0.25	0.35	2.23	0.08	50.8

- Globally, salient object detection models generate more performed saliency maps than fixation prediction models.
- The RBD model outperforms the other models of object detection since it is based on an optimized framework that efficiently and effectively combines different saliency cues.
- The SIM salient fixation model is the worst model.

The processing time is as important as the saliency map accuracy. Obtained results using 20 images acquired by our NA_System, (typical image resolution: 480×640), presented in Table 1, show that:

- The SR model is the fastest model (around 0.08 ms).
- The AC model has close time processing to RBD but its saliency maps are not very informative to be processed.
- The COV model can be considered as the lowest model.

To summarize, the RBD model gives satisfactory efficient scene simplification/time processing rate compromise.

Figures 10, 11 and 12 present the output results of the proposed method applied on different real-indoor scenes. The output images have fewer details than input images

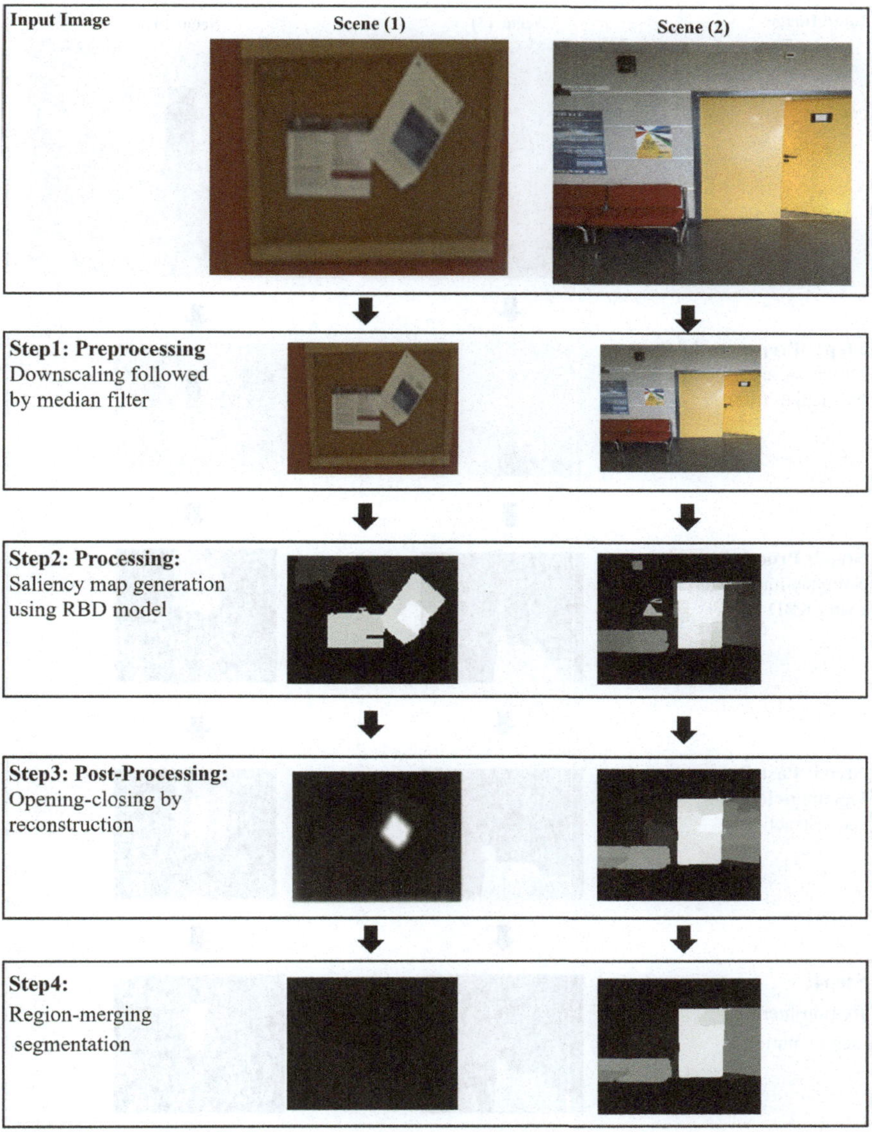

Fig. 10 Output images for scene (1), and scene (2)

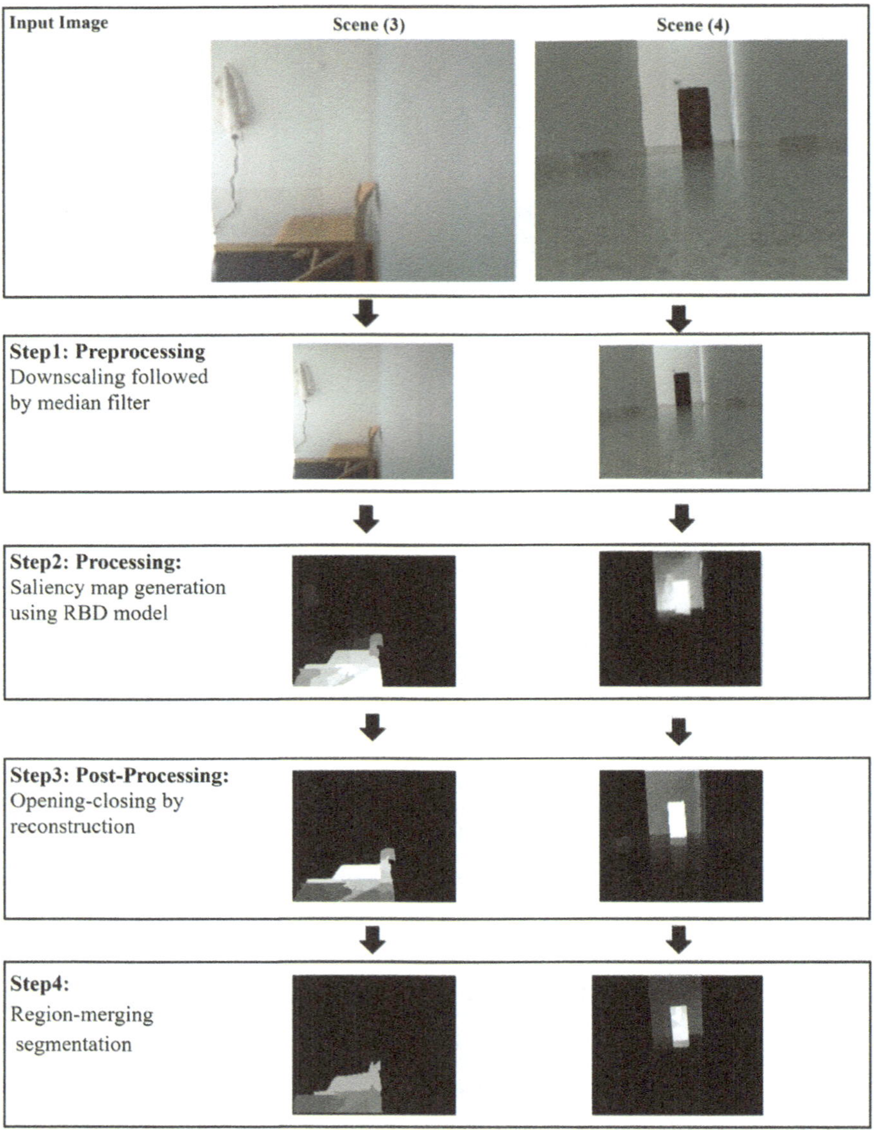

Fig. 11 Output images for scene (3), and scene (4)

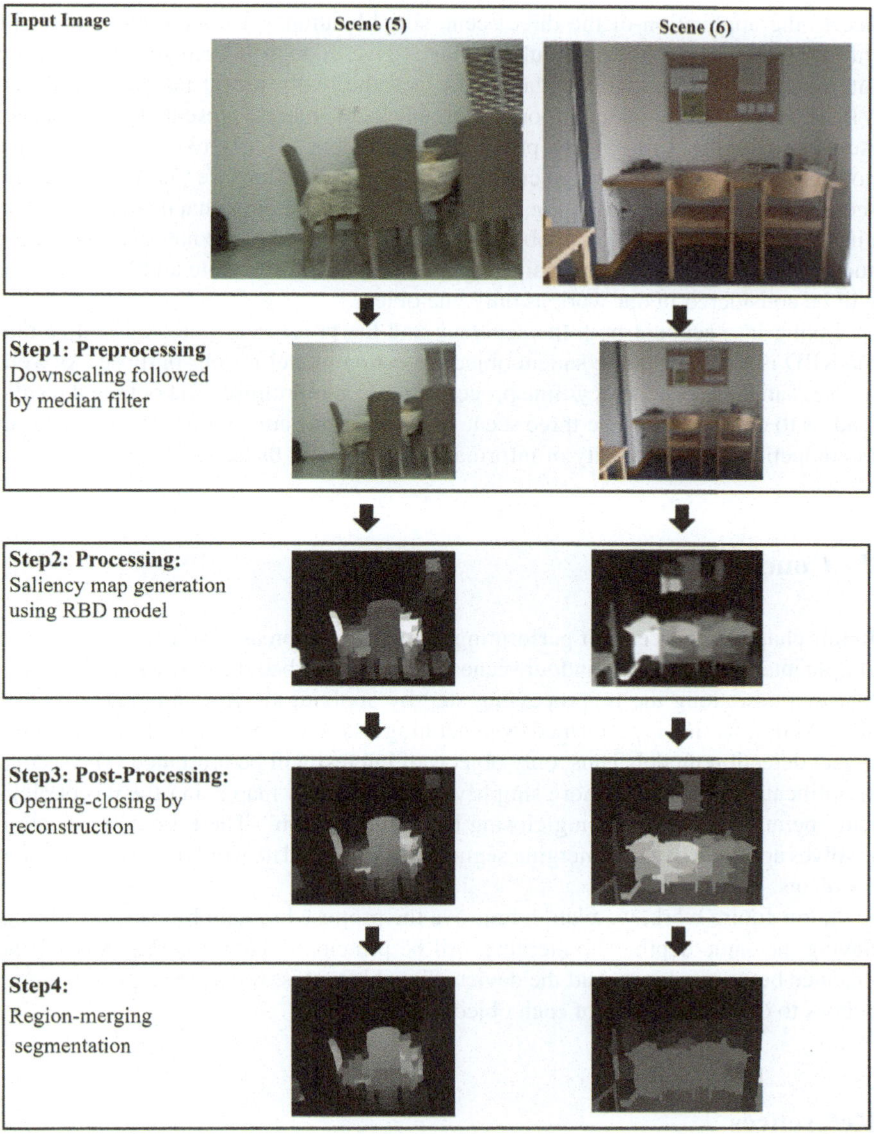

Fig. 12 Output images for scene (5), and scene (6)

due to the application of the three scene simplifications. Although the amount of information is reduced, the resulted images give enough information that can be interpreted to ensure safe navigation. In fact, the really useful information that a VIP needs to navigate is incorporated in objects of interest presented in the scene. Actually, hanging wall-objects, presented in the scene (1), (2), (3), (5), and (6), are not fundamental information to ensure safe navigation. Moreover, in the case of the scene (5) and (6) presented in Fig. 12, the essential information that the VIP has to be informed is the existence of an object in front of him and not the nature of the object such as whether it is a table, chairs, etc. In other words, the table and the chairs set will be considered in our work as only one object.

Taking into account both the accuracy and the processing time, we realize that the RBD is the appropriate salient object detection model for our purpose. As well, it gives satisfactory accuracy/time processing rate compromise. The obtained results lead us to conclude that the three scene simplifications are significant and useful in the reduction of the intensity of information involved in the scene.

7 Conclusion

In this chapter, a process for performing a safe navigation assessment of VIP with a simple interpretation of an indoor scene has been described. Firstly, we reduce time-consumption along the preprocessing step by applying downscaling. Then, a first scene simplification is performed by generating a saliency map using the RBD salient object detection model. Thus, only objects of interest will be presented. The second simplification consists of more simplifying the saliency map using the morphological operation called "opening-closing by reconstruction". The third simplification involves applying a region merging segmentation method to avoid over-segmentation problems.

In our future work, we plan to improve the proposed system by merging objects having the same depth. The merging will be principally based on the value of the distance between objects and the device. To achieve this, we will use the ultrasonic echoes to define the depth of each object.

References

Achanta, R., Hemami, S., Estrada, F., & Susstrunk, S. (2009). Frequency-tuned salient region detection. *2009 IEEE Conference on Computer Vision and Pattern Recognition* (pp. 1597–1604), Miami, FL, USA.

Aulinas, J., Petillot, Y., Salvi, J., & Lladó, X. (2008). The slam problem: A survey. In *Proceedings of the 2008 Conference on Artificial Intelligence Research and Development: Proceedings of the 11th International Conference of the Catalan Association for Artificial Intelligence* (pp. 363–371), Amsterdam, The Netherlands, The Netherlands.

Bay, H., Tuytelaars, T., & Van Gool, L. (2006). Surf: Speeded up robust features. In Leonardis, Alešand Bischof, H. & Pinz, A. (Eds.), *Computer Vision – ECCV 2006* (pp. 404–417), Berlin, Heidelberg.

Becker, S., Peter, M., & Fritsch, D. (2015). Grammar-supported 3d indoor reconstruction from point clouds for "as-built" bim. *ISPRS Annals of Photogrammetry, Remote Sensing and Spatial Information Sciences, II-3/W4*, 17–24.

Beevers, K. R., & Huang, W. H. (2007). An embedded implementation of slam with sparse sensing. In *2008 IEEE International Conference on Robotics & Automation (ICRA 2008)* (pp. 1–6).

Besbes, B., Ammar, S., Kessentini, Y., Rogozan, A., & Bensrhair, A. (2011). Evidential combination of svm road obstacle classifiers in visible and far infrared images. In *2011 IEEE Intelligent Vehicles Symposium (IV)* (pp. 1074–1079).

Bharath, R., Jian Nicholas, L. Z., & Cheng, X. (2013). Scalable scene understanding using saliency-guided object localization. In *2013 10th IEEE International Conference on Control and Automation (ICCA)* (pp. 1503–1508), China, Hangzhou.

Bonato, V., Peron, R., Wolf, D., Mascagni de Holanda, J. A., Marques, E., & Cardoso, J. (2007). An fpga implementation for a kalman filter with application to mobile robotics. In *2007 International Symposium on Industrial Embedded Systems*, Lisbon, Portugal.

Budroni, A., & Boehm, J. (2010). Automated 3d reconstruction of interiors from point clouds. *International Journal of Architectural Computing, 8*, 55–73.

Burgard, W., & Hebert, M. (2008). World modeling. *Springer Handbook of Robotics* (pp. 853–869). Berlin: Springer.

Cadena, C., Carlone, L., Carrillo, H., Latif, Y., Scaramuzza, D., Neira, J., et al. (2016). Past, present, and future of simultaneous localization and mapping: Toward the robust-perception age. *IEEE Transactions on Robotics, 32*(6), 1309–1332.

Chen, L., Guo, B.-L., & Sun, W. (2011). Obstacle detection system for visually impaired people based on stereo vision. In *2010 Fourth International Conference on Genetic and Evolutionary Computing* (pp. 723–726), China, Shenzhen.

Cheng, M., Mitra, N. J., Huang, X., Torr, P. H. S., & Hu, S. (2015). Global contrast based salient region detection. *IEEE Transactions on Pattern Analysis and Machine Intelligence, 37*(3), 569–582.

Cheng, M., Warrell, J., Lin, W., Zheng, S., Vineet, V., & Crook, N. (2013). Efficient salient region detection with soft image abstraction. In *2013 IEEE International Conference on Computer Vision* (pp. 1529–1536), Sydney, NSW, Australia.

Choi, S., Zhou, Q., & Koltun, V. (2015). Robust reconstruction of indoor scenes. In *2015 IEEE Conference on Computer Vision and Pattern Recognition (CVPR)* (pp. 5556–5565), Boston, MA, USA.

Díaz Vilariño, L., Boguslawski, P., Khoshelham, K., Lorenzo, H., & Mahdjoubi, L. (2016). Indoor navigation from point clouds: 3d modelling and obstacle detection. *ISPRS - International Archives of the Photogrammetry, Remote Sensing and Spatial Information Sciences, XLI-B4*, 275–281.

Díaz Vilariño, L., Martínez-Sánchez, J., Lagüela, S., Armesto, J., & Khoshelham, K. (2014). Door recognition in cluttered building interiors using imagery and lidar data. *ISPRS - International Archives of the Photogrammetry, Remote Sensing and Spatial Information Sciences, 40*, 203–209.

Eade, E., Fong, P., & Munich, M. E. (2010). Monocular graph slam with complexity reduction. In *2010 IEEE/RSJ International Conference on Intelligent Robots and Systems* (pp. 3017–3024), Taiwan, Taipei.

Erdem, E., & Erdem, A. (2013). Visual saliency estimation by nonlinearly integrating features using region covariances. *Journal of Vision, 13*, 1–20.

Gifford, C. M., Webb, R., Bley, J., Leung, D., Calnon, M., & Makarewicz, J., et al. (2008). Low-cost multi-robot exploration and mapping. In *2008 IEEE International Conference on Technologies for Practical Robot Applications* (pp. 74–79), Woburn, MA, USA.

Gond, L., Pham, Q. C., Begard, J., Allezard, N., & Sayd, P. (2008). Imagerie infrarouge pour la surveillance de foules. *Reconnaissance des Formes et Intelligence Artificielle* (pp. 750–758).

Grisettiyz, G., Stachniss, C., & Burgard, W. (2005). Improving grid-based slam with rao-blackwellized particle filters by adaptive proposals and selective resampling. In *Proceedings of the 2005 IEEE International Conference on Robotics and Automation* (pp. 2432–2437), Barcelona, Spain.

Hou, X., Harel, J., & Koch, C. (2012). Image signature: Highlighting sparse salient regions. *IEEE Transactions on Pattern Analysis and Machine Intelligence, 34*(1), 194–201.

Hudelot, C. (2005). *Towards a cognitive vision platform for semantic image interpretation; application to the recognition of biological organisms*. PhD thesis, université de Nice -Sophia Antipolis, France.

Ismail, I. (2019). Survey of slam in low-resourced hardware. *IJAIT (International Journal of Applied Information Technology), 3*(01), 1–9.

Jiang, H., Wang, J., Yuan, Z., Wu, Y., Zheng, N., & Li, S. (2013). Salient object detection: A discriminative regional feature integration approach. In *2013 IEEE Conference on Computer Vision and Pattern Recognition* (pp. 2083–2090), Portland, OR, USA.

Khoshelham, K., & Díaz Vilariño, L. (2014). 3d modelling of interior spaces: Learning the language of indoor architecture. *ISPRS - International Archives of the Photogrammetry, Remote Sensing and Spatial Information Sciences, 45*, 321–326.

Koch, C., & Ullman, S. (1985). Shifts in selective visual attention: Towards the underlying neural circuitry. *Human Neurbiol, 4*, 219–227.

Lehtola, V. V., Kaartinen, H., Nüchter, A., Kaijaluoto, R., Kukko, A., Litkey, P., et al. (2017). Comparison of the selected state-of-the-art 3d indoor scanning and point cloud generation methods. *Remote Sensing, 9*(08), 1–26.

Lemonde, V. (2005). *Stéréovision embarquée sur véhicule : de l'auto-calibrage à la détection d'obstacles*. PhD thesis, INSA de Toulouse, France.

Lin, Z., Xu, Z., Hu, D., Hu, Q., & Li, W. (2017). Hybrid spatial data model for indoor space: Combined topology and grid. *ISPRS International Journal of Geo-Information, 6*, 343.

Liu, T., Sun, J., Zheng, N., Tang, X., & Shum, H. (2007). Learning to detect a salient object. In *2007 IEEE Conference on Computer Vision and Pattern Recognition* (pp. 1–8), Minneapolis, MN, USA.

Ma, Y.-F., & Zhang, H.-J. (2003). Contrast-based image attention analysis by using fuzzy growing. In *Proceedings of the Eleventh ACM International Conference on Multimedia* (pp. 374–381), New York, NY, USA.

Mannan, S., Kennard, C., & Husain, M. (2009). The role of visual salience in directing eye movements in visual object agnosia. *Current Biology : CB, 19*(06), R247–8.

Marr, D. (1982). *Vision: A Computational Investigation into the Human Representation and Processing of Visual Information*. New York: Henry Holt and Co. Inc.

Matthies, L., Maimone, M., Johnson, A., Cheng, Y., Willson, R., Villalpando, C., et al. (2007). Computer vision on mars. *International Journal of Computer Vision, 75*, 67–92.

Michailidis, G.-T., & Pajarola, R. (2017). Bayesian graph-cut optimization for wall surfaces reconstruction in indoor environments. *Visual Computer, 33*(10), 1347–1355.

Murray, N., Vanrell, M., Otazu, X., & Parraga, C. A. (2011). Saliency estimation using a non-parametric low-level vision model. In *CVPR 2011* (pp. 433–440), Colorado Springs, CO, USA.

Muthulakshmi, L., & Ganesh, A. (2012). Bimodal based environmental awareness system for visually impaired people. *Procedia Engineering, 38*, 1132–1137.

Nikolic, J., Rehder, J., Burri, M., Gohl, P., Leutenegger, S., Furgale, P. T., & Siegwart, R. (2014). A synchronized visual-inertial sensor system with fpga pre-processing for accurate real-time slam. In *2014 IEEE International Conference on Robotics and Automation (ICRA)* (pp. 431–437).

Oesau, S., Lafarge, F., & Alliez, P. (2014). Indoor scene reconstruction using feature sensitive primitive extraction and graph-cut. *ISPRS Journal of Photogrammetry and Remote Sensing, 90*, 68–82.

Okorn, B., Xiong, X., and Akinci, B. (2010). Toward automated modeling of floor plans. In *Symposium on 3D Data Processing, Visualization and Transmission, Proceedings, 3DPVT* (pp. 1–8).

Pang, Y., Zhang, C., Zhou, L., Lin, B., & Lü, G. (2018). Extracting indoor space information in complex building environments. *ISPRS International Journal of Geo-Information*, *7*, 321.
Rahtu, E., Kannala, J., Salo, M., & Heikkilä, J. (2010). Segmenting salient objects from images and videos. In *Computer Vision – ECCV 2010* (pp. 366–379). Springer: Berlin.
Rosati, G., Faccio, M., Carli, A., & Rossi, A. (2011). Convenience analysis and validation of a fully flexible assembly system. In *ETFA2011* (pp. 1–8). Toulouse, France.
Sanchez, V., & Zakhor, A. (2012). Planar 3d modeling of building interiors from point cloud data. In *2012 19th IEEE International Conference on Image Processing* (pp. 1777–1780), Orlando. FL, USA.
Serra, J. (1983). *Image Analysis and Mathematical Morphology*. Orlando: Academic Press Inc.
SolÃ, J. (2007). *Towards visual localization, mapping and moving objects tracking by a mobile robot: A geometric and probabilistic approach*. PhD thesis, LAAS-CNRS, Toulouse, France.
Spampinato, G., Lidholm, J., Ahlberg, C., Ekstrand, F., Ekström, M., & Asplund, L. (2011). An embedded stereo vision module for 6d pose estimation and mapping. In *2011 IEEE/RSJ International Conference on Intelligent Robots and Systems* (pp. 1626–1631).
Terol-Villalobos, I. R. (2001). Morphological image enhancement and segmentation. *Advances in Imaging and Electron Physics*, *118*, 207–273.
Vincent, L. M., & Dougherty, E. R. (1994). Morphological segmentation for textures and particles. In *Digital Image Processing Methods*.
Vincke, B., Elouardi, A., & Lambert, A. (2010). Design and evaluation of an embedded system based slam applications. In *2010 IEEE/SICE International Symposium on System Integration* (pp. 224–229), Sendai, Japan.
Vincke, B., Elouardi, A., Lambert, A., & Merigot, A. (2012). Efficient implementation of ekf-slam on a multi-core embedded system. In *IECON 2012–38th Annual Conference on IEEE Industrial Electronics Society* (pp. 3049–3054), Montreal, QC, Canada.
Wang, R., Xie, L., & Dong, C. (2017). Modeling indoor spaces using decomposition and reconstruction of structural elements. *Photogrammetric Engineering and Remote Sensing*, *83*, 827–841.
Wang, S., Yang, X., & Tian, Y. (2013). Detecting signage and doors for blind navigation and wayfinding. *Network Modeling Analysis in Health Informatics and Bioinformatics*, *2*(2), 81–93.
Wolfe, J., & Horowitz, T. (2004). What attributes guide the deployment of visual attention and how do they do it? Nature reviews. *Neuroscience*, *5*, 495–501.
Yan, Q., Xu, L., Shi, J., & Jia, J. (2013). Hierarchical saliency detection. In *2013 IEEE Conference on Computer Vision and Pattern Recognition* (pp. 1155–1162), Portland, OR, USA.
Yang, C., Zhang, L., Lu, H., Ruan, X., & Yang, M. (2013). Saliency detection via graph-based manifold ranking. In *2013 IEEE Conference on Computer Vision and Pattern Recognition* (pp. 3166–3173). Portland, OR, USA.
Yap, T. N., & Shelton, C. R. (2009). Slam in large indoor environments with low-cost, noisy, and sparse sonars. In *2009 IEEE International Conference on Robotics and Automation* (pp. 1395–1401), Kobe, Japan.
Zaffar, M., Ehsan, S., Stolkin, R., & Maier, K. (2018). Sensors, slam and long-term autonomy: A review (pp. 285–290).
Zhang, J., & Sclaroff, S. (2013). Saliency detection: A boolean map approach. In *2013 IEEE International Conference on Computer Vision* (pp. 153–160), Sydney, NSW, Australia.
Zhu, W., Liang, S., Wei, Y., & Sun, J. (2014). Saliency optimization from robust background detection. In *2014 IEEE Conference on Computer Vision and Pattern Recognition* (pp. 2814–2821), Columbus, OH, USA.

Towards Intelligent Control of Electric Wheelchairs for Physically Challenged People

Kasim M. Al-Aubidy and Mokhles M. Abdulghani

Abstract The chapter deals with the use of soft computing techniques in solving the mobility problems of physically handicapped people using available signals such as face directional gesture, voice, brain and electromyogram (EMG) signals. These signals, depending on the type and degree of handicap, are used to classify commands required to drive a wheelchair. The user's intention is transferred to the wheelchair controller through the human-computer interface (HCI), and then the wheelchair is guided to the intended direction. Additionally, the wheelchair can perform safe and reliable motions by detecting and avoiding obstacles autonomously. Several detection methods and commands classification algorithms will be discussed. For smooth and reliable operation, an intelligent controller will be proposed to drive wheelchair motors. An adaptive Neuro-fuzzy inference system (ANFIS) technique will be used in the controller. The chapter introduces a modified method to design multiple-input, multiple-output (MIMO) ANFIS using only MATLAB. This controller relies on real data received from obstacle avoidance sensors and the HCI unit. The implemented wheelchair will be equipped with path detection sensors, GPS tracking and battery level monitoring to guaranty more safety for the user. It has been tested on 3D simulation software, and the obtained results from the wheelchair prototype and 3D simulation model demonstrated the performance of the proposed real-time controller in dealing with user requirements and working environment constraints. The cost of the proposed smart wheelchair is suitable with user case. By combining the concepts of soft computing and mechatronics, the implemented wheelchair will be more sophisticated and gives people more mobility. The obtained results show that the proposed intelligent wheelchair is feasible for the disabled and the elderly with severe mobility disabilities.

K. M. Al-Aubidy (✉) · M. M. Abdulghani
Faculty of Engineering & Technology, Philadelphia University, Amman, Jordan
e-mail: kma@philadelphia.edu.jo

O. Kanoun and N. Derbel (eds.), *Advanced Systems for Biomedical Applications*,
Smart Sensors, Measurement and Instrumentation 39,
https://doi.org/10.1007/978-3-030-71221-1_11

Keywords Smart wheelchair control · Brain-computer interface · Human-computer interface · Voice recognition · Intelligent control · ANFIS · V-REP mechatronics

1 Introduction

According to the statistics of the World Health Organization (WHO), there are currently more than 75 million people (1% of the world's population) who need a wheelchair on a daily basis (WHO 2019). Without a doubt, this number will increase as a result of wars, conflicts and accidents.People with disabilities who suffer from some permanent disabilities due to accidents, paralysis or old age cannot live a normal life (although their brains were not affected) without using wheelchairs.The people affected with any disability must deal with a new lifestyle by using wheelchairs to move around from one place to another. Most of them cannot be independent and thus find it difficult to practice their daily lives. Many systems have been developed to address this problem, but there is not an integrated system that allows the patient to move autonomously. These restrictions contributed to encouraging several researchers and specialized research centers to adopt modern technologies in computers, communications, and mechatronics systems to build smart systems that suit their needs to practice the requirements of daily life naturally.People with disabilities need comfortable wheelchairs that are compatible with their disability and can be easily controlled to improve their lives (Terashima et al. 2008). Such wheelchairs must be smart and equipped with intelligent real-time controller and a set of sensors for navigation and obstacle avoidance tasks.

The operation of a wheelchair mainly depends on the type of disability and the nature of the signals that can be obtained using special sensors such as head and eye movement, sound, muscle movement sensors and brain signals. By applying special algorithms, the generated signals from sensors are used to generate the required commands to drive a wheelchair. An embedded microcomputer uses these commands together with obstacle avoidance sensors to generate driving signals for wheelchair motors.With the technological developments in computers and microelectronics, it became possible to deal with signals obtained from sensors to generate commands to drive the wheelchair. Several research projects have been carried out in many countries of the world, such as wheelchair control using; voice (Krishnamurthy and Ghovanloo 2006), brain signals (Abdulghani and Al-Aubidy 2019; Carlson and Millán 2013), eye movement (Bigras et al. 2019), head movement (Priandani et al. 2017); and electromyography (EMG) signals (Reaz et al. 2006).Recently, human brain signals and voice commands have gained increasing importance in wheelchair systems. Therefore, the given literature review will focus on these two technologies.

Human-brain-based control of wheelchairs has attracted great attention due to advances in sensors, computers and communication technologies (Abiyev et al. 2016; Prince et al. 2015). New research has emerged by understanding different functions of the brain to serve people suffering from disabilities. An important problem that constraints the design of an interactive human-computer interface is the availability of appropriate sensors and the computer abilities to process signals to achieve the desired results at minimum time. In addition, it is bounded by the inherent abilities of humans to analyze, store, and interpret information to create behavior; and by limitations in the ability of computers to predict human intentions for mobility and communications (Prashant et al. 2015).

A Brain Computer Interface (BCI) provides electronic communication between the computer and the human brain. This interface is mainly based on brain activities during muscular movements or changes in the rhythms of brain signals (Velasco-Alvarez and Ron-Angevin 2017). BCI transforms the electroencephalography (EEG) signals produced by brain activities into commands, which can be used to generate a sequence of control signals to guide the wheelchair (Abdulghani and Al-Aubidy 2019). The history of BCI began with the discovery of human brain electrical activities and the development of brain EEG signals. The detection and processing of human brain signals to generate commands is the domain of many researchers over the past two decades, where many published papers and BCI products are available (Abdulghani and Al-Aubidy 2019; Barbosa et al. 2013; Prashant et al. 2015). A typical BCI system comprises a data acquisition system, pre-processing of the acquired signals, feature extraction process, classification of commands, and finally the control interface and device controller.

The electric wheelchair can be controlled directly by voice commands of a person with a disability. In this case, the applied voice recognition technique presents a direct human machine interaction method with whilechair. In fact, voice recognition is not an easy task for a computer, especially when applied in real-time applications. However, developments in computers and signal processing technologies have been invested in employing voice recognition in wheelchair applications. An electric wheelchair can be driven directly using a smartphone, as it has voice recognition features (Malik et al. 2017).

Incorporating fuzzy logic and artificial neural network (ANN) as intelligent tools in voice recognition and real-time control makes it very attractive for engineers to design and implement smart wheelchairs (Abdulghani and Al-Aubidy 2019; Kaur and Tanwar 2015). A fuzzy controller, embedded on a field programmable gate array (FPGA) hardware, has been used for driving motors of a wheelchair. Several obstacle avoidance sensors were equipped around the wheelchair to provide the controller with feedback signals for navigation decisions (Rojas et al. 2018). Another research work (Mazo et al. 1995) proposed a voice-controlled wheelchair system uses both PID and fuzzy controllers. The PID controller is used for position and speed control, while the fuzzy controller is used for obstacle avoidance.

In this chapter, two main issues will be considered; the human-machine interface and real-time control of a smart wheelchair. The rest of the chapter is organized as follows; the main elements of an intelligent wheelchair are given in Sect. 2. The human-machine sensing methods are discussed in Sect. 3. The command generation techniques based on voice recognition and brain signals are explained in Sects. 4 and 5 respectively. The hardware design of the embedded controller and related devices is given in Sect. 6. The ANFIS-based real-time controller design is explained in Sect. 7.

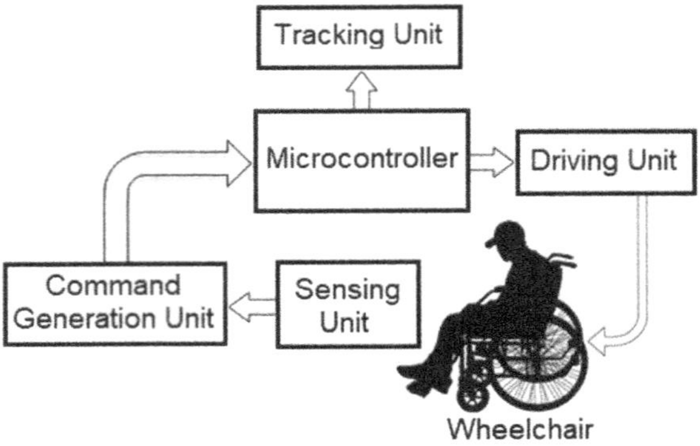

Fig. 1 Main elements of the proposed wheelchair system

A modified method has been developed to interface the MATLAB-Simulink with the open source 3D simulation engine is given in Sect. 8. Finally, experimental and simulation results are discussed in Sect. 9.

2 Wheelchair System

The main elements of the wheelchair system are illustrated in Fig. 1, where the electrical wheelchair is equipped with five units. These units are; sensing unit, human-computer interface unit, driving unit, tracking unit and an embedded microcontroller.

2.1 Electrical Wheelchair

Despite the rapid technological developments in medical devices, especially devices for persons with disabilities, there has been no significant progress in the design of wheelchairs. Studies indicate that the medical folding wheelchair was cleaned in 1933, while electric wheelchairs were developed in the early 1970s (Wallam and Asif 2011).

In this study, an electrical wheelchair prototype with two geared DC motors is considered. The motor actuating unit has a gear ratio of 1:48 and an electronic drive unit type (LD298). The wheelchair is equipped with special sensors to aid the user in avoiding static obstacles such as walls, and dynamic obstacles like people and other wheelchairs.

2.2 Sensing Unit

Two types of sensing units are considered; the first detects the movement that the disabled person means to guide the wheelchair. The second sensing unit represents set of sensors to detect any obstacle in the movement path.

- Human command sensors: Special sensors are used to detect the operation required by the disabled person. These sensors generate analog signals, which are translated by the command generation unit and the microcontroller to control the wheelchair.
- Safety sensors: Six ultrasonic sensors (type HC-SR04) are mounted on the wheelchair body to detect any obstacle and to give the movement more safety. Two sensors were mounted on the front, two on the back, and one for each side of the wheelchair, as shown in Fig. 2. These sensors provide 2 cm to a 400 cm non-contact measurement function and cover a good range accuracy (about 2 mm) with stable readings.

2.3 Human-Computer Interface (HCI)

A smart wheelchair requires user-friendly human-computer interface unit. The main function of this unit is to transfer the user's intention into commands for wheelchair driving. The computing element of the HCI is used to convert signals generated from human sensing unit into driving commands; stop, forward, back, right, and left. These commands are serially connected with the main microcontroller through a Bluetooth module (type HC-06). Two methods of HCI design will be discussed in this study; the first using voice signals, and the second using brain signals.

For the purpose of verifying the wheelchair driving orders, the disabled person is required to confirm the required command again within a period of time (30 s), otherwise the command will be ignored. For further assistance, a verification screen can be added to show the current commands or direction performed on the wheelchair.

2.4 Control Unit

A single-chip microcontroller has been used to implement all tasks of real-time control according to the generated command from the HCI and signals received from safety sensors. A single chip microcontroller type (ARDUINO MEGA-2560) has been chosen, since it has enough input/output pins andmemory size to implement different tasks of the proposed control and tracking operations. The main microcontroller is connected directly with motors driving unit and tracking unit, while it is connected serially with HCI unit. The navigation and steering of the wheelchair has been controlled using an Adaptive Neuro-Fuzzy Inference System (ANFIS) implemented by the main microcontroller.

2.5 Tracking and Safety Unit

The wheelchair prototype system has been equipped with a number of sensors to obtain additional tracking and safety features that enable the disabled person comfortably and safely use (Hamadi et al. 2020). These includes wheelchair;

- Tracking through an emergency message sent to the owner containing the current location of the wheelchair.
- Battery level monitoring and alarming,
- and wheelchair level monitoring to regulate the speed of the wheelchair while moving down or up on tilted ground, and also to prevent falls in the holes.

3 Sensing Methods

Electric wheelchair users suffer from weakness that may affect their ability to move safely, so there is an urgent need to design a smart wheelchair system that fits the type and degree of disability. The designer needs to know the disabled person's ability to generate wheelchair control signals. There are situations that do not allow a person with a disability to use body parts, and in other cases people with disabilities cannot use their voice.Therefore, advances in sensing techniques and signal processing algorithms have been adopted to develop suitable detection method suitable for each disability. Different sensing methods are used in wheelchair systems including; joystick, head movement, voice recognition, brain signals, muscle signals, and others.By using the appropriate sensing method, embedded microcontroller, and intelligent control, traditional wheelchair systems can be replaced by competitive developed wheelchair systems.

3.1 Finger Movement Detection

Special sensors are used to track the movement of the disabled person's fingers. Most of these sensors generate analog signals that are translated by a microcomputer for the smooth and accurate control of a wheelchair.

3.1.1 Using Flex Sensors

The wheelchair can be controlled using a group of fingers. These fingers allow control of the wheelchair, where one finger controls speed while another finger controls the direction of movement. Flexible sensors change their resistance when bent, and used to track finger movement. These sensors can be attached to the finger position on the gloves as shown in Fig. 2.

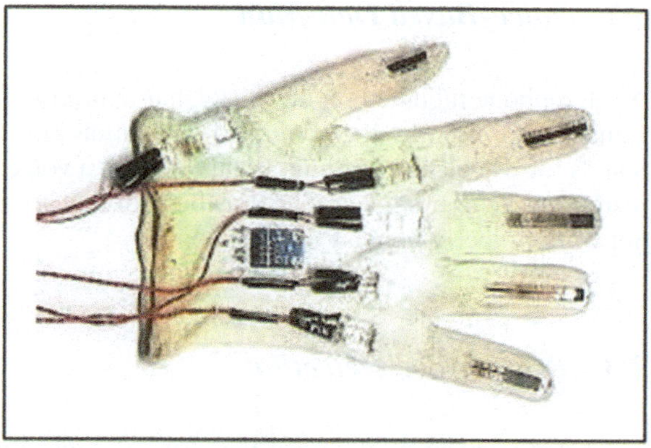

Fig. 2 Flex sensors attached to a hand glove

Fig. 3 Joystick for wheelchair guiding

3.1.2 Using a Joystick

The joystick can be used by disabled people to control wheelchair movement, but there are others that are unable to use it efficiently (Fig. 3). In the joystick-based wheelchair control, the joystick tool is considered as the main control interface for the wheelchairs guiding. It has the advantage that the detailed movement commands, such as direction and speed control, are used friendly. However, it has drawbacks as it requires complex wrist movement that may become difficult for some disable people and the elderly, and in some cases it may even lead to accidents (Yokota et al. 2010).

3.2 Voice-Based Detection

A microphone mounted on the wheelchair converts the voice signals into electrical signals to be used by the command generation unit to generate the required commands for wheelchair guiding. In this system, a voice recognition module is used to recognize the user's voice. The drawback of this sensing is that the wheelchair will not work properly in noisy environment.

3.3 Brain Wave Detection

Several studies have addressed the issue of controlling wheelchairs by brainwave without the need for any physical responses from the user. The best way to record brain activity is by electroencephalogram (EEG). In this case, a brain computer interface (BCI) is required to enable direct communication between the brain and the wheelchair controller. This interface includes a brain headset to capture the EEG signal. Then, by applying classification algorithms, the EEG signals are processed and converted into a mental commands.

3.4 Muscle Wave Detection

Electromyography (EMG) is an electro diagnostic technique for evaluating and recording the electrical activity produced by skeletal muscles (Sairam et al. 2018). EMG detects the electric potential generated by muscle cells when these cells are electrically or neurologically activated. The signals can be analyzed to detect activation level or to analyze the biomechanics of human or animal movement. These encourage researchers to use EMG as an input interface for electric wheelchair system. In this case, EMG signals recorded from muscle.

3.5 Head Motion Controller

The idea is to design a wheelchair system based on head movements to help disabled persons that cannot move any of the body parts, except of the head. Such a system requires head motion sensors and recognition techniques. Normally, a head-operated wheelchair employs tilt sensors placed in the headset to determine head position. A novel head motion recognition technique based on Gyroscope sensor (Priandani et al. 2017) or accelerometer attached to the patient head (Pajkanović and Dokić 2013) can be used for head movement detection. Furthermore, eye ball can be detected by using a head-mounted camera focused on the person eye and images are sent to the embedded microcontroller for further processing to generate the required commands for wheelchair driving.

4 Voice Computer Interface

In the early 1960s, a number of studies addressed the subject of speech recognition, but the rapid advances in computer technology and digital signal processing contributed effectively to improving speech recognition capabilities. Speech can be a useful interface for real-time interacting with devices such as wheelchairs. However, this still faces a lot of problems, due to the difference that occurred in the speaker because of age, gender, speed of signal, different pronunciation, surroundings noise etc. (Gevaert et al. 2010; Malik et al. 2017). Recently, there has been a huge growth in voice control applications especially after the launch of smartphones, where sophisticated hardware and software products have been developed for voice recognition and applications.Voice recognition approach is mainly classified into;

- speaker dependent approach; is based on training the person,
- and speaker independent approach: is based on training the system to respond to a word regardless of who speaks.

In fact, the first approach has high accuracy for word recognition; therefore, it is recommended for voice-based wheelchair control.The general layout of the voice controlled wheelchair system is given in Fig. 4. It contains two primary components; voice-based command generation unit and real-time controller. The command generation unit consists of three stages; signal analysis, feature extraction, and pattern recognition.

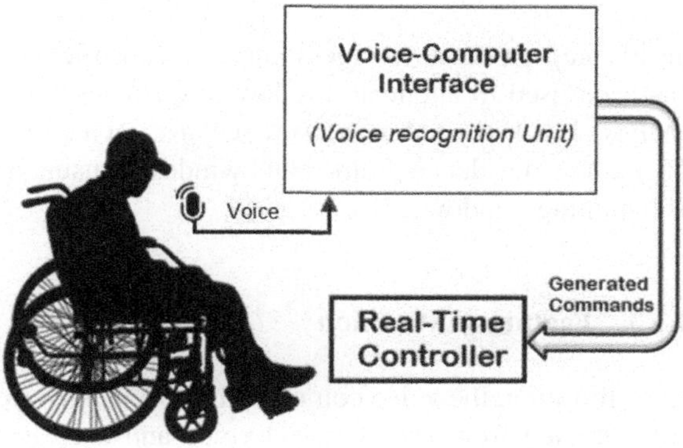

Fig. 4 Elements of the proposed system

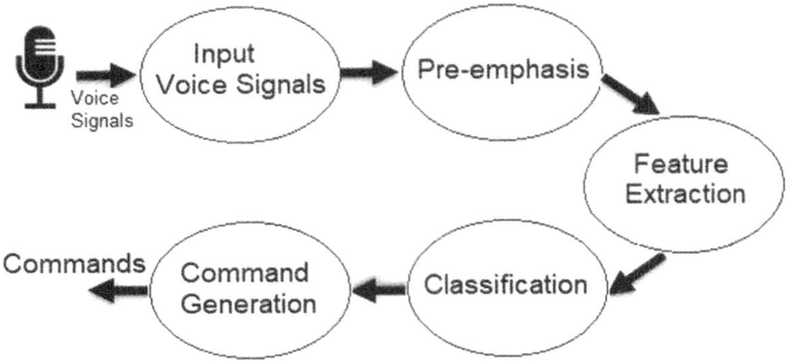

Fig. 5 Voice recognition process

4.1 Voice Recognition Process

Voice recognition is mainly done in two stages named as training and testing. But before this, some basic procedures are necessary applied to speech signals. The basic process of speech recognition is shown in Fig. 5.

4.1.1 Input Voice Signals

The different voice signals come from a microphone are preprocessed using suitable techniques like filtering. The regarding useful features are extracted to distinguish between different signals (Rani et al. 2015).

4.1.2 Pre-processing

In this step an equal loudness curve is constructed. Finite impulse response filter has been used to attenuate the low frequencies and reduce the noise. The overlap analysis block is used to convert scalar samples to a frame output at a lower rate. Then, the voice data is framed and windowed using available window function such as hamming window.

4.1.3 Features Extraction

After filtration, the voice command has been converted to a feature vector. Features are extracted from preprocessed voice and used to represent the voice signal. In general, the voice commands have been divided to 80 frequency frames for each action; Stop, Forward, Back, Right, and Left. Seven statistic features (Mean, Median,

Minimum, Mode, Peak-to-Peak, RMS, and Standard Deviation) have been calculated for each voice command and the result was a 1×80 matrix for each feature.

4.1.4 Classification

In this step, two data sets (one for training and the other for validation and testing) are chosen based on above statistics features. The dimension of the training and testing input matrices is of (7×400) each. While the target data is a matrix of (5×400) dimension.

4.1.5 Command Generation

The final step is dedicated to convert the trained and classified sound commands to control commands. In this study an adaptive neuro-fuzzy inference system model was evaluated to predict five control commands yield on the basis of inputs.

4.2 Neural Network-Based Classification

The classification has been done using neural network tool on MATLAB version R20116a workspace. The implemented neural network topology was of (7-25-10-5). It has 7-node linear input layer, two sigmoidal nonlinear hidden layers of 25, 10 units respectively, and 5-node linear output layer, as shown in Fig. 6.

The developed ANN was a multilayer perceptron (MLP) with seven neurons in the input layer, two hidden layers with 25 and 10 neurons respectively, and five neurons in the output layer. For the best ANN model, error-back propagation learning algorithm has been applied with learning rate of 0.05 and stopping criterion of mean error square less than 0.005. As illustrated in Fig. 7, after 197 iterations the neural network has been learnt effectively (Abdulghani et al. 2020).

The tested data is used to confirm learning with an output of the required action signal is perfectly achieved, as shown in Tables 1 and 2 as a sample. The seven selected features of each voice commands given in Table 1 are used for network training to recognize each command. While, Table 2 shows the five required outputs

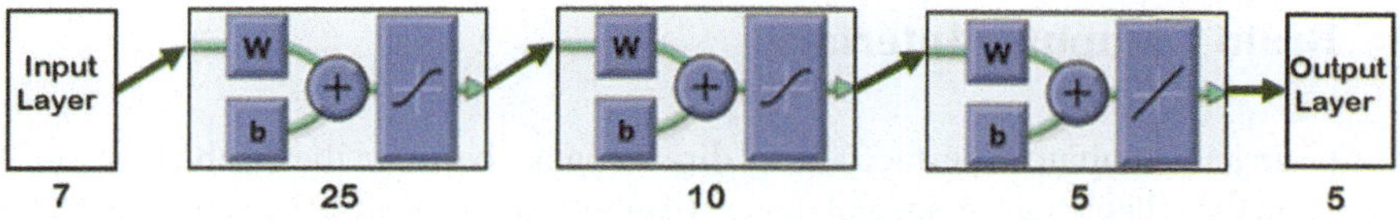

Fig. 6 Topology of the implemented neural network

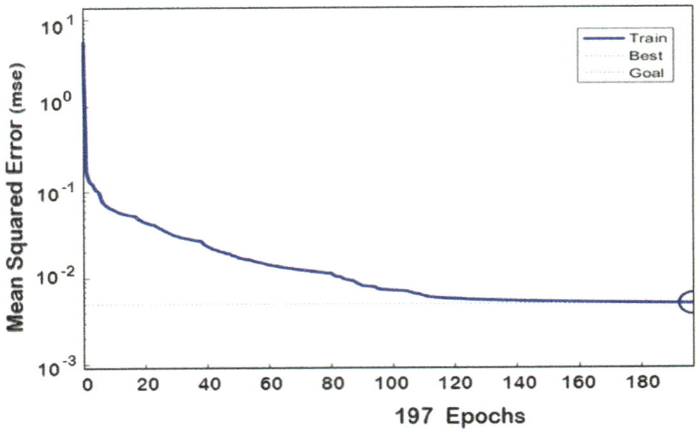

Fig. 7 Performance of neural networks

Table 1 Testing input pattern

Features	Stop	Forward	Back	Right	Left
Mean	−0.008	−0.0075	0.0056	−0.0006	0.0048
Median	−0.002	0.00015	−0.00006	0.00018	0
Minimum	−0.21	−0.3176	−0.05	−0.1021	−0.0313
Mode	0	−0.0002	−0.00006	0	0
Peak to peak	0.4356	0.4716	0.2078	0.2809	0.286
RMS	0.0468	0.0615	0.0284	0.024	0.03
STD	0.0463	0.0614	0.028	0.024	0.0302

for each voice commands which will be implemented by the main microcontroller. These five outputs represent:

- Single output to activate a timer counter as a safety feature to indicate if the rider is in sleep situation or not.
- Four outputs for the drive unit (two for each motor) to control the rotation direction of the wheels.

5 Brain Computer Interface

The brain and computer interface allows direct contact between the disabled person's brain and the wheelchair. A special device (Emotiv EPOC headset) has been used for capturing the EEG signal. This headset is able to transmit the EEG signal wirelessly via Bluetooth to the PC (personal computer). The EEG headset, together with its

Table 2 Target and actual NNs output for given input pattern

Output pattern		
Action	Target	Actual NNs output
Stop	1	1.012
	0	−0.002
	0	0.018
	0	−0.025
	0	−0.002
Forward	0	−0.0008
	1	1.0014
	0	−0.0005
	0	−0.00002
	0	−0.00007
Back	0	0.0058
	0	−0.0002
	1	1.0262
	0	−0.004
	0	−0.0283
Right	0	−0.0117
	0	0.0001
	0	0.0308
	1	1.0313
	0	−0.0507
Left	0	−0.01
	0	0.0002
	0	0.018
	0	0.0014
	1	0.99

software, represents the BCI module to process detected signals and convert them to commands (Stop, Forward, Back, Right and Left). Figure 8 shows the general layout of a brain-controlled wheelchair, where the BCI consists of three parts;

- Signal acquisition: reading EEG signals from the user brain and converting into the digital representations.
- Feature extraction: extracting features of the brain activities and translate them to commands.
- Training algorithm: training the brain activities and finally outputs their commands.

The selected headset comes with a set of software to classify and train the brain EEG signals as follows.

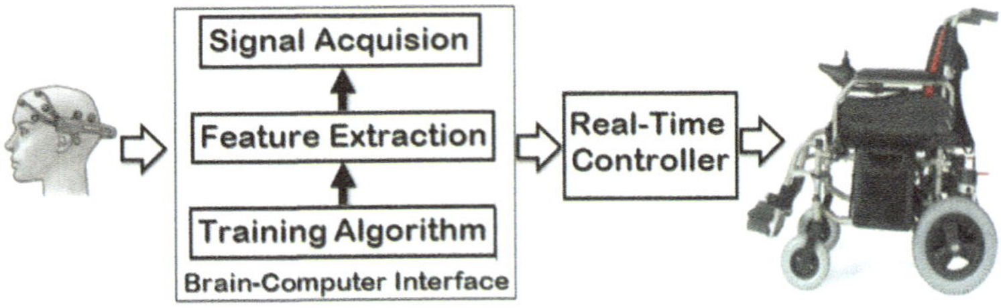

Fig. 8 Brain-controlled wheelchair system

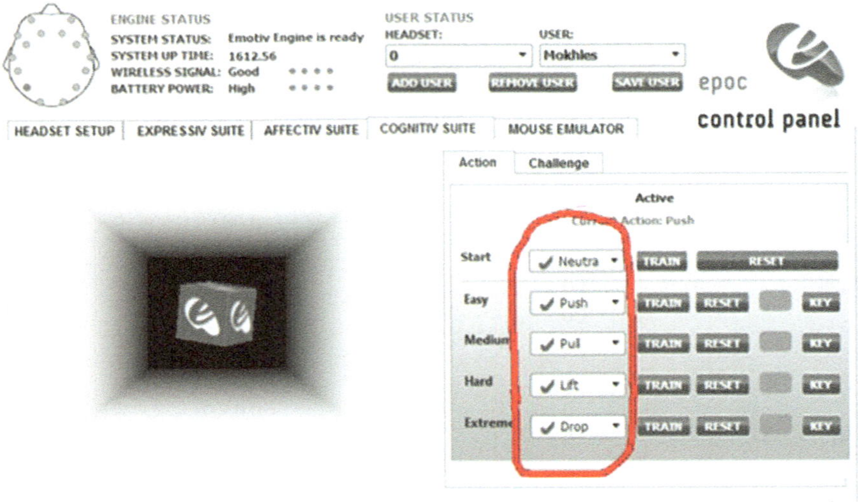

Fig. 9 Brain activities training

5.1 *Brain Activities Training*

The EEG headset starts reading the EEG signals produced by the patient brain and sends them as a set of 14 signals to the Emotive control panel software to be classified (Abdulghani and Al-Aubidy 2019). To start training these sets of signals, the neutral situation of the brain should be firstly trained, then start training a sequence of activities, as illustrated in Fig. 9.

5.2 *Generating Control Commands*

The next step is to convert the trained and classified activities to control commands using the EmoKey software, as shown in Fig. 10. Five control commands are consid-

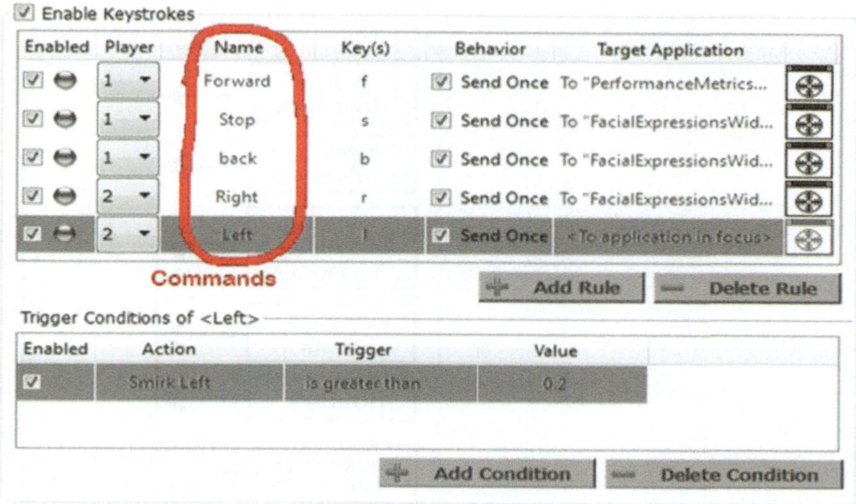

Fig. 10 Converting the trained brain activities to control commands

ered; forward (F), backward (B), stop (S), turn right (R), and turn left (L). Through the EmoKey software module, the list of the trained brain activities are imported from the Emotive control panel software. Each activity will be connected with a certain command to be sent to the micro-controller when this activity is active. For example, when the forward activity is active, the letter "F" will be sent serially to the micro-controller, and then the real-time control algorithm will implement the forward movement of the wheelchair (Abdulghani and Al-Aubidy 2019).

6 Hardware Design

The overall layout of the hardware design of the implemented wheelchair prototype is shown in Fig. 11; it has two microcontrollers, motors drive unit, HCI unit, and set of sensors for safety and tracking. Two microcontrollers are used, the ATmega2560 microcontroller is used as the main controller, while the second microcontroller is used for tracking purposes. The HCI unit is connected serially to the main micro-controller via a Bluetooth module (type HC-06), while all other units are connected directly as shown in Fig. 12. The main microcontroller generates the triggering signals for the eight ultrasonic sensors; while the output signals for these sensors are used together with HCI output command to generate the appropriate control signals.

An electronic drive unit (type L298N) is used to drive each DC motor using two control signals; direction, and duty cycle of the pulse width modulated (PWM) signal, for both right and left DC motors. The total number of required input/out lines is 30

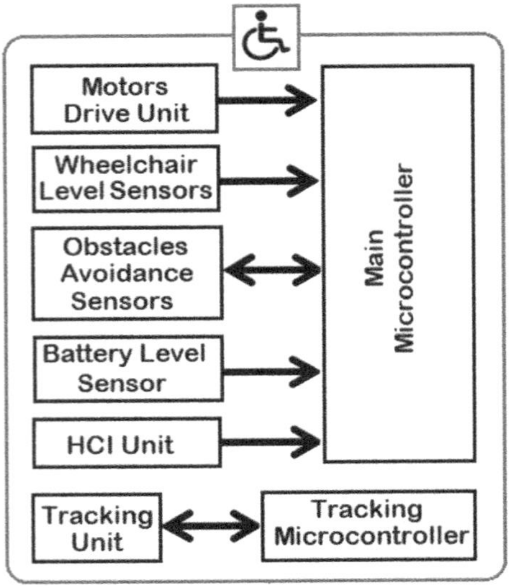

Fig. 11 General layout of the microcontroller-based design

lines digital and a single line analog for battery level sensing. The selected micro controller has 54 digital input/output pins (of which 15 can be used as PWM outputs) and 16 analog inputs.

The second microcontroller (type ARDUINO UNO) is connected directly to the GSM/GPS module (type SIM808). It is responsible for position-tracking task and equipped with an independent power source to keep it working 24 h (Al-Aubidy et al. 2015). The position tracking task will be managed by sending an SMS with the "track" command from the owner's cell phone to the GSM unit. The position tracking algorithm in the UNO microcontroller responds directly by resending and texting to the owner's cell phone with a Google Map link showing the latitude and longitude of the exact current position of the wheelchair according to the reading data of the GPS chip.

7 ANFIS-Based Controller Design

For the safe mobility and smooth driving of the wheelchair, an intelligent controller will be designed and implemented. Soft computing tools, such as fuzzy logic and neural networks, will be applied to designing real-time control algorithms in a smart wheelchair without the need for a mathematical model (Abdulghani and Al-Aubidy 2019; Turnip et al. 2015). In this research work, the MATLAB-based ANFIS has been used to control the speed and direction for each wheel. The speed control of each motor is achieved by calculating the accurate duty cycle of the PWM signal according to the feedback signals generated from sensors and required command generated by the HCI unit.

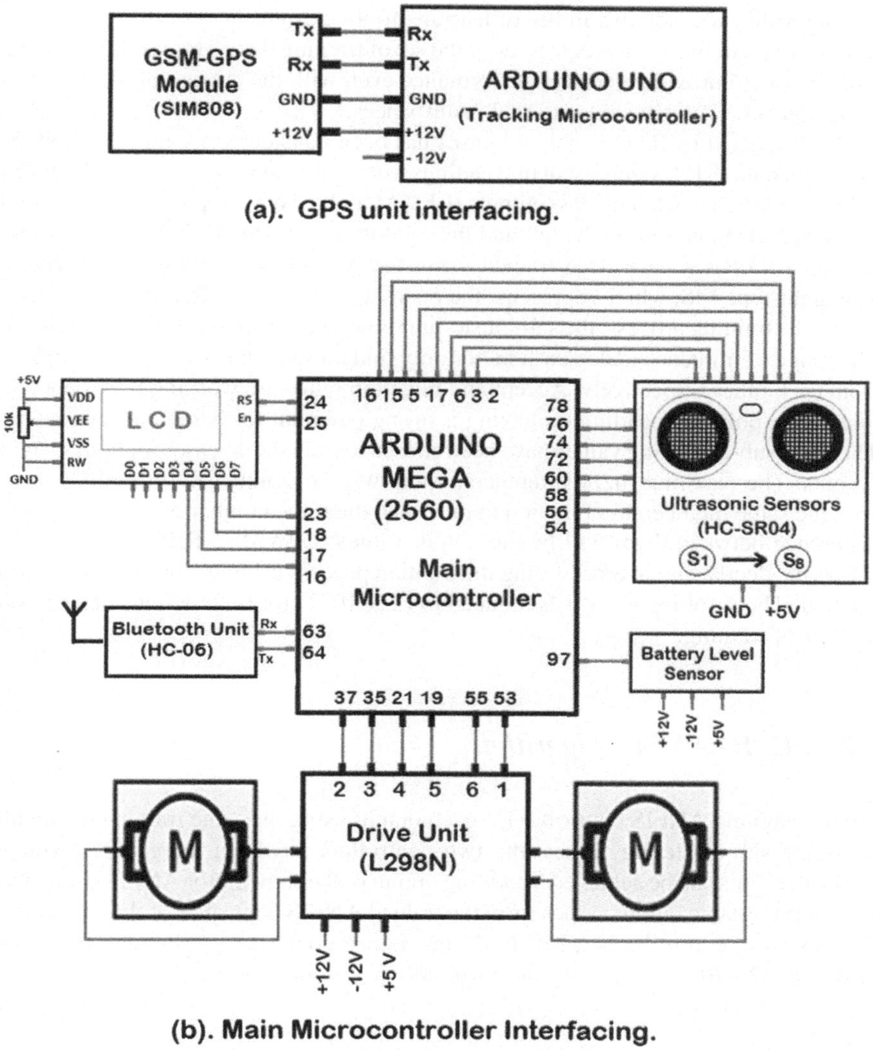

Fig. 12 Hardware design of the wheelchair controller

7.1 *MIMO ANFIS Design*

A MATLAB-based MIMO ANFIS algorithm has been designed without the need for extra software or advanced programming skills. Forty-four cases were chosen to train the algorithm to achieve good performance by ensuring obstacle avoidance and operating the wheelchair smoothly and safely.The overall error obtained in the designed controller is about (2×10^{-5}), which is accepted for such application.Using this algorithm will add the ability of learning to the designed controller, which will lead to cover all the in-between cases in the set of training data. The resulting system will be able to provide a perfect performance even with the change of the system's condition and overcome any external disturbances.

As illustrated in Table 3, only 44 cases has been considered to train the ANFIS. The input data (S1–S8) represent the readings from 8 ultrasonic sensors distributed on the body of the wheelchair. The outputs (LPWM and RPWM) represent the required PWM signals to be sent to the left and the right motors respectively.For each output the same 44 cases have been trained using the ANFIS application to construct a Mamdani-type FIS, which represents the resulting system after training.

For the training process, the sub-clustering values have been set as given in Table 4. The range of influence and squash factor only find clusters that are approach and far from each other respectively. Accept and rejection ratios are working on accept and reject data points depending on having a strong potential for being cluster centers. The same sub-clustering values have been chosen to train the 44 cases for both system outputs. The resulting fuzzy inference system will be symmetric for both of them with the same membership function number and the same number of rules. The only difference between them will be the output values (LPWM & RPWM). Figure 13 shows the decreasing in error during the training process using the back-propagation method. The resulting error was reduced to (2×10^{-5}) for both outputs during 200 Epochs of training.

7.2 *MIMO ANFIS Algorithm*

After designing ANFIS controller for each output using the same training data with the same sub-clustering values, the two controllers has been merged in a single controller. This can be achieved by adding output of the right motor ANFIS controller and inter the same output values form the trained ANFIS controller of the left motor. The resulting controller is a MIMO Neuro-Fuzzy controller works with the same error vale (2×10^{-5}) to produce the required PWM signal for each motor.

7.3 *ANFIS Performance*

The ANFIS controller performance has been compared with the classical Proportional, Integral and Derivative (PID) controller.The responses of the two DC motors of the wheelchair have been checked during its movement and during obstacle avoiding mission. To achieve this, the PID and ANFIS controllers were tested with the mathematical model of the wheelchair and with the implemented prototype. The mean square error (MSE) performance index has been calculated for both the ANFIS and PID controllers using this formula:

Table 3 Dataset used for training the real-time controller

Case	S_1	S_2	S_3	S_4	S_5	S_6	S_7	S_8	LPWM	RPWM
1	40	40	20	20	20	20	3	2	80	80
2	35	40	20	20	20	20	3	2	75	80
3	40	35	20	20	20	20	3	2	80	75
4	30	40	20	20	20	20	3	2	60	80
5	40	30	20	20	20	20	3	2	80	60
6	25	40	20	20	20	20	3	2	50	80
7	40	25	20	20	20	20	3	2	80	50
8	20	40	20	20	20	20	3	2	40	80
9	40	20	20	20	20	20	3	2	80	40
10	15	40	20	20	20	20	3	2	30	80
11	40	15	20	20	20	20	3	2	80	30
12	10	40	20	20	20	20	3	2	20	80
13	40	10	20	20	20	20	3	2	80	20
14	10	10	20	20	20	20	3	2	0	0
15	40	40	15	20	20	20	3	2	70	80
16	40	40	20	15	20	20	3	2	80	70
17	40	40	10	20	20	20	3	2	50	80
18	40	40	20	10	20	20	3	2	80	50
19	40	40	5	20	20	20	3	2	20	80
20	40	40	20	5	20	20	3	2	80	20
21	10	10	10	20	20	20	3	2	30	80
22	10	10	20	10	20	20	3	2	80	30
23	10	10	5	20	20	20	3	2	30	80
24	10	10	20	5	20	20	3	2	80	30
25	40	40	20	20	15	20	3	2	70	80
26	40	40	20	20	20	15	3	2	80	70
27	40	40	20	20	10	20	3	2	50	80
28	40	40	20	20	20	10	3	2	80	50
29	40	40	20	20	5	20	3	2	20	80
30	40	40	20	20	20	5	3	2	80	20
31	40	40	20	20	20	20	4	1	75	75
32	40	40	20	20	20	20	2	1	80	80
33	40	40	20	20	20	20	5	1	65	65
34	40	40	20	20	20	20	3	3	70	70
35	40	40	20	20	20	20	3	4	0	0
36	40	40	20	20	20	20	5	2	65	65
37	40	40	20	20	20	20	3	0	80	80

(continued)

Table 3 (continued)

Case	S_1	S_2	S_3	S_4	S_5	S_6	S_7	S_8	LPWM	RPWM
38	40	40	20	20	20	20	7	2	0	0
39	30	40	15	20	15	20	3	2	80	80
40	40	30	20	15	20	15	3	2	80	80
41	20	40	10	20	10	20	3	2	40	80
42	40	20	20	10	20	10	3	2	80	40
43	10	40	5	20	5	20	3	2	30	80
44	40	10	20	5	20	5	3	2	80	30

Table 4 Sub-clustering values

Range of influence	Squash factor	Accept ratio	Rejection ratio
0.4	1	0.3	0.1

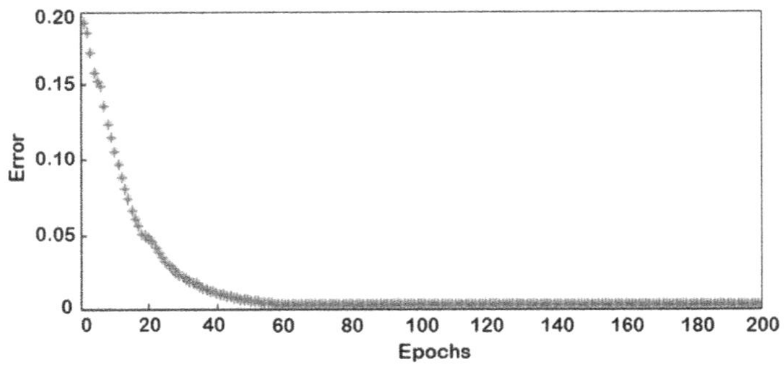

Fig. 13 ANFIS error after the back propagation training of the data set

$$MSE = \frac{1}{n}\sum_{i=1}^{n}(\theta_{desired} - \theta_{actual})^2 \tag{1}$$

where n is the number of samples in the time period and θ is the steering angle.

8 Direct Interface Between SIMULINK and V-REP

The Virtual-Robot Experimentation Platform (V-REP) is a 3D robot simulation software with an integrated development environment that allows modeling, editing,program, and simulates any robotic system such as KUKA, PUMA and PIONEER robots (V-REP 2020).A modified method has been developed to interface the

Fig. 14 Interface between SIMULINK and V-REP

MATLAB-Simulink with the open source 3D simulation engine V-REP software.The MATLAB lacks an easy to use 3D simulation options, especially to robotics systems. Therefore, V-REPcould use as a valid alternative, it is an opensource and freely available to research applications.The 3D simulation has become more and more widespread in robotics systems.The V-REP offers many possibilities to include its own code or to interface external software.

A direct interface between SIMULINK and V-REP software, as illustrated in Fig. 14, has been successfully achieved through these steps;

- Preparing a V-REP model to test the designed ANFIS controller.
- Preparing a SIMULIMK model for the ANFIS controller.
- Writing a MATLAB script code has been designed to send and receive data between SIMULINK & V-REP.

8.1 Preparing the V-REP Module

The Pioneer robot model, shown in Fig. 15, has been used to simulate the designed ANFIS controller and implement the obstacle avoidance job. The Pioneer model is equipped with 16 ultrasonic sensors; six of them are used to test the resulting ANFIS controller. The other two sensors are used to monitor the height level in the back and front side of the wheelchair have been set to be constant in the V-REP and have been tested on the wheelchair prototype only.

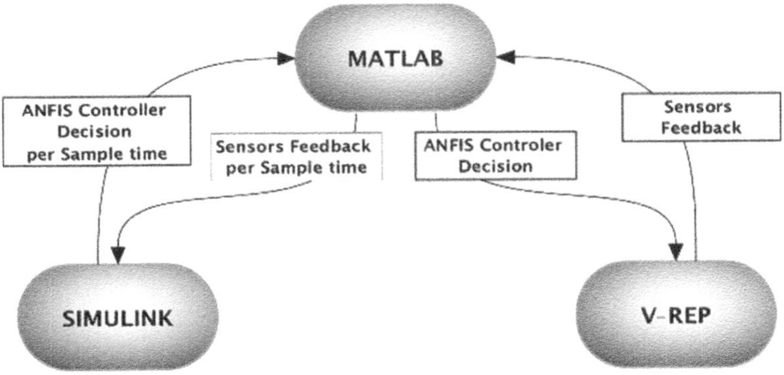

Fig. 15 Pioneer robot model

8.2 SIMULINK Model for the ANFIS Controller

The layout of the designed ANFIS controller of the wheelchair system using SIMULINK is illustrated in Fig. 16. The ANFIS for the left and right motors of the wheelchair will receive the eight ultrasonic readings from the V-REP Pioneer robot and make the required decision according to the training data used to train the ANFIS controller.

The sensors reading received in meters from the V-REP are converted to centimeters in the SIMULINK model using Gain and MATLAB function blocks for each input.The ANFIS controller produce the output as a PWM signal then it has been converted to a rotation speed to be send to the V-REP Pioneer model.A set of 44 cases, given in Table 5, has been used to cover the wheelchair steering and driving to avoid obstacles and move in a safe way.

The 44 cases used to train the ANFIS have been tested, and the obtained results show good performance. The real-time algorithm has been converted to a low-level language code to be uploaded to ARDUINO microcontroller memory. The resulted system has been tested on a wheelchair prototype and the obtained results show acceptable performance in the real environment. Figure 17 shows the result of testing the implementation case 23 of the training data (the 44 cases).

9 Results and Discussion

The resulting MIMO ANFIS algorithm has been tested on the simulation model and the real prototype. The performance of the resulting MIMO ANFIS algorithm was perfect and all the cases has been covered, even the in-between cases has been covered extremely perfect. Pioneer mobile robot model on the 3D simulator (V-REP) has been used to test the resulting system. It has been tested on a wheelchair prototype after

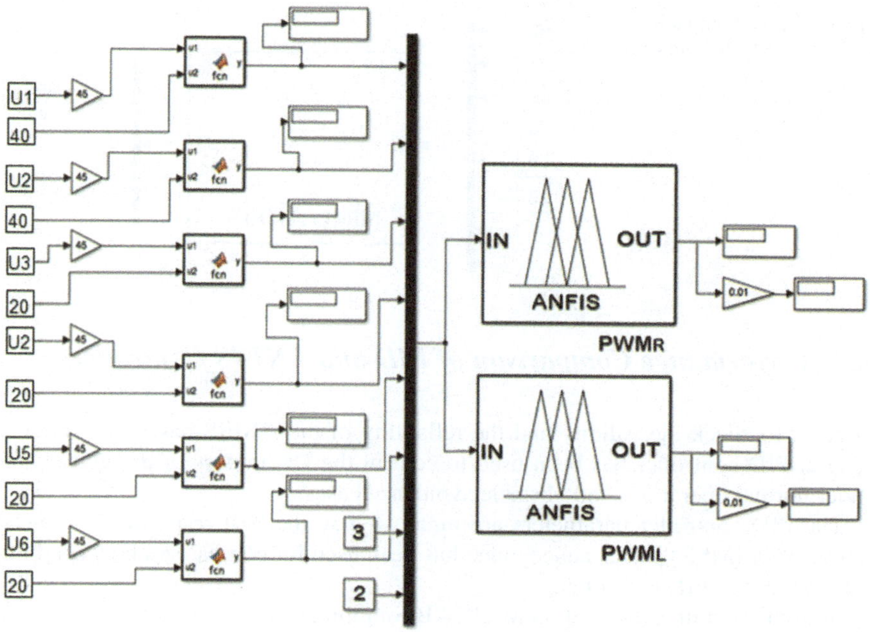

Fig. 16 SIMULINK model of the ANFIS controller

Table 5 Description of the selected 44 cases

Cases	Description
1–14	Avoiding dynamic and static obstacles on the front left and front right side
15–20	Avoiding dynamic and static obstacles on the right and left side
21–22	Driving wheelchair autonomously to the opened way. If the front side and the left side are facing obstacle, but the right side is clear then turn right the wheelchair, and vice versa
23–24	Avoiding dynamic and static obstacles on the right and left side
25–30	Avoiding dynamic and static obstacles on the back left and back right side
31–38	Regulating the wheelchair speed during up and down stairs movement
39–44	Making decision in the case of multiple obstacles surrounding the wheelchair

converting the MIMO ANFIS algorithm to a C++ code to be uploaded on the main microcontroller memory. Table 6 shows the output values of the resulting MIMO ANFIS controller for both left and right motors, according to the 44 cases used in the learning process.

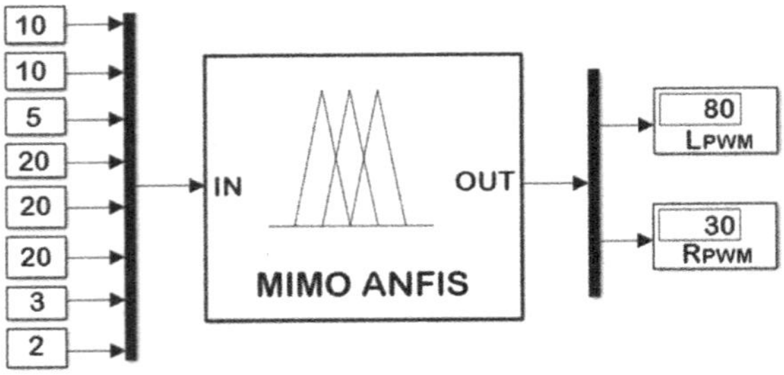

Fig. 17 Testing output of the MIMO ANFIS controller

9.1 Performance Comparison of PID and ANFIS Controllers

In order to check the validity and the reliability of the ANFIS-based controller, a general PID controller has been used to control the DC motors of the wheelchair under normal operations and obstacle avoidance cases.

The PID controller parameters are tuned to give the best response. The mean square error (MSE) performance index has been used for comparison between both the ANFIS and PID controllers.

A simulation model based on MATLAB Simulink has been used to test the performance of the PID and ANFIS controllers. As illustrated in Fig. 18, the inputs have been set according to case 23 from Table 6. The controller should respond to the change in reading of the ultrasonic sensors (S_1, S_2 and S_3) and regulates the speed of left and right motors accordingly to turn the wheelchair to the left direction. The operation of the PID controller shows lowest smoothness while changing in the motor speed.

Figure 19 shows the time response of the wheelchair when the steering angle is changed to avoid obstacles using both PID and ANFIS controllers.

It is clear that the ANFIS controller shows better performance with (MSE = 0.017) against the PID controller with (MSE = 0.049). Although, the overall MSE values in the PID and the ANFIS controllers are acceptable for low speed wheelchair system, the ANFIS performance is smoother than the PID performance in the transient area as shown in Fig. 20.

As shown in Figs. 21, 22, 23 and 24, the speed of left and right motors have been decreased to (500 RPM) and (2400 RPM) respectively to turn the wheelchair to the left. Figures 21 and 22 shows the performance of the PID controller in adjusting speed for both left and right motors, while Figs. 23 and 24 shows the performance of the ANFIS controller with the same tests. It is clear from time response parameters of each speed test, given in Table 7, that the ANFIS controller has better performance than the PID.

Table 6 Resulting outputs of the dataset used for controller training

Case	S_1	S_2	S_3	S_4	S_5	S_6	S_7	S_8	LPWM	RPWM
1	40	40	20	20	20	20	3	2	80	80
2	35	40	20	20	20	20	3	2	75	80
3	40	35	20	20	20	20	3	2	80	75
4	30	40	20	20	20	20	3	2	60	80
5	40	30	20	20	20	20	3	2	80	60
6	25	40	20	20	20	20	3	2	50	80
7	40	25	20	20	20	20	3	2	80	50
8	20	40	20	20	20	20	3	2	40	80
9	40	20	20	20	20	20	3	2	80	40
10	15	40	20	20	20	20	3	2	30	80
11	40	15	20	20	20	20	3	2	80	30
12	10	40	20	20	20	20	3	2	20	80
13	40	10	20	20	20	20	3	2	80	20
14	10	10	20	20	20	20	3	2	−0.0001	0
15	40	40	15	20	20	20	3	2	70	80
16	40	40	20	15	20	20	3	2	80	70
17	40	40	10	20	20	20	3	2	50	80
18	40	40	20	10	20	20	3	2	80	50
19	40	40	5	20	20	20	3	2	20	79.99
20	40	40	20	5	20	20	3	2	80	20
21	10	10	10	20	20	20	3	2	30	80
22	10	10	20	10	20	20	3	2	80	30
23	10	10	5	20	20	20	3	2	30	80
24	10	10	20	5	20	20	3	2	80	30
25	40	40	20	20	15	20	3	2	70	80
26	40	40	20	20	20	15	3	2	80	70
27	40	40	20	20	10	20	3	2	50	80
28	40	40	20	20	20	10	3	2	80	50
29	40	40	20	20	5	20	3	2	20	80
30	40	40	20	20	20	5	3	2	79.999	20
31	40	40	20	20	20	20	4	1	75	75
32	40	40	20	20	20	20	2	1	80	80
33	40	40	20	20	20	20	5	1	65	65
34	40	40	20	20	20	20	3	3	70	70
35	40	40	20	20	20	20	3	4	−0.0001	0
36	40	40	20	20	20	20	5	2	65	65

(continued)

Table 6 (continued)

Case	S_1	S_2	S_3	S_4	S_5	S_6	S_7	S_8	LPWM	RPWM
37	40	40	20	20	20	20	3	0	80	80
38	40	40	20	20	20	20	7	2	0	0
39	30	40	15	20	15	20	3	2	80	80
40	40	30	20	15	20	15	3	2	80	80
41	20	40	10	20	10	20	3	2	40	80
42	40	20	20	10	20	10	3	2	80	40
43	10	40	5	20	5	20	3	2	30	80
44	40	10	20	5	20	5	3	2	80	30

9.2 V-REP Test

The direct interface between MATLAB Simulink, and the V-REP 3D simulator is an interesting approach to simulate any mechatronic system. With simple programming skills on the MATLAB scripting and using the remote API commands of the V-REP, it was easy to test the resulting system on a 3D model. The following 4 points illustrate the behavior of the 3D simulation model during the implementation of the resulting MIMO ANFIS algorithm.

9.2.1 Avoiding Static Obstacle

The sequence of movements of the Pioneer robot model to avoid static obstacles is illustrated in Fig. 25.It is clear that the robot can avoid static and dynamic obstacles on the left and right front sides. In fact, the MIMO ANFIS controller is able to make the required decision even with obstacle distance not included in the training data (the 44 cases).

9.2.2 Avoiding Dynamic Obstacles

Figure 26 shows the sequence of movements of the Pioneer model to avoid dynamic obstacles. The Pioneer robot model shows good performance to avoid unexpected dynamic obstacles during its movement. The interface between MATLAB Simulink and the V-REP software has been well synchronized, where the Pioneer robot model shows a fast response even with dynamic obstacles.

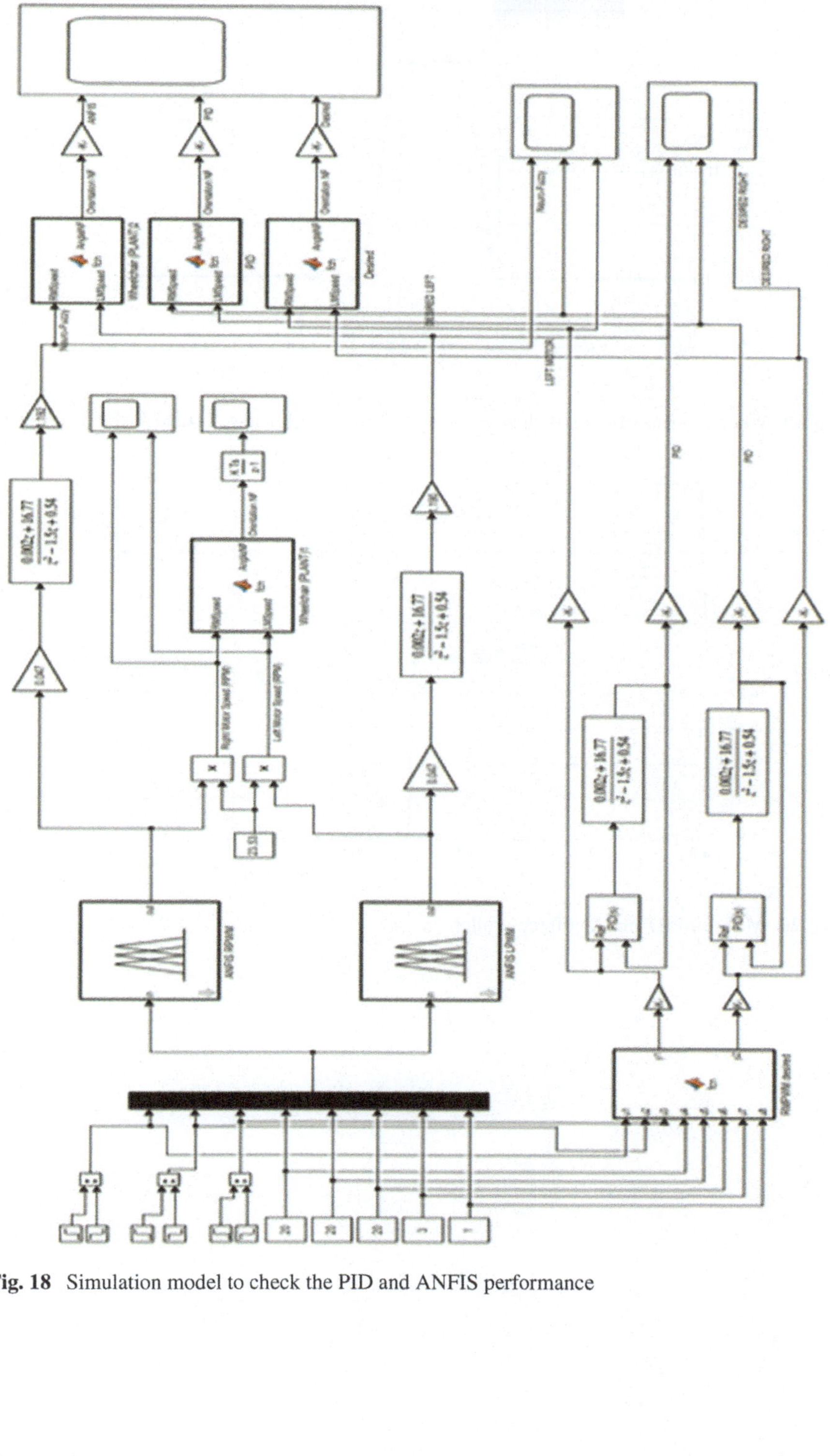

Fig. 18 Simulation model to check the PID and ANFIS performance

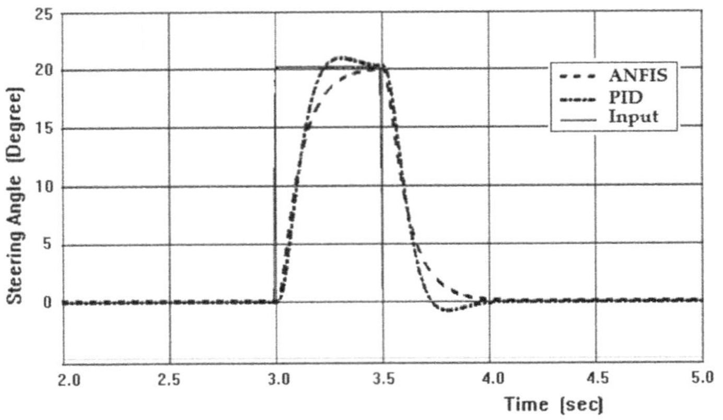

Fig. 19 Wheelchair steering angle change while avoiding obstacle using PID and ANFIS controllers

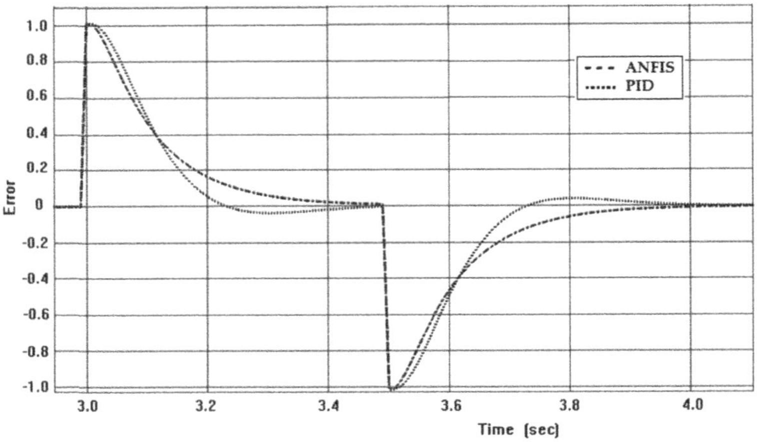

Fig. 20 MSE for PID and ANFIS controllers

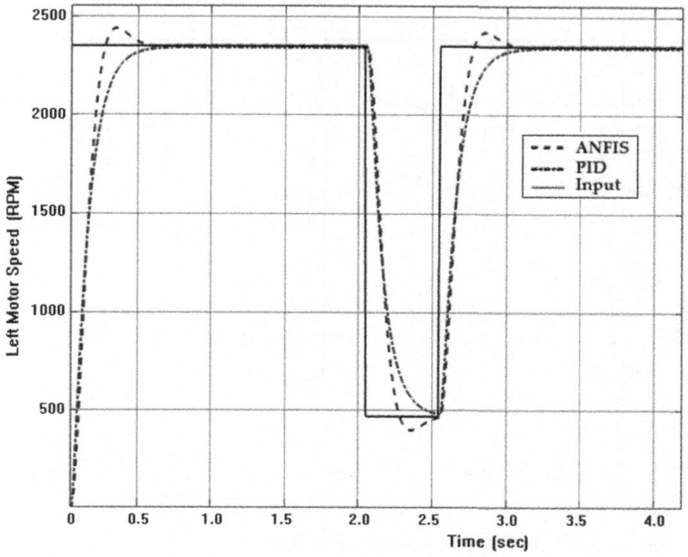

Fig. 21 Left motor speed response with PID controller

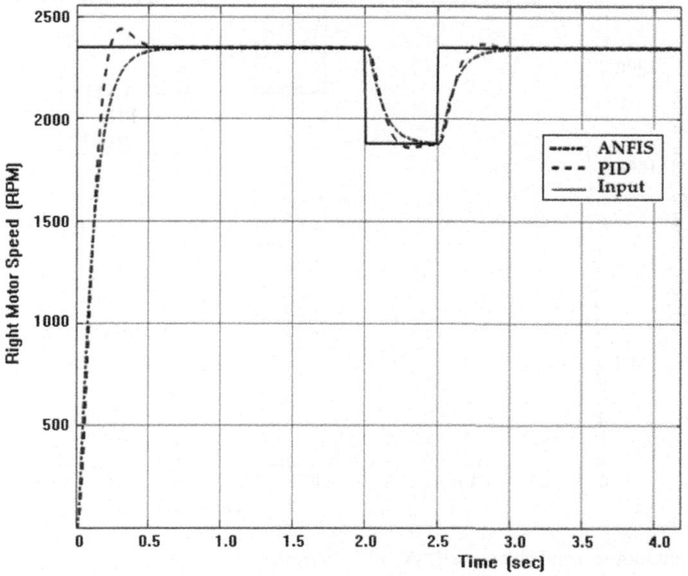

Fig. 22 Right motor speed change with PID controller

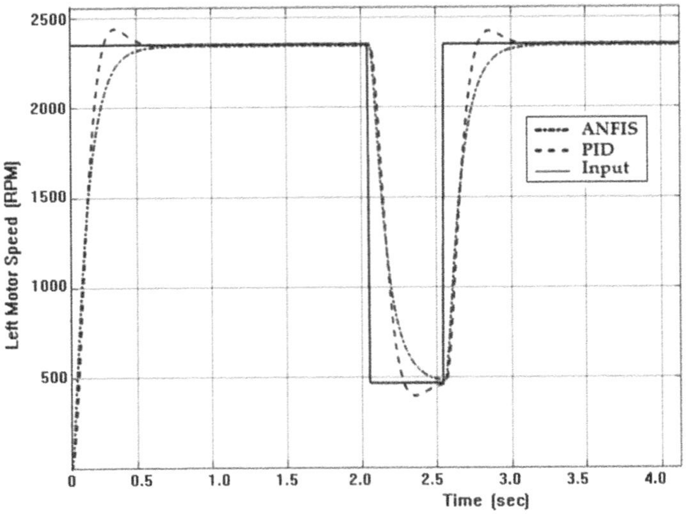

Fig. 23 Left motor speed change with ANFIS controller

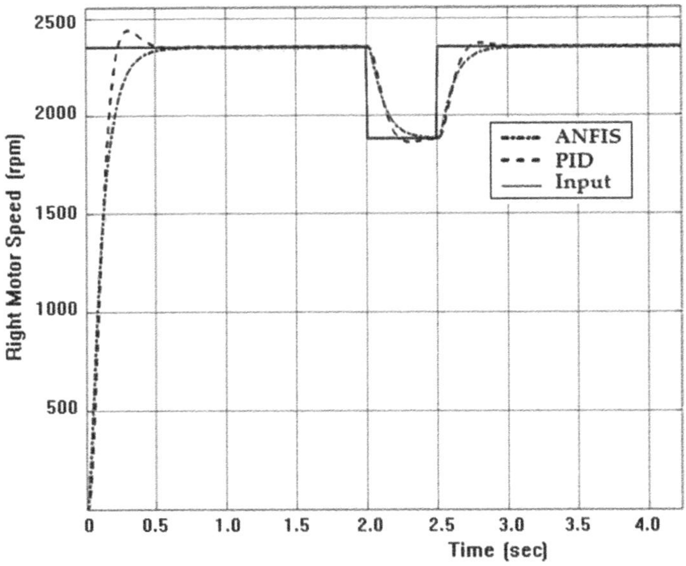

Fig. 24 Right motor speed change with ANFIS controller

Table 7 Time response characteristics for the given tests

Figure no.	Minimum	Maximum	Peak-to-peak	Mean	Median	RMS
21	0	2.441×10^3	2.441×10^3	2.115×10^3	2.352×10^3	2.197×10^3
22	0	2.440×10^3	2.441×10^3	2.256×10^3	2.352×10^3	2.278×10^3
23	0	2.345×10^3	2.345×10^3	2.099×10^3	2.345×10^3	2.171×10^3
24	0	2.349×10^3	2.349×10^3	2.244×10^3	2.349×10^3	2.264×10^3

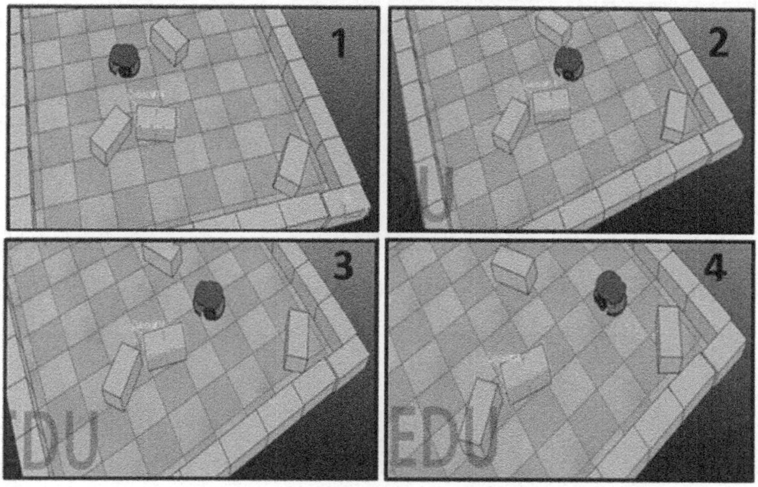

Fig. 25 V-REP simulation during facing static obstacle

9.2.3 Driving the Wheelchair Through a Narrow Way

Figure 27 shows the movement sequence to drive the Pioneer model through a narrow path with obstacles on the left and right side.The MIMO ANFIS controller is able to regulate the left and right motor speeds according to the received feedback from the ultrasonic sensors. The MIMO ANFIS controller actuates the Pioneer robot model successfully through a narrow path.

9.2.4 Autonomously Steering the Wheelchair to the Opened Way

Figure 28 shows the sequence of movements of the Pioneer model to turn autonomously to the opened way. The readings of the ultrasonic sensors on the Pioneer body demonstrate that the way is closed on the front and right side of the Pioneer robot. As shown, the MIMO ANFIS controller is able to autonomously turn left the Pioneer robot model when the way was closed in the front and right side.

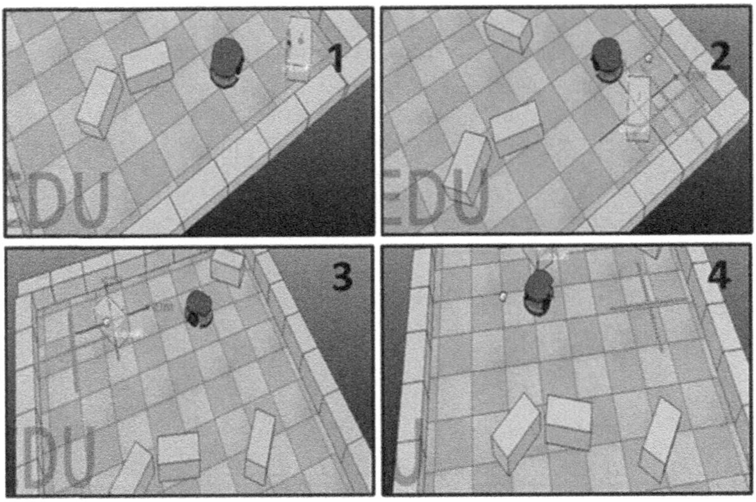

Fig. 26 V-REP simulation during facing dynamic obstacle

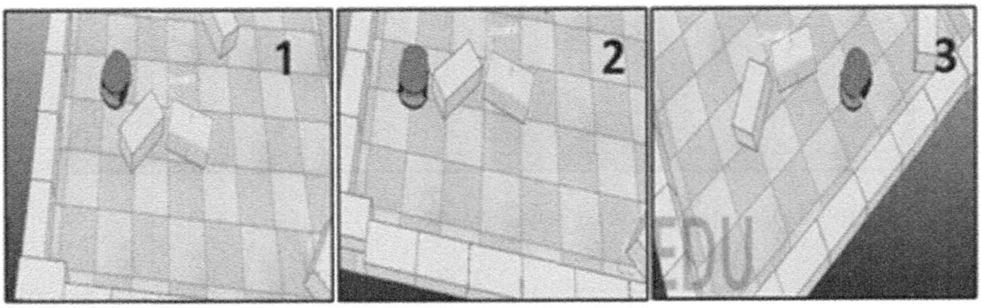

Fig. 27 V-REP simulation during drives through a narrow path

9.3 Wheelchair Prototype Test

The resulting ANFIS algorithm has been generated as a C++ code using the code generation tool in MATLAB Simulink. Then the code has been uploaded to the main micro-controller in the wheelchair prototype. The Emotive Epoc headset has been used as a BCI unit to generate the control commands. Figure 29 shows the prototype and the BCI unit during the real-time test of the implemented wheelchair prototype.

Although all the parts used in the wheelchair prototype implementation are simple, cheap, and low quality, the resulting system shows a good performance. The implementation of the BCI commands was accurate (almost 100%). The wheelchair prototype has been tested on all cases, such as moving downstairs and upstairs, avoiding static obstacles, avoiding dynamic obstacles, and stopping when facing a wall. Figure 30 shows the prototype test during regulating the motors' speed to drive it

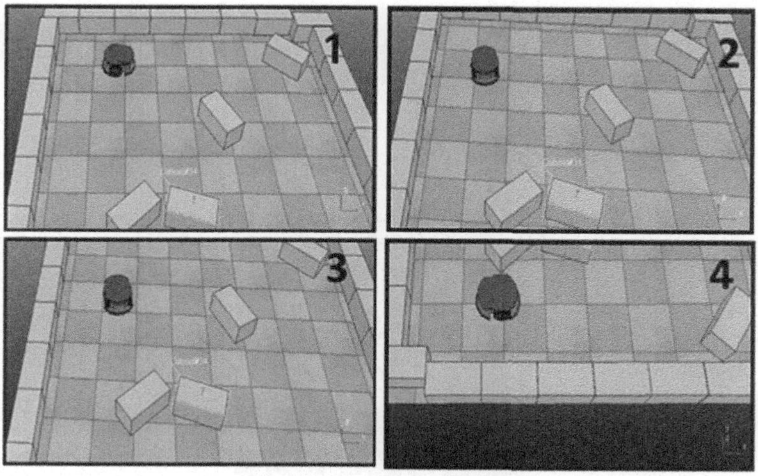

Fig. 28 V-REP simulation during autonomously turns to the opened way

Fig. 29 Wheelchair prototype controlled by the BCI

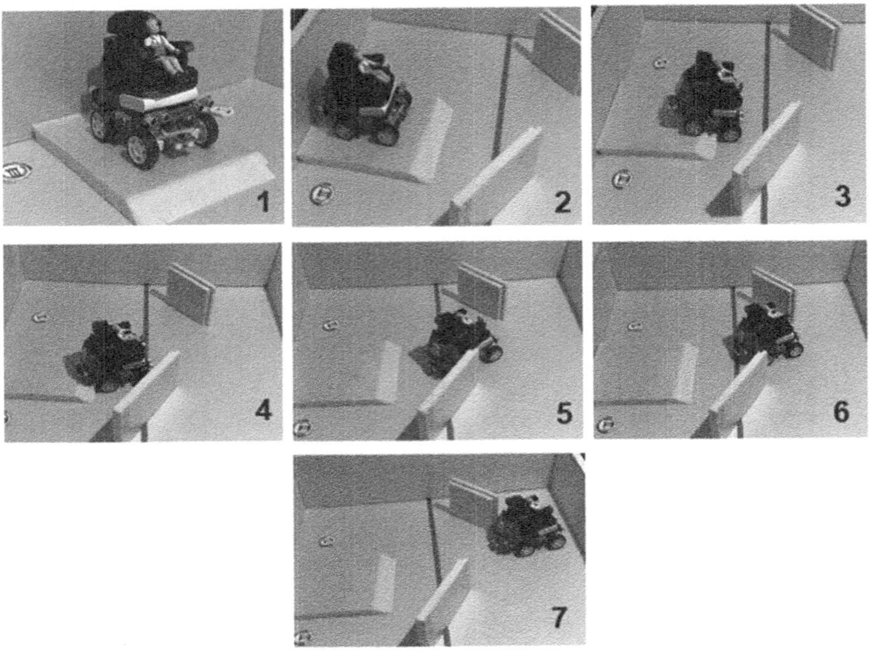

Fig. 30 Testing the resulting system on the wheelchair prototype

downstairs and to avoid obstacles. To overcome the signal interference, moving forward command has been implemented using metal brain activities (using thoughts only). The other 4 commands has been implemented with the brain activities produced with the face expressions, such as left and right eyes wink to turn left and right, smiling to go back and fully eyes closed to stop. If the patient or the user slept, the wheelchair would implement the stopping command responding to the closed eyes of the patient.

Moreover, the GSM/GPS unit has been successfully tested to send emergency SMS with the location and battery level. The owner of the wheelchair could send SMS to the wheelchair at any time to request a location and system information. If the patient stooped the wheelchair more than 5 minutes, this could be an indicator that the patient is sleeping, so that the wheelchair will send emergency SMS to the owner.

References

Abdulghani, M. M., & Al-Aubidy, K.M. (2019). Wheelchair neuro fuzzy control using brain computer interface. In *The 12th International Conference on the Developments in E-systems Engineering "DeSE2019"*, Kazan, Russia, 7–10 October.

Abdulghani, M. M., Al-Aubidy, K. M., Ali, M. M., & Hamarsheh, Q. J. (2020). Wheelchair neuro fuzzy control and tracking system based on voice recognition. *Sensors*, *20*, 2872.

Abiyev, R. H., Akkaya, N., Aytac, E., Günsel, I., & ÇaLman, A., (2016). Brain-computer interface for control of wheelchair using fuzzy neural networks. *BioMed Research International*, *2016*, Article ID–9359868, 1–9.

Al-Aubidy, K. M., Ali, M. M., & Derbas, A. M. (2015). Multi-robot task scheduling and routing using neuro-fuzzy control. In *12th IEEE International Multi-Conference on Systems, Signals, Devices (SSD15)*, Tunisia, 16–19 March.

Barbosa, A. O. G., Freitas, D. Z., Guedes, J. Q. M., & Meggiolaro, M. A. (2013). Implementation of a wheelchair control using a four-command brain computer interface. In *22nd International Congress of Mechanical Engineering (COBEM-2013)*, Ribeirão Preto, São Paulo, 3–7 November.

Bigras, C., Kairy D., & Archambault, P.S. (2019). Augmented feedback for powered wheelchair training in a virtual environment. *Journal of NeuroEngineering and Rehabilitation*, *16*(12), 1–12.

Carlson, T., & Millán, J. d. R. (2013). Brain-controlled wheelchairs: A robotic architecture. *IEEE Robotics and Automation Magazine*, *20*, no. EPFL-ARTICLE-181698, 65–73.

Gevaert, W., Tsenav, G., & Mladenov, V. (2010). Neural network used for speech recognition. *Journal of Automatic Control*, *20*, 1–7.

Hamadi, H., Suhendro, B., Alamsyah, M. S., & Ibrahim, M. (2020). Human tracking control system using Kinect sensors on wheelchair based on Arduino. *Journal of Physics: Conference Series*, *1436*. https://doi.org/10.1088/1742-6596/1436/1/012003.

Kaur, & Tanwar, P. (2015). Developing brain computer interface using fuzzy logic. *International Journal of Information Technology and Knowledge Management* 2, No. 2, 429–434, October 29–30.

Krishnamurthy, G., & Ghovanloo, M. (2006). Tongue drive: A tongue operated magnetic sensor based wireless assistive technology for people with severe disabilities. In *IEEE International Symposium on Circuits and Systems (ISCAS 2006)* (pp. 5551–5554).

Malik, M.I., Bashir, T., & Khan, O.F. (2017). Voice controlled wheel chair system. *International Journal of Computer Science and Mobile Computing*, *6*, No. 6, 411–419.

Mazo, M., Rodriguez, F. J., Lazaro, J. L., Urena, J., Garcia, J. C., Santiso, E., Revenga, P., & Garcia, J. J. (1995). Wheelchair for physically disabled people with voice ultrasonic and infrared sensor control. *Autonomous Robots*, No. 2, 203–224.

Pajkanović, A. & Dokić, B. (2013). Wheelchair control by head motion. *Serbian Journal of Electrical Engineering*, *10*, No. 1, 135–151.

Prashant, P., Joshi, A., & Gandhi, V. (2015). Brain-computer interface: A review. In *2015 5th Nirma University International Conference on Engineering (NUiCONE)*, Ahmedabad, India, 26–28 November, pp. 1–6.

Priandani, N. D., Tolle, H., & Utaminingrum, F. (2017). Real time advanced head movement recognition for application controller based on android internal gyroscope sensor. *International Journal of Advances in Soft Computing and its Applications*, *9*, No. 1, 70–87, ISSN 2074-8523.

Prince, D., Edmonds, M., Sutter, A., Cusumano, M., Lu, W., & Asari, V. (2015). Brain machine interface using emotiv EPOC to control robai cyton robotic arm. In *IEEE National Aerospace and Electronics Conference*. Department of Electrical and Computer Engineering University of Dayton, USA, paper 376.

Rani, P., Kakkar, S., & Rani, S. (2015). Speech recognition using neural network. In *Proceedings of the International Conference on Advancement in Engineering and Technology, ICAET 2015*, Incheon, South Korea 11–13 December, pp. 11–14.

Reaz, M. B. I., Hussain, M. S., & Mohd-Yasin, F. (2006). Techniques of EMG signal analysis: Detection, processing, classification and applications. *Biological Procedures Online*, *8*, 11–35.

Rojas, M., Ponce, P., & Molina, A. (2018). A fuzzy logic navigation controller implemented in hardware for an electric wheelchair. *International Journal of Advanced Robotic Systems*, 1–12. https://doi.org/10.1177/1729881418755768.

Sairam, B.V., Vijaykumar, N., Addanki, S., & Parapanch, S. (2018). Smartphone and wheelchair control for bedridden and semi-paralyzed people using brain-control interface (BCI). *International Journal of Advance Research, Ideas and Innovations in Technology*, *4*, No. 1, 820–825.

Terashima, K., Urbano, J., Kitagawa, H., Miyoshi, T. (2008). Development of a human-friendly omnidirectional wheelchair with safety, comfort and operability using a smart interface (chapter 13). In Zemliak, A. (Ed.), *Frontiers in Robotics, Automation and Control*. www.Intechopen.com.

Turnip, M., Dharma, A., Pasaribu, H. H. S., Harahap, M., Amri, M. F., Suhendra, M. A., & Turnip, A. (2015). An application of online ANFIS classifier for wheelchair based brain computer interface. In *2015 International Conference on Automation, Cognitive Science, Optics, Micro Electro-Mechanical System, and Information Technology, (ICACOMIT)*, Indonesia, October 29–30, pp. 134–137.

Velasco-Alvarez, F., & Ron-Angevin, R. (2017). Asynchronous brain-computer interface to navigate in virtual environments using one motor imagery. In *International Work Conference on Artificial Neural Networks*, Spain, 14–16 June, pp. 698–705.

V-REP, (2020). 3D simulation software. Available online at 10th May 2020, http://www.coppeliarobotics.com

Wallam, F., & Asif, M. (2011). Dynamic finger movement tracking and voice commands based smart wheelchair. *International Journal of Computer and Electrical Engineering*, *3*, No. 4, 497–502.

WHO. (2019), Disabled people in the world in 2019: Facts and figures. World Health Organization, Available on line at 7th April 2020, https://www.inclusivecitymaker.com/disabled-people-in-the-world-in-2019-facts-and-figures/.

Yokota, S., Hashimoto, H., Ohyama, Y., & She, J. (2010). Electric wheelchair controlled by human body motion. *Journal of Robotics and Mechatronics*, *22*(4), 439–446.

Fuzzy Control of an Intelligent Electric Wheelchair Using an EMOTIV EPOC Headset

Rabeb Abid, Firas Hamden, Mohamed Amine Matmati, and Nabil Derbel

Abstract This chapter involves the design and the implementation of an intelligent control method of an electric wheelchair for several disables persons using technologies derived from mobile robotics. The main objective of this work is to improve the performance of the wheelchair which is controlled by the brain without requirements of any physical feedback from the user. A detection system and obstacle avoidance based on the ultrasonic sensors with the integration of fuzzy logic controller is established. Intelligent control, via brain signal based on EMOTIV EPOC headset which can control the wheelchair with expressive, affective and cognitive data using a graphical interface, is achieved.

Keywords Electric wheelchair · Intelligent control · Ultrasonic sensors · Fuzzy controller · Emotiv EPOC headset

R. Abid (✉) · N. Derbel
Control & Energy Management Laboratory (CEMLab), Sfax Engineering School, Digital Research Center of Sfax (CRNS), University of Sfax, Sfax, Tunisia
e-mail: abid.rabeb.bah@gmail.com

N. Derbel
e-mail: nabil.derbel@enis.tn

F. Hamden · M. A. Matmati
Computer & Embedded Systems laboratory (CES-Lab), Sfax Engineering School, University of Sfax, Sfax, Tunisia
e-mail: firashamden14@gmail.com

M. A. Matmati
e-mail: med.amine.matmati@gmail.com

O. Kanoun and N. Derbel (eds.), *Advanced Systems for Biomedical Applications*, Smart Sensors, Measurement and Instrumentation 39,
https://doi.org/10.1007/978-3-030-71221-1_12

1 Introduction

A wheelchair is the most common assistive device used to improve a person's mobility. Mobility is a necessary prerequisite for enjoying human rights and enabling people with disabilities to become productive members. For many people, a suitable and well-designed wheelchair can be the first step towards inclusion and social participation. However, a big number of people with disabilities cannot use a standard electric wheelchair or have difficulty for driving it. For this reason, several researchers have been interested to improve the human-machine interface by implementing new control interfaces. Gajwani et al. (2010) used eye tracking and eye blinking obtained by a camera mounted on a cap to control a wheelchair. Gray et al. (2007) developed a visual based HMI for controlling a wheelchair by head gestures which were recognized by detecting the position of the nose on user's face. In the other hand, some researchers have used the tongue movements to operate the wheelchair, in which the movement data was obtained from a magnetic tracer on the tongue (Huo and Ghovanloo 2009).

The method employed in this project is the Brain-computer Interface, which enables direct communication between the brain and the electrical wheelchair. The best method for recording the brain's activity is electroencephalogram (EEG) (Carrino et al. 2012; Swee and You 2016). The used device for capturing the EEG signal is the Emotiv EPOC headset which transmits this signal wirelessly to the PC (Swee et al. 2016). Using the PC software, EEG signals are processed and converted into mental commands. According to the obtained mental commands (e.g. right, left, forward...), the output electrical signal is transmitted to the electrical wheelchair to perform the desired movement.

This chapter consists to improve the performance of an electric wheelchair by fixing an obstacle detection and an avoidance system based on ultrasonic sensors. In addition, the implementation and the realization of an intelligent command through an EMOTIV EPOC headset is presented. Emotiv EPOC, is available on the market to provide potential applications on hands-free HMIs. It has three suites: 'cognitive suite' to detect thoughts, 'expressive suite' to detect facial expressions and 'affective suite' to detect emotions (Rechy et al. 2014).

Section 2 presents the historical evolution of wheelchairs, their different types and the famous projects on smart wheelchairs. Section 3 present the considered wheelchair project and their specifications. Section 4 is devoted to the presentation of different parts of the project: the control card (Arduino Mega), the ultrasonic sensors, and the EMOTIV EPOC headset. Finally, Sect. 5 presents the software and the hardware implementation of the system specifying the choice of materials and components as well as practical required steps in order to achieve the complement system.

2 State of the Art

2.1 Historical Evolution of Wheelchairs

In order to provide the well-being of disabilities of people, the conventional electric wheelchair is becoming increasingly insufficient. So, improving the traditional wheelchair has become a necessity. The growth of science and technology has led to many inventions which have benefited men in many ways. One of such inventions is the wheelchair which has been a major break through in the history of mankind. In this chapter, we try to make a state of the art and present the wheelchair specifications which will be used later in the conceptual choices. There are different types of wheelchairs which are listed as follows (Herman 1969; Medola et al. 2012; Sidik et al. 2018; Simpson 2005):

- *Goutteux's chair:* It is the oldest wheelchair manufactured around 1595 by J. Lhermitte, for the King of Spain, Philippe II.
- *Wheelchair—John Dawson:* It was invented by the Englishman John Dawson in 1783. This chair was created with two large rear wheels and a small wheel forward.
- *Manual wheelchair:* It was invented in 1933, consisting of a foldable frame, two large rear wheels, two small front wheels, an adjustable seat, armrests and footplates. Three main modes of propulsion have been listed. The first one is the most used (90%) called the handrail. It consists of a metallic ring attached to the outside rims of the rear wheels. Sometimes both handrails are on the same side, allowing people with hemiplegia to propel and steer the chair with one hand. The second system uses a lever arm driving a wheel through a link. The lever is used to propel, steer and brake the chair. Finally, the last system is an adaptation of the transmission of a bicycle. This system can either be fixed on a chair with handrails, or constituting a full-fledged chair, generally for sports.
- *Electric wheelchair:* The first electric wheelchair was invented by Georges Kleinien in 1950 to facilitate the movement of disabled veterans of the second World War. For its manufacture, it used steel tubes and mechanical components, with a joystick. An electric propulsion system has been invented by increasing the voltage of the electric motor. The electric wheelchair is easier to use and allows disabled people to live more independently.
- *Intelligent wheelchair:* An intelligent wheelchair is actually an electric wheelchair with a human-machine interaction module. The aim interest is to provide automatic functionalities allowing a set of predefined tasks to be carried out with a completely assistance of the user in his daily life. For example, the intelligent wheelchair should allow the user to avoid hitting the obstacles that are around.

2.2 Electric Wheelchairs Types

No single model of the wheelchair can be adapted to the needs of all users, and the diversity of users implies the need for different types of wheelchairs. Those who select wheelchairs should understand the needs of the user and know how they will use the wheelchair. The electric wheelchair is for people who do not have the ability to use a manual wheelchair. It can be used indoors as well as outdoors. There are different types of wheelchairs as listed as follows (Frank 2010; Goher 2016):

- *Inside wheelchair*: It is a small motorized wheelchair model with small wheels and limited size, usually foldable with a speed of up to 6 km/h.
- *Outside wheelchair*: It is a large armchair with a speed of up to 10 km/h and thicker wheels to facilitate the movements outdoors.
- *The stand-up chair*: Electric stand-up wheelchairs are sophisticated models that allow the user to stand upright in order to keep them in a position close to standing. These models are especially suitable for quadriplegics, or paraplegics, who need their autonomy to reach anything high up.
- *Height-adjustable wheelchair*: It allows the user to be lifted into a seated position. It allows a better understanding of the environment and can be important for social integration. Some electric wheelchairs can combine verticalization and variable height when seated.

3 Presentation of the Selected Electric Wheelchair

In this chapter, the selected wheelchair is an indoor chair referenced by KY 110 AD produced by the Chinese company, Galaxy Medical Equipments & Instruments Co.
Feature of the Implemented Wheelchair
In choosing an electric wheelchair, the patient's first needs are:

- Moving needs (indoor, outdoor, etc.).
- The needs for installation, comfort, positioning, pain prevention, respect for fatigue.

The technical characteristics of the electric wheelchair are shown in Table 1.
In the next part, we will give an idea of the motors installed on the wheelchair.

3.1 Gear Motors

Generally, DC continuous motors are built to operate in a rapidity range close to their no-load speed. This speed range is generally too high for the majority of applications. So, to reduce this speed, a reduction gearbox is added to the motor. The whole system

Table 1 Characteristics of the project's electric wheelchair

Height	89 cm
Width	61 cm
Seat width	14/46 cm
Weight	90 kg
Authorized weight	110 kg
Maximum forward speed	6 km
Descending speed	0–4 km
Stopping distance	1.5 m
Maximum operating distance	20 km
Motor power	168 W × 2
Battery	20 AH 12V × 2
Charging time	8–16 h

is called a geared motor. The DC continuous motor turns the reducer, the energy flows from the input to the output shaft.

3.2 *Power Card*

The power part is an interface between the control card and the motor section. It allows to control the DC motors according to the orders provided by the command card to permit the moving of the electric wheelchair.

A controller for two direct current motors was produced. The detailed work of this power card is presented in Samet (2014). It is based on the principle of PWM (Pulse Width Modulation) to regulating the speed motor and the operation of the load H-bridge (motor) to choose the rotation direction. This power card consists essentially of (Fig. 1):

- Two complete H-bridges for controlling the two motors of the chair.
- MOSFET IR2110 drivers.
- The supply part.
- Relays for controlling the brakes.

3.3 *System Design*

To promote the mobility of a disabled person using an electric wheelchair, several researchers have used various technologies developed for mobile robots to create "the intelligent wheelchairs". A system to help piloting electric wheelchairs using

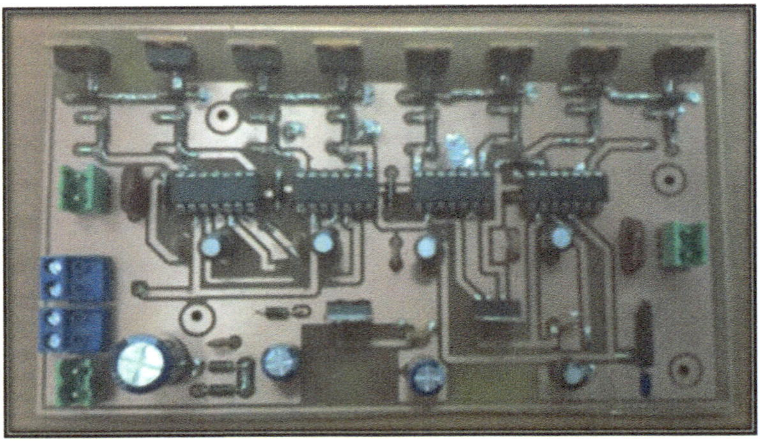

Fig. 1 Power card

methods and technologies from mobile robotics and adding an autonomous and intelligent aspect to these chairs is presented in this chapter. The methodologies are:

- The implementation of an obstacle detection and an avoidance system using ultrasonic sensors.
- The intelligent control through the EMOTIV EPOC helmet.

4 Sub-parts of the Wheelchair

After defining the methodology, the presentation of different parts is detailed. We start by the control card, then we present different types of ultrasonic sensors and their operating principle, and finally we close this section by the presentation of the EMOTIV EPOC headset.

4.1 Control Card

Usually, the control part of a system is its heart. It gives orders to the operational parts and receives their reports. The control card communicates between different control inputs and the power card, using the Arduino board as a control card.

General Information About the Arduino Board

The Arduino system is an open-source platform based on a simple micro-controller board (from the AVR family), and software to write, compile and transfer the program to the micro-controller card. This system gives the possibility of combining the performances of programming with those of electronics and the programming of electronic systems. The big advantage of programmed electronics that simplifies the electronic diagrams and therefore the cost of the realization. Indeed, the main advantages that encourages to use the ARDUINO card are:

- Inexpensive Hardware: Arduino boards are relatively inexpensive compared to other development platforms.
- Open source and extensible hardware: The diagrams of the modules are published under a common creative license, and it is possible to make our own versions of the Arduino boards.
- Free and Open Source Software: Arduino software and Arduino language are released under an open source license, available to be completed by experienced programmers. The language can also be extended using C^{++} libraries.
- Cross-platform software: Arduino software, written in Java, runs on Windows, Macintosh and Linux operating systems.
- "Shield" or "extension modules": Additional cards connecting to the Arduino module to increase the possibilities: color graphic display, GPS, etc.

4.2 Ultrasonic Sensors

To provide an obstacle detection, a distance sensor is required. There are several types of distance sensors such as the ultrasonic sensors. In this part, the operating principle and a study of different types of ultrasonic sensors is presented and detailed.

4.2.1 Operating Principle

An ultrasonic sensor emits short pulses of sound at regular intervals. These impulses propagate through the air at the speed of sound. When they detect an object, they are reflected and return in the form of an echo to the sensor which then calculates the distance separating it from the target on the time basis elapsed between the transmission of the signal and the reception echo. Generally, any sound reflecting material can be detected, regardless of its color. Even transparent objects or thin films are previewed for an ultrasonic sensor. Ultrasonic sensors are available for ranges from 20 mm to 10 m and give the measured value to the nearest millimeter.

4.2.2 Ultrasonic Sensor Selection

There are several types of ultrasonic sensors. In this part, we present the most used types in mobile robotics.

Lego Mindstorms NXT

This sensor is characterized by a detection range between 0 and 255 cm with a difference of $\pm$ 3 cm. Its resolution is about 6 cm, it communicates with the micro-controllers via an I2C bus. Measurements smaller than 3 cm cannot be made. This corresponds to the problem of the necessary return time of the wave. Measurements up to 20 cm are relatively precise enough at an angle between 8° and 30° . Indeed, the receiver is on the left and the transmitter is on the right, which explains why measurements on the left are less precise than measurements on the right. This sensor is very expensive (around $ 300).

VexRobotics

This sensor is characterized by its 40 kHz transmission frequency. It measures a distance between 3 cm and 3 m and its pulse duration is 250 milliseconds. This sensor requires the Robot C programming kit which is included in the kit sold on Generation Robots. It is programmed only by the VEX microcontroller. This sensor is priced at around $ 35

Paralax

The technical characteristics of the sensor are the distance measurement between 2 cm and 3.3 m, the average error is less than 0.5 cm.

Maxbotix "MaxSonar-EZ1"

This sensor gives a scanned distance of 0 to 6.45 m.

Ultrasonic Sensor SRF08

This module is designed for applications related to telemetry and robotics. Generally, the SRF08 sensor is able to determine the distance to an obstacle in front of it between 3 cm and 6 m. It interfaces with an I2C bus and controlled with an EEPROM memory.

SRF04 Ultrasonic Sensor

The SRF04 sensor is composed of a transmitter and a receiver. By measuring the time between the emission and the reception of the wave and it's speed of propagation, we can deduce the distance from the obstacle. The SRF04 needs a 10 μs pulse on the Trigger input to start the detection. Therefore, it generates eight ultrasound pulses at 40 kHz and waits for the echo.

The SRF04 sends to the Echo output a signal proportional to the distance separating it from the obstacle. To calculate the distance, it measures the width of the signal returned by the SRF04 in μs and to divide it by 58 to have a distance in cm (or 148 to have it in inch).

HC-SR04 Ultrasonic Sensor

This sensor detects all obstacles within a distance up to 4.5 m. This module presents four output pins: VCC, TRIGGER, ECHO and GND. The HC-SR04 ultrasonic sensor is adaptable to an Arduino board. A high level pulse (5V) for at least 10 μs must be applied to the "TRIGGER" pin to start reading. If an obstacle is detected, the "ECHO" pin passes to high level (5V). The distance where the obstacle is located is proportional to the duration of the pulse.

Ultrasonic sensors are effective means of obstacle detection requiring simple implementations. A general idea on their operating principle as well as the famous ultrasonic sensors used in mobile robotics have been presented. In the following section, an intelligent command with the EMOTIV EPOC headset is detailed.

4.3 Casque EMOTIV EPOC

In this part, we focus our attention on the EMOTIV EPOC helmet (design presentation and its simulators).

4.3.1 EMOTIV EPOC Headset

Reading people's minds is a dream that researchers are now celebrating. Some researchers have managed to extract certain pictures from the brain of their patients, others are interested in emotions. Emotiv Systems are specialized in this field. His EMOTIV EPOC helmet allows to analyze what is happening in the head.

Once the headset is placed on the head of the patient, the device picks up brain waves and transmits them to the software, which analyzes them in detail to detect emotions.

The headset applications are endless. The EMOTIV EPOC allows the greatest handicapped to interact much more easily with their environment. Simulators with which the EMOTIV EPOC headset works is presented in the next section.

4.3.2 EmoComposer Simulator

The EmoComposer allows sending the user's emotion states to the Emotive Control Panel, EmoKey or any other application that uses an Emotive application.

4.3.3 Emotiv Control Panel Supervisor

Emotiv Control Panel is a supervisor that visualizes the command result sent from the headset or the EmoComposer.

Expressive Suite

The Expressive Suite (Fig. 2) details the facial expressions and the non-verbal communication skills of the Emotiv EPOC headset. To show the facial expressions, we need to perform them while wearing the Emotiv EPOC helmet. These expressions can be visualized on the avatar (virtual face in the expressive suite)

Affective Suite

The affective suite (Fig. 3) monitors the user's emotional states in real time. It offers an additional dimension by allowing the computer to respond to the user's emotions. The affective suite can be used to monitor the user's state of mind. It allows developers to adjust the difficulty to suit each situation.

Cognitive Suite

The cognitive suite (Fig. 4) reads and interprets the thoughts and the intentions of the user. Users can manipulate virtual or real objects using the power of their thoughts.

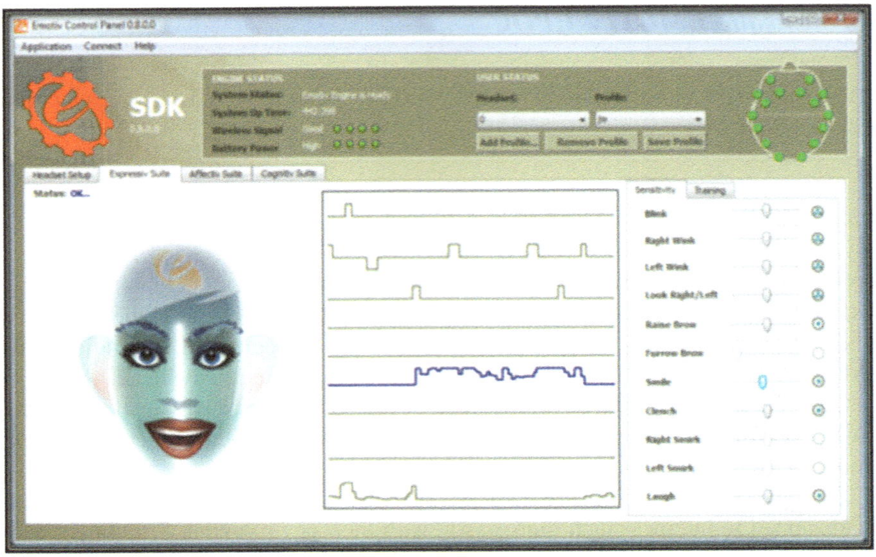

Fig. 2 Expressive suite on the EMO CONTROL PANEL interface

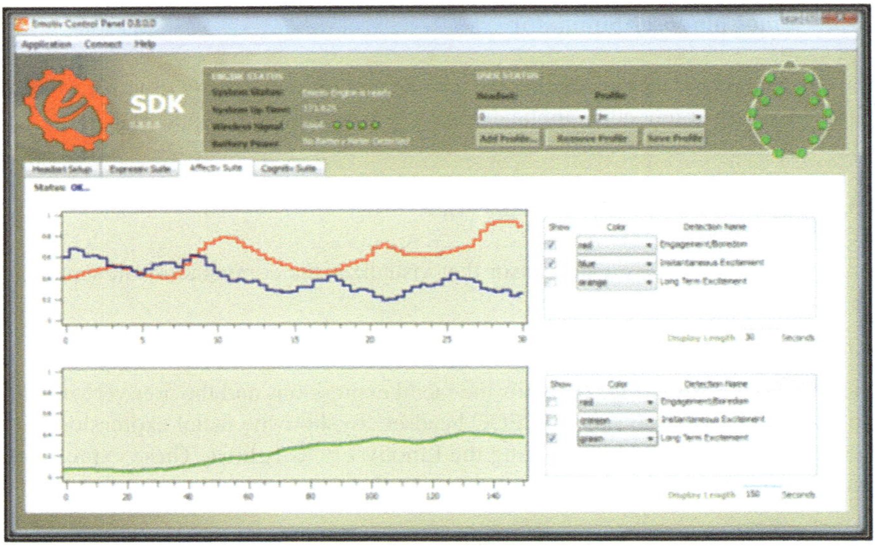

Fig. 3 Affective suite on the EMO CONTROL PANEL interface

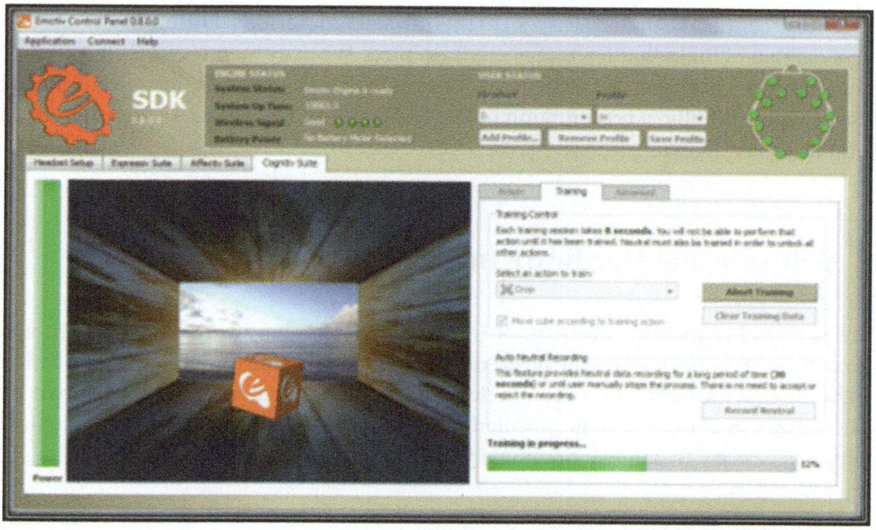

Fig. 4 Cognitive suite on the EMO CONTROL PANEL interface

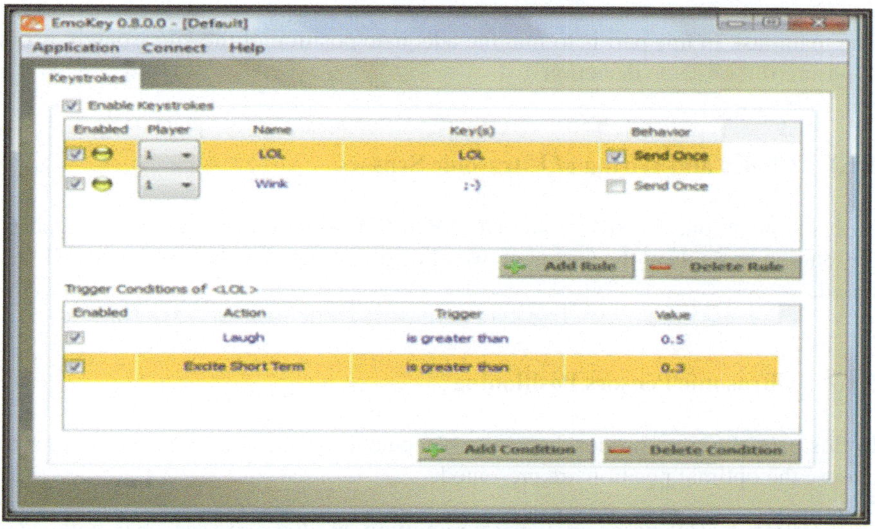

Fig. 5 EMOKEY interface

4.3.4 EmoKey

The EmoKey (Fig. 5) is used to encode the detection results received from the Emotiv Control Panel according to an easy logical rules. It also provides a mechanism to integrate the Emotiv EPOC headset with a pre-existing application with the keyboard.

In this section, the material aspect of this project is described. Different elements of the system and their characteristics are presented. We start by presenting the control board used to manage the system and then we give an idea about the ultrasonic sensor types. Finally, a presentation of the EMOTIV EPOC headset and its different simulators are shown.

5 Software and Hardware Implementation

The practical realization consists of implementing the hardware, the software and development tools to obtain the desired system that meets the functional specification set in the previous part. This section deals with the practical implementation part by specifying the choice of materials and components as well as the various practical phases necessary to achieve the final system.

5.1 *Ultrasonic Sensors*

The main task in this part is to find the adequate position of the Ultrasonic sensors to ensure full obstacle detection.

5.1.1 Choice and Wiring of Ultrasonic Sensors

In this application, the SRF04 and HC-SR04 ultrasonic sensors are used. Figure 6, presents an example of the wiring of two ultrasonic sensors with the Arduino MEGA board.

5.1.2 Ultrasonic Pensors Positioning

In this part, different proposals for the placement of ultrasonic sensors in order to choose the optimal position are presented.

Proposition 1 *This proposal consists in putting an ultrasonic sensor in both sides in front of the wheelchair. Two sensors are used. Each one has a 50° angle detection. The problem with this method is that the dead zone is very large, so that if there is an obstacle near the chair, it cannot be detected (Fig. 7).*

Proposition 2 *The second proposition consists of placing three ultrasonic sensors in front of the chair (two sensors on both sides and a sensor in the middle) so that the dead zone becomes smaller but it is not completely eliminated (Fig. 8).*

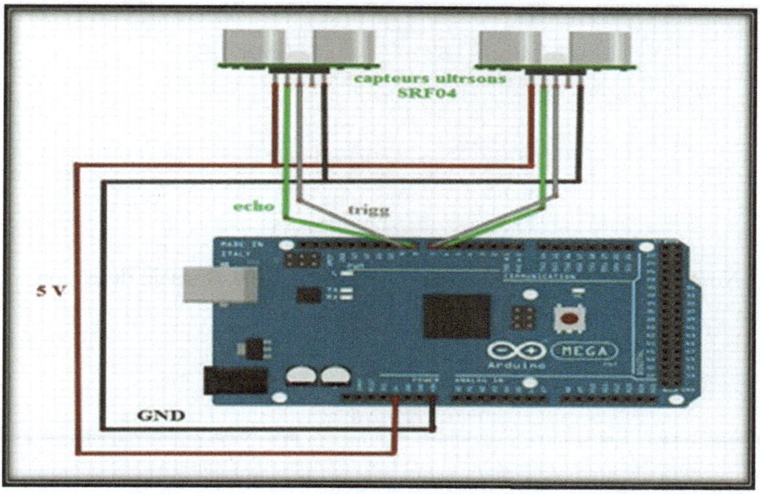

Fig. 6 Connection of two SRF04 ultrasonic sensors with the Arduino MEGA board

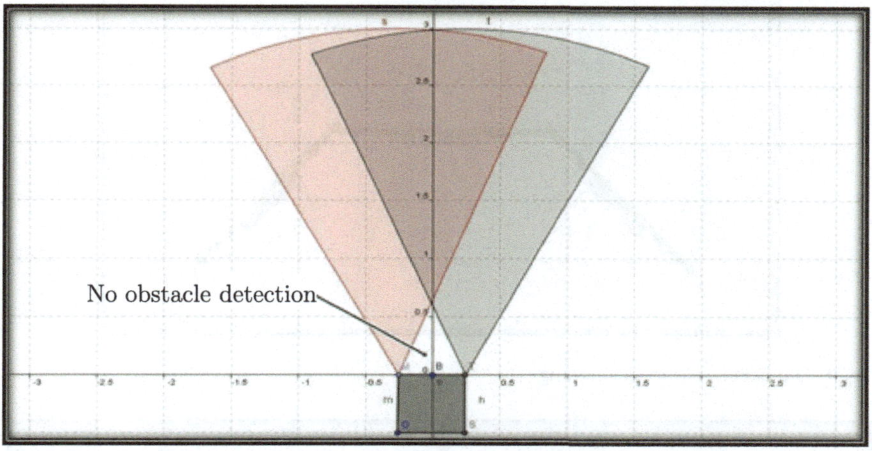

Fig. 7 Simplified diagram of sensor positioning (Proposition 1)

Proposition 3 *The aim of this proposal is to eliminate all the dead zone so that we have the possibility of detecting any obstacle that may arise in front of the chair. The idea consists in putting three ultrasonic sensors practically in a single point as shown in Fig. 9.*

This positioning of the sensors makes it possible to completely eliminate the dead zone with a maximum detection angle equal to 150° as shown in Fig. 10.

The fixing of the sensors according to the chosen position is carried out using a plexiglass piece (Fig. 11):

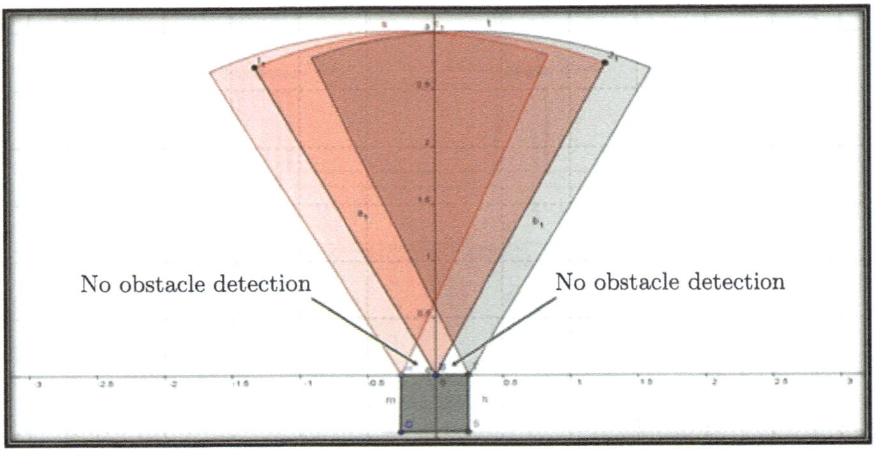

Fig. 8 Simplified diagram of sensor positioning (Proposition 2)

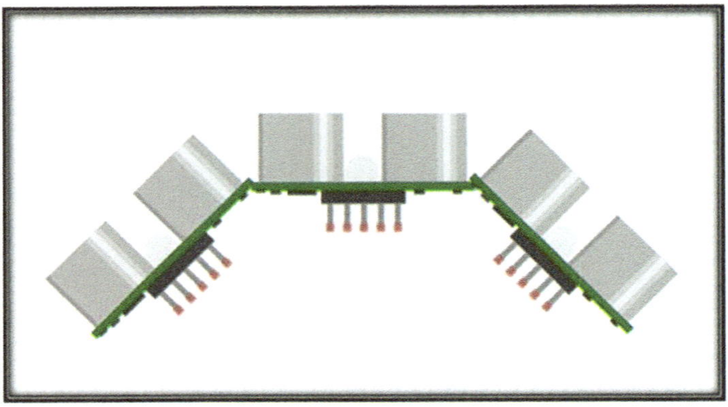

Fig. 9 Ultrasonic sensors positioning

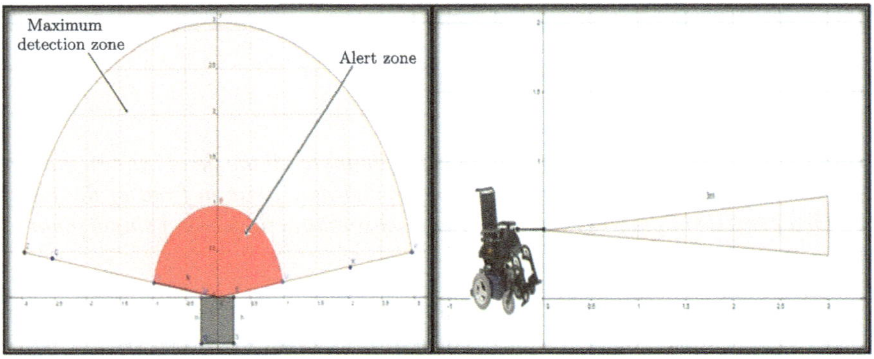

Fig. 10 Simplified diagram of sensor positioning (Proposition 3)

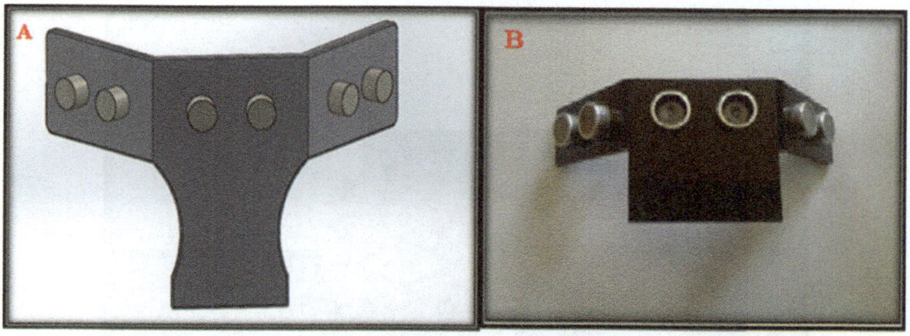

Fig. 11 Attachment part of the ultrasonic sensor, A: SOLIDWORKS picture, B: Real picture

Fig. 12 Photo of the wheelchair with the fixation of the ultrasonic sensors (in SOLIDWORKS)

This part is attached to a movable metal arm fixed to the wheelchair as shown in Fig. 12:

Proposal 3 is implemented in the practical realization, which is more efficient and allows to eliminate the dead zones detecting any obstacle in front of the chair. An attachment device and a mobile arm on which the ultrasonic sensors are fixed are manufactured (Fig. 13).

5.2 *Obstacle Avoidance, Fuzzy Controller*

A fuzzy controller system is implemented to determine the speed of the right and the left wheels according to the distance from obstacles around the wheelchair.

5.2.1 Linguistic Variables

The first step is to determine the linguistic input and output variables. The front, the left and the right distances between obstacles and the wheelchair are the inputs of

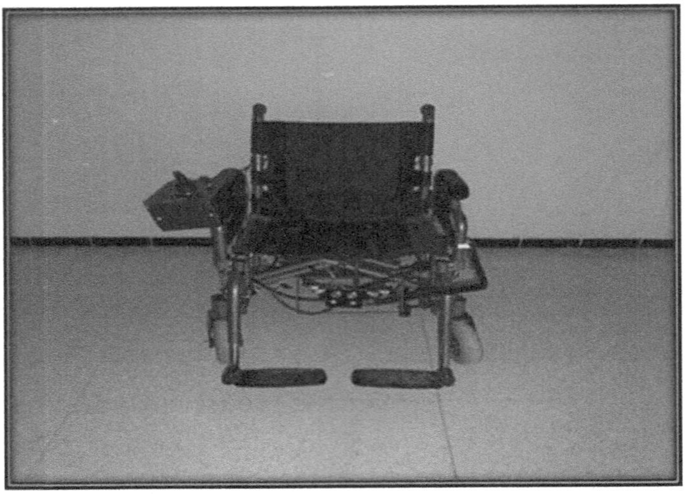

Fig. 13 Real picture of the wheelchair with fixing the ultrasonic sensors

the fuzzy controller. Two linguistic variables are used in order to present the distance data, which are:

- small distance,
- large distance.

The output variables are the speeds of the wheels. A speed of a wheel is represented by a PWM signal (duty cycle between 0 and 1). To qualify this duty ratio the chosen linguistic variables are:

- small speed,
- medium speed,
- high speed.

After setting the linguistic input and output variables, the membership functions must be chosen (Fig. 14). A small distance is less than 1 m. It is considered as a high distance if it is larger than 3 m. The duty cycle is: (i) small if it is around zero, (ii) medium if it is around 0.5, and (iii) large if it is around 1. Membership functions are presented in Fig. 14.

5.2.2 Inference Rules

Three forward sensors have been used. Each sensor measurement is quantified into two linguistic variables: small and large. Therefore, we have at most 8 inference rules.

- **Case 1**: If no obstacle is detected, so we can go straight at full speed.

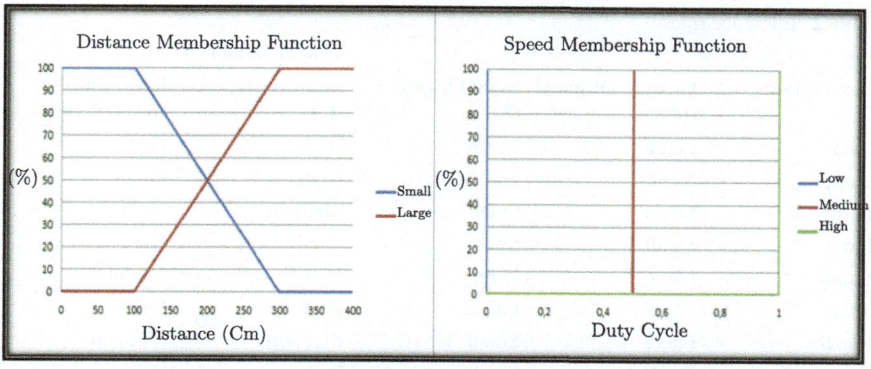

Fig. 14 Distance and speed membership functions

- **Case 2**: If an obstacle in front of the wheelchair is detected but nothing on the sides. Then, we have to turn in the right direction (high speed is needed on the left wheel and low on the right ones).
- **Case 3**: If an obstacle is present only on the right side, then the wheelchair must turn slightly to the left (speed of the right wheel is high while that of the left one is medium).
- **Case 4**: If an obstacle at the right side and in front of the wheelchair is detected, then it must turn strongly to the left (speed of the right wheel is large and that of the left one is small).
- **Case 5**: If an obstacle is detected on the left, the wheelchair must turn slightly to the right (speed of the left wheel is high while that of the right one is medium).
- **Case 6**: If an obstacle in the left and in front of the chair is detected, it must turn strongly to the right (speed of the left wheel is large and the right one is small).
- **Case 7**: If the way is clear in front of the chair, but there are two obstacles on each side of the robot, then it has must advance carefully (speed of the two wheels is medium).
- **Case 8**: Finally, if the robot detects obstacles on all the sensors, then the motors stop.

The inference rules are presented in Table 2:

5.2.3 Defuzzification

There is three linguistic variables per wheel (six linguistic variables in total).

To calculate the duty cycles of the right and left wheels, we must apply a barycenter calculation on each of the linguistic variables :

Table 2 Inference rules

Input variables			Output variable	
Left Distance (LD)	Forward Distance (FD)	Right Distance (RD)	Left wheels speed (LS)	Right wheels speed (RS)
Large	Large	Large	High	High
Large	Small	Large	High	Small
Large	Large	Small	Medium	High
Large	Small	Small	Small	High
Small	Large	Large	High	Medium
Small	Small	Large	High	Small
Small	Large	Small	Medium	Medium
Small	Small	Small	Zero	Zero

$$Left_{speed} = \frac{(large_{left_{speed}} \times 1) + (meduim_{left_{speed}} \times 0.5) + (small_{left_{speed}} \times 0)}{large_{left_{speed}} + meduim_{left_{speed}} + small_{left_{speed}}}$$

$$Right_{speed} = \frac{(large_{right_{speed}} \times 1) + (meduim_{right_{speed}} \times 0.5) + (small_{right_{speed}} \times 0)}{large_{right_{speed}} + meduim_{right_{speed}} + small_{right_{speed}}}$$

5.3 *EMOTIV EPOC Helmet*

The EMOTIV EPOC headset runs with an SDK installed on the PC. The headset generates signals, which are sent to the PC through wireless transmissions. The SDK (or EMOCONTROL PANEL) analyzes these signals and gives the planned actions (BLINK, WINK, SMILE ...) to the PC bord. Then, through EMOKEY, these actions are coded and then sent to a PC/ARDUINO interface to control the wheelchair. In the following, different steps of connecting the headset with the ARDUINO card are presented.

Figure 15 shows the communication between different blocks.

5.3.1 Actions Coding with EMOKEY

EmoKey emulates a compatible keyboard on Windows. It sends the keystrokes (such as characters or numbers) in the system queue. The application in the foreground on Windows (in our case is the PC/ARDUINO interface) receives these emu-

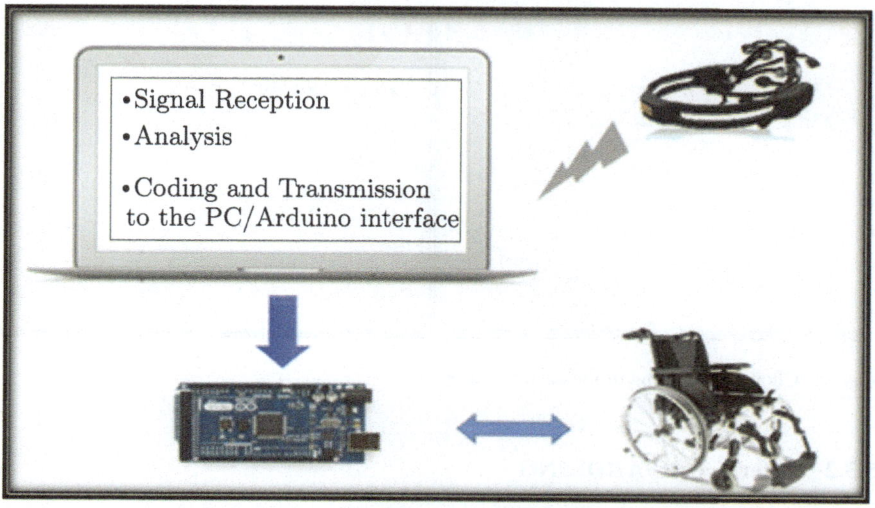

Fig. 15 Communication between different parts

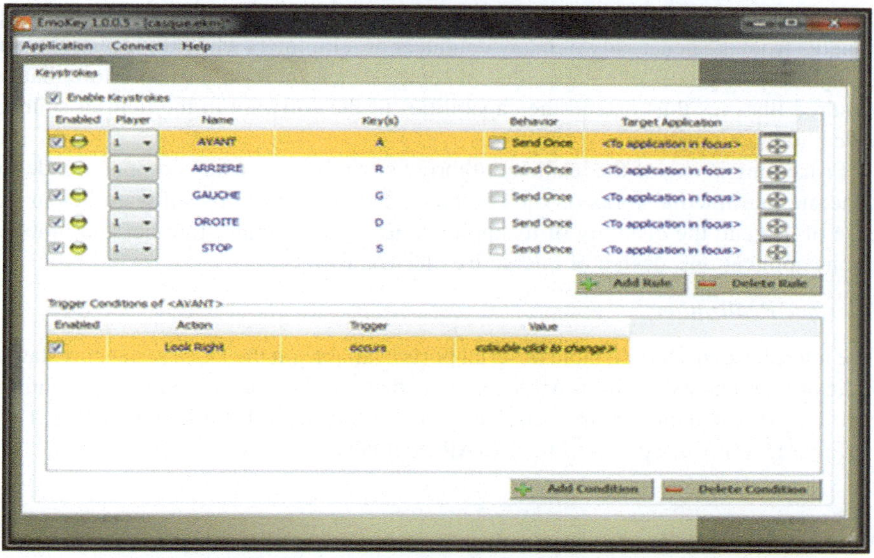

Fig. 16 Actions coding with characters

lated keystrokes and sends orders to the Arduino board according to the characters received. Figure 16 shows the coding actions through characters.

Figure 17 shows the choice and the coding of the desired action through a word or a character.

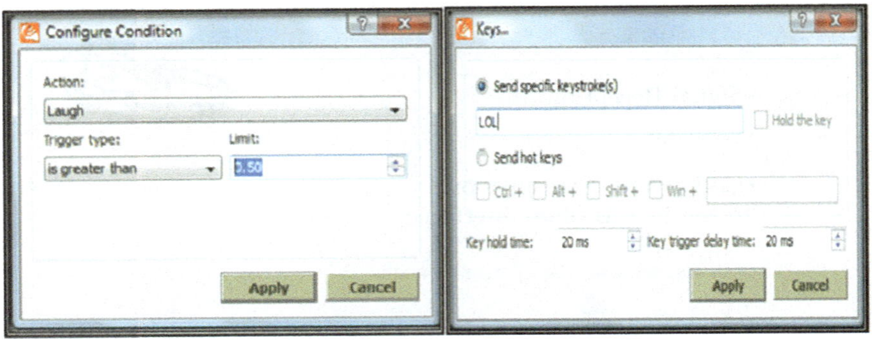

Fig. 17 Choice and coding of the desired action

5.3.2 Interface PC/ARDUINO

Introduction to PROCESSING Software

Processing is an open source software. It runs on Windows, Linux, Mac (and any other platform that can run software designed in Java). As free software, Processing benefits from the generosity of many volunteer programmers who provide users with easily reusable pieces of code (called in computer jargon for libraries). More than a hundred libraries thus extend the capabilities of the software in the field of sound, video, interaction, etc.

Processing is one of the leading authoring environments using computer code to generate multimedia works on the computer. The appeal of this software lies in its ease of use and the diversity of its applications: image, sound, Internet and mobile phone applications, design of interactive electronic objects.

Interface Realization

The interface consists of a table of serial ports available on the PC, a serial port status indicator, and an exit button. All the actions that will be transmitted to the Arduino board are encoded through this interface program written in JAVA language (Fig. 18).

Figure 19 shows a photo of the PC/ARDUIONO interface.

5.3.3 EMOTIV Headset Operating Flowchart

After analyzing different elements that ensure the operation of the headset, we present at Fig. 20 the operating flowchart of the headset mode.

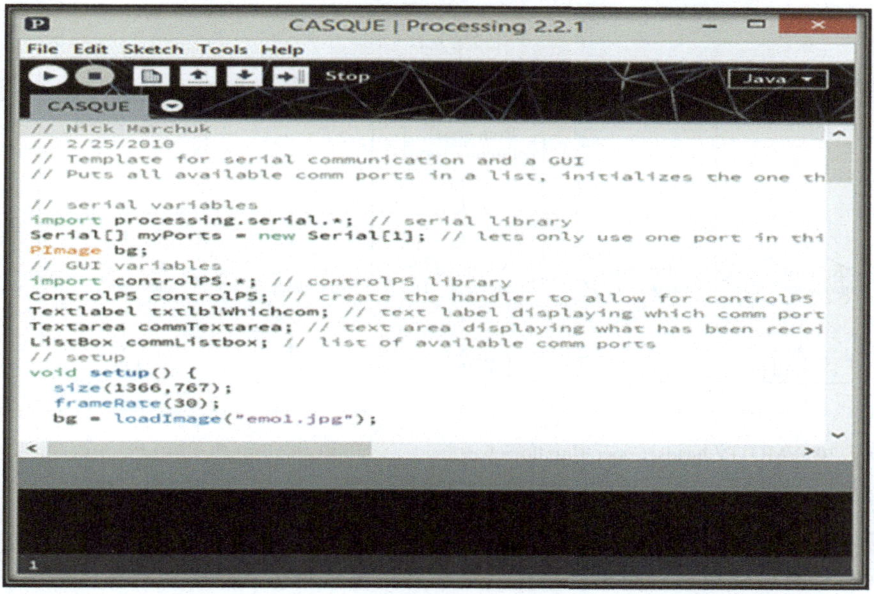

Fig. 18 "Processing" program

Fig. 19 PC/ARDUINO interface

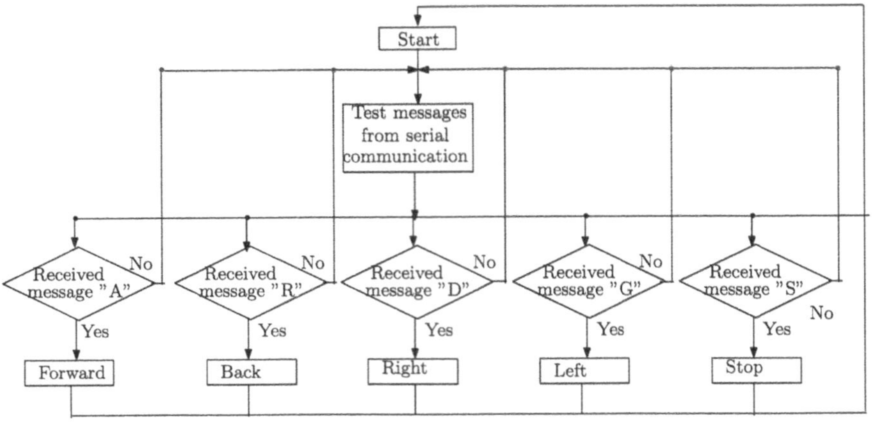

Fig. 20 EMOTIV headset operating flowchart

5.4 *Wheelchair Operation Flowchart*

In this section we present the realization of different parts of the project. We start with the fixing and positioning the method of the ultrasonic sensors to improve the functioning of the wheelchair. Then, we detail the use of the Emotiv EPOC headset as well as the creation of the graphic interface that communicate the headset with the Arduino board (Fig. 21).

5.5 *Experimental Results*

Two tests on the wheelchair are carried out to validate the theoretical work.

5.5.1 Validation of the Ultrasonic Sensors Placement

In this mode, the Ultrasonic sensors are placed on the wheelchair to detect any obstacle and allow to move away.

Figure 22 shows a series of photos of the Ultrasonic sensors test of the intelligent wheelchair.

From Fig. 22, we can see that if the wheelchair is in front of a wall (about 1m), he has turned at the right side without any intervention from the user.

Fig. 21 Wheelchair flowchart

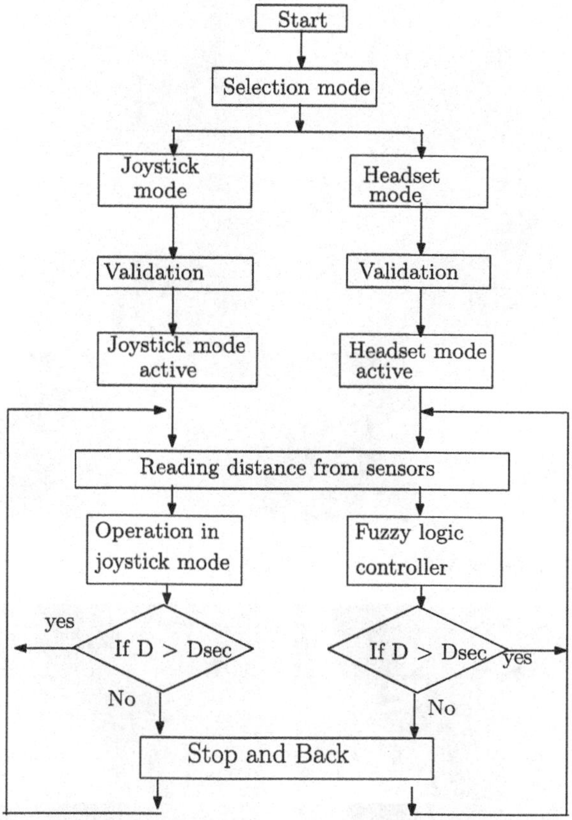

5.5.2 Validation of the EMOTIV Headset Operation with the Wheelchair

In this mode, the user place the EMOTIV EPOC helmet in his head and drives the wheelchair without any intervention of hands. Only the face and the eyes grimace are uses to make the wheelchair move forward or turn in the left or the right side. Figure 23 shows a series of photos of the EMOTIV headset test with the intelligent wheelchair.

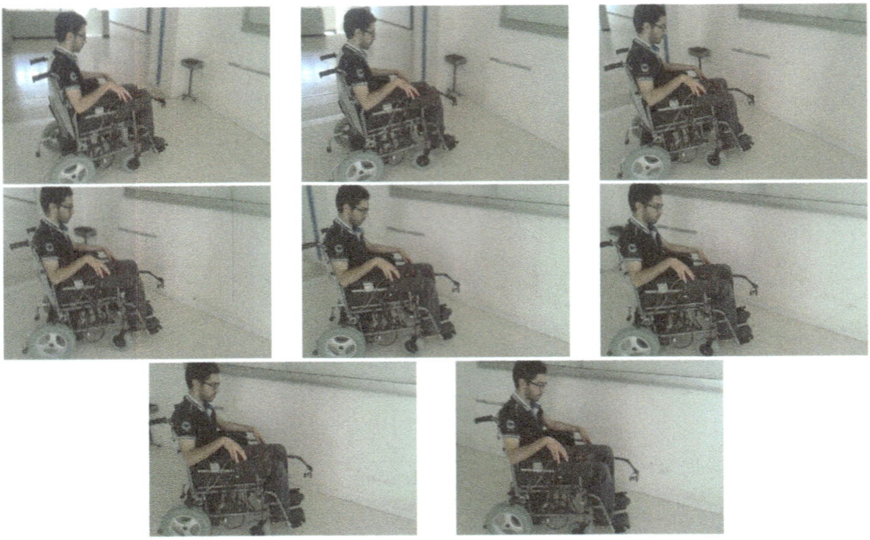

Fig. 22 Sequence 1

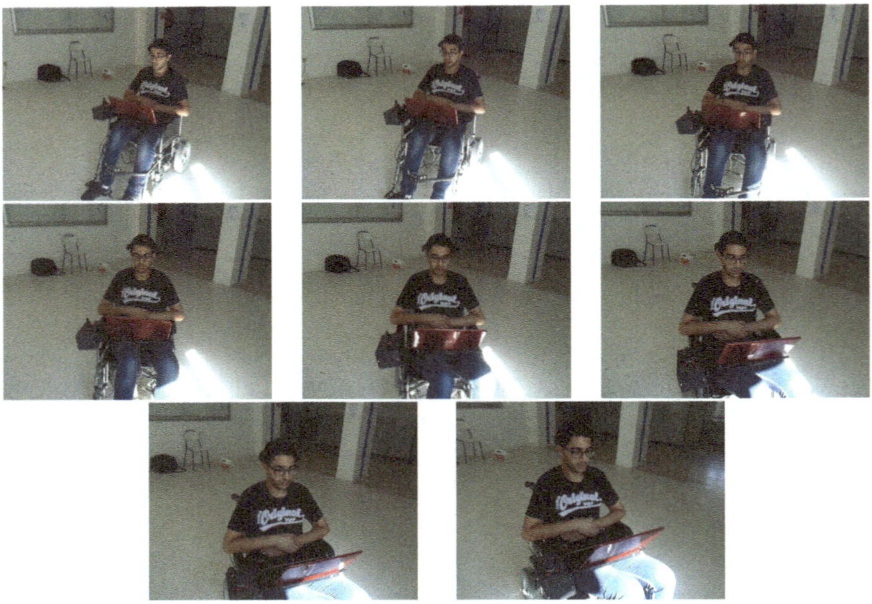

Fig. 23 Sequence 2

6 Conclusion

In this chapter, the control system of an intelligent wheelchair has been presented. As a solution to the operating problem of an electric wheelchair for several types of disabilities, a new system for controlling the chair is presented in this chapter. New driving modules have been added, the aim of which is to provide new functionalities to perform a set of predefined tasks. A new smart control options have been incorporated to facilitate the wheelchair command. To implement an obstacle detection and avoidance system a detailed steps based on ultrasonic sensors established. In addition, we have been managed to control the wheelchair through the EMOTIV EPOC helmet which consists of controlling the chair by a facial expressions using a graphical interface.

References

Carrino, F., Dumoulin, J., Mugellini, E., Abou, K. O., & Ingold, R., (2012). A self-paced BCI system to control an electric wheelchair: Evaluation of a commercial, low-cost EEG device. In *ISSNIP Biosignals and Biorobotics Conference: Biosignals and Robotics for Better and Safer Living (BRC)* (pp. 1–6), IEEE Organization.

Frank, A., Neophytou, C., Frank, J., & Souza, L. (2010). Electric-powered indoor/outdoor wheelchairs (EPIOCs): Users' views of influence on family, friends and carers. *Journal of Disability and Rehabilitation: Assistive Technology*, *5*, no. 5, 327–338, Publisher Taylor & Francis.

Gajwani, S., & Chhabria, S. A. (2010). Eye motion tracking for wheelchair control. *International Journal of Information Technology*, *2*(2), 185–187.

Goher, K. M. (2016). A reconfigurable wheelchair for mobility and rehabilitation: Design and development. *Journal Homepage of Cogent Engineering*, Publisher Taylor & Francis.

Gray, J., Pei, H., Huosheng, H., Lu, T., & Yuan, K. (2007). Head gesture recognition for hands-free control of an intelligent wheelchair. *International Journal of Industrial Robot*, Publisher Emerald Group Publishing Limited.

Herman L. K., (1969). A brief history of the wheelchair. *Journal of the History of Medicine and Allied Sciences*, *XXIV*, 205–2010.

Huo, X. & Ghovanloo, M., (2009). Using unconstrained tongue motion as an alternative control mechanism for wheeled mobility. *International Journal of IEEE Transactions on Biomedical Engineering*, *56*, no. 6, 1719–1726, IEEE Publisher.

Medola, F. O., Fortulan, C. A., da Silva S. C., & Elui, V. M. (2012). Reaching for independence: challenges for a new concept of wheelchair design. *International Journal of Physical Therapy Perspectives in the 21st Century–Challenges and Possibilities*, p. 371.

Rechy, R., Ericka, J., & Hu, H. (2014). Flexible bi-modal control modes for hands-free operation of a wheelchair by head movements and facial expressions. *New Trends in Medical and Service Robots*, (pp. 109–123). Berlin: Springer.

Samet, K. (2014). Conception et réalisation d'un système de commande pour les fauteuils roulants électriques. Mémoire du Diplôme National d'Ingénieur en Génie électrique, Ecole Nationale d'Ingénieurs de Sfax, Tunisia.

Sidik, M. M., Ghani, C. S., Ishak, M., Harun, W. S. W., & Daud, N. (2018). A review on electric wheelchair innovation to ease mobility and as a rehabilitation tool for spinal cord impairment patient. *International Journal of Engineering and Technology*, 803–815.

Simpson, R. C. (2005). Smart wheelchairs: A literature review. *Journal of Rehabilitation Research and Development*, *42*, no. 4, 423, Publisher Superintendent of Documents.

Swee, S. K., Kiang, K. D., & You, L. Z. (2016). EEG controlled wheelchair. *MATEC web of conferences* (vol. 51, pp. 02011).

Swee, S. K., & You, L. Z. (2016). Fast fourier analysis and EEG classification brainwave controlled wheelchair. In *2nd International Conference on Control Science and Systems Engineering (ICC-SSE)* (pp. 20–23), IEEE Transiction.

The manufacturer's authorised representative in the EU is Springer Nature Customer Service Centre GmbH, Europaplatz 3, 69115 Heidelberg, Germany. If you have any concerns regarding our products, please contact ProductSafety@springernature.com

Printed and bound by CPI Group (UK) Ltd, Croydon, CR0 4YY

30/09/2026

02225294-0011